Problems of Living

Perspectives from Philosophy, Psychiatry, and Cognitive-Affective Science

Problems of Living

Perspectives from Philosophy, Psychiatry, and Cognitive-Affective Science

Dan J. Stein, MD, PhD, DPhil
Department of Psychiatry, University of Cape Town and Groote Schuur Hospital, Cape Town, South Africa

Academic Press is an imprint of Elsevier
125 London Wall, London EC2Y 5AS, United Kingdom
525 B Street, Suite 1650, San Diego, CA 92101, United States
50 Hampshire Street, 5th Floor, Cambridge, MA 02139, United States
The Boulevard, Langford Lane, Kidlington, Oxford OX5 1GB, United Kingdom

Library of Congress Cataloging-in-Publication Data
A catalog record for this book is available from the Library of Congress

British Library Cataloguing-in-Publication Data
A catalogue record for this book is available from the British Library

ISBN 978-0-323-90239-7

For information on all Academic Press publications
visit our website at https://www.elsevier.com/books-and-journals

Publisher: Nikki Levy
Acquisitions Editor: Natalie Farra
Editorial Project Manager: Pat Gonzalez
Production Project Manager: Punithavathy Govindaradjane
Cover Designer: Matthew Limbert

Typeset by SPi Global, India

Dedication

To Solly, Fanny, Adam, David, Naomi, Jonathan,
Heather, Gabriella, Joshua, and Sarah,
with immense gratitude and love.

Excerpt from "The Labyrinth"

by W.H. Auden

Anthropos apteros for days
Walked whistling round and round the Maze,
Relying happily upon
His temperament for getting on.

The hundredth time he sighted, though,
A bush he left an hour ago,
He halted where four alleys crossed,
And recognized that he was lost.

Anthropos apteros, perplexed
To know which turning to take next,
Looked up and wished he were the bird
To whom such doubts must seem absurd.

Contents

Preface

What sorts of things are humans? What sort of life should we lead? How do we know these things? Philosophers have asked and answered these 'big questions' for thousands of years. Socrates was reputedly one of the first to engage with others as a philosopher, debating with the citizens of Athens, and reminding them of the inscription 'know thyself'. Psychiatrists and clinical psychologists have long emphasized the value of self-insight, and more recently have often emphasized the views of other classical philosophers, the Stoics, who argued that although life often leads to circumstances over which we have little control, we always have power over our own responses.

'The person on the street' has also long debated these kinds of questions. Oftentimes, though, people are not so much seeking abstract conceptual clarification, but rather wanting practical solutions to the day-to-day problems that life invariably brings. For millennia, shamans have responded to people who find life meaningless, or who feel so burdened by life that they see no viable way forwards. To some extent, this role has today been taken over by psychiatrists and clinical psychologists, who provide insight-oriented and cognitive-behavioural psychotherapy not only to those with severe mental disorders, but also to those whom they refer to as having 'problems in living'.

Children start to grapple with the big questions of life as well as problems of living early on. Like many, I became particularly intrigued with the big questions as a teenager, perhaps because that is when abstract thought becomes increasingly sophisticated, and perhaps because that is a time when day-to-day problems of living seem particularly acute. My parents had a wonderful library, allowing me to search for answers across the sciences and humanities. Knowing that my search had just begun, I thought that training in medicine, and then in psychiatry, would allow me to keep reading across a range of fields. As an adult working in psychiatry, though, the big questions and hard problems continued to plague me, at home and at work: including the really big one, *the meaning of life*: what was life all about, really?

In philosophy, key approaches include those of analytic philosophy and of continental philosophy. Analytic philosophy provides many rigorous arguments about ourselves and the world, but at the same time it has often regarded questions about the meaning of life as meaningless. Continental philosophy also provides rich intellectual resources, but it often sees answers to the big questions as just one more narrative, emerging from particular lived experiences.

Furthermore, oftentimes, writing from analytic philosophy is highly technical, and that from continental philosophy is highly abstruse; this sort of writing hasn't helped me to find good answers to the big questions, and I would think it also hasn't helped others to do so either.

In psychiatry, key approaches include those of biological psychiatry and psychoanalytic psychiatry. Biological psychiatry sees psychiatry as a clinical neuroscience; this has led to useful contributions to understanding specific psychiatric disorders, but it is often focused on narrow questions about brain circuits and molecules, rather than on bigger questions about who we are and should be. Psychoanalysis has emphasized the importance of the unconscious; this has led to a range of productive ideas in the clinic, but its conceptual framework seems increasingly dated. I have been fascinated but also at times disappointed by the literature in these areas. Reading molecular neuroscience can be fairly dull, while psychoanalytic writing is often filled with meaningless jargon, and it's not clear that either addresses problems of living in a helpful way.

Worryingly, while these debates in philosophy and psychiatry have played out, the big questions and hard problems only seem to have gained urgency. The digital age has perhaps altered aspects of our psychological make-up, and the genetics age leads to the possibility that our neurobiology will be edited in the future. To cope with the current digital age, and the future genetics age, we need to have a clear idea of who we are, and who we want to be. I thought that training in philosophy might help, but this training at times raised more questions than answers for me. I would imagine that for many, strident debates in philosophy and psychiatry may shed more darkness than light.

Significantly, while these debates in philosophy and psychiatry have continued, science has been gradually advancing. In particular, the field of cognitive science, which includes disciplines such as neuroscience and psychology, and which I will refer to as cognitive-affective science, or the cognitive-affective sciences, has provided us with many answers to important questions, ranging from how genes and environments influence brain development, through to understanding how our thinking and feeling is often based on metaphors. We've learned a great deal about our brain–minds: it turns out that our 'wetware' is quite different from the hardware and software that have so often been used to conceptualize the brain and the mind. These advances in the cognitive-affective sciences seem crucially important for philosophy, for psychiatry, and for answering the big questions and hard problems.

At the same time, it is important to be aware how much remains to be learned. The new and growing field of neurophilosophy has argued that philosophy has already ceded territory in metaphysics to science, and must now look to neuroscience to help address questions in epistemology. In parallel, neuroethics has explored both the neuroscience of ethics, as well as at the ethics of neuroscience. And certainly, *it has become difficult in the age of neurophilosophy and neuropsychiatry to think about the big questions and hard problems without some reference to what science in general, and the cognitive-affective sciences in particular, have to say about the brain and the mind.* But these fields are addressing an enormously complex subject; right now what they have to say is often underwhelming.

Differences between those who regard the cognitive-affective sciences with high regard and great optimism, and those who are more cautious and perhaps more pessimistic about their status, resonate with deeply different responses to the big questions and hard problems, including the meaning of life. There is a famous view in philosophy that ours is the best of all possible worlds. Science, for example, has shown us just how intricate and complex the workings of nature are. There is also, however, the view that our current world is a dreadful one. Science has shown us not only that the laws of replication promote the survival of the fittest, but also that nature is in many ways deeply immoral, necessarily causing great suffering. These contrasting approaches to answering the big questions and hard problems have further contributed to my and others' perplexity.

Polar views on philosophy, on psychiatry, on science, and on the meaning of life, and the perplexity they exacerbate, are the impetus for this volume. The volume aims to develop an approach to resolve these contrasting views, and so to provide an appealing way forwards for responding to the big questions of life and the hard problems of living.

This way forwards is based on several key ideas.

First, it is possible to draw on the best aspects of opposing positions in philosophy and psychiatry, to develop strong integrated approaches that help resolve a number of debates. I'll suggest that there are useful ways of bringing together aspects of analytic and continental philosophy, as well as of biological and psychoanalytic psychiatry. In order to facilitate this sort of resolution, we need a multidimensional strategy (that brings together different concepts and findings) as well as a multidisciplinary one (that brings together different fields and methods). Such a strategy allows us to address the big questions, as well as a range of more focused issues in philosophy and psychiatry, with the breadth and depth that they deserve.

Second, some work in philosophy and psychiatry has been particularly prescient, foreshadowing many current findings in cognitive-affective science. It turns out that particular philosophical positions taken by authors such as Aristotle, Spinoza, Hume, Dewey, and Jaspers are supported by recent research in cognitive-affective science. These authors may be especially helpful in formulating an integrated approach, and their work clearly remains pertinent to addressing a range of questions. These include not only the big questions (e.g. 'how should we live our lives?'), but also more focused but related problems of life (e.g. 'how do I best express anger?').

Third, advances in the cognitive-affective sciences provide a unique opportunity to rethink debates in philosophy and psychiatry, and to provide sharper answers to the big questions and hard problems. Consider, for example, the finding that similar neurocircuitry is involved in mediating anxiety across different mammalian species, and that very similar neurocircuitry is involved in mediating anxiety across different primate species. These findings may mean that particular positions taken by philosophy and by psychiatry in the past need updating; these include positions that bear on the big questions (such as 'who are we?'), but also on a range of subsidiary questions (such as the question of how we should treat other animals).

Fourth, one of the most important contributions of modern cognitive-affective science is its understanding of how the mind is embodied in the brain, and embedded in society, and of how thoughts and feelings often rely crucially on metaphors. Aspects of this view have long been foreshadowed in philosophy and in psychiatry, are useful for supporting integrated positions in each field, and are helpful in bringing these fields together to address the big questions and hard problems. As we search for a 'way forwards', we'll find ourselves turning back (pun intended) to a number of key metaphors, such as 'balance', 'parenting', and 'journeys'.

Fifth, the cognitive-affective sciences help provide us with key insights into the big questions and hard problems, but they also have crucial limitations. When it comes to considering how much the sciences have advanced, how meaningful life is, and whether optimism or pessimism is more justified, this volume will often take an intermediate position. Advances in the cognitive-affective sciences have been really impressive, and nature is truly wonderful; but at the same time it's impossible to simply dismiss the many gaps and errors in our scientific knowledge and moral decision-making, and the opacity and suffering that life brings. We need to find a middle path forwards.

While none of these five ideas is entirely new, it is novel for a dialogue between philosophy, psychiatry, and the cognitive-affective sciences to encompass all of them. Furthermore, given recent advances in the cognitive-affective sciences, we perhaps have the potential to see further by standing on the shoulders of our predecessors. Finally, because I live and work in Africa, I will bring some concepts and discoveries from Africa into the discussion. I hope this synthesis of ideas and findings is comprehensive, stimulating, and useful for a broad academic audience.

To appeal to a broad academic audience I have tried to write as clearly and nontechnically as possible, and perhaps the volume will therefore also appeal to the interested 'person in the street'. More particularly, I hope that a wide-ranging synthesis of different voices and views in philosophy, psychiatry, psychology, and neuroscience is appealing to those with more in-depth training in any of these fields. For those more specialized interests, I've included more detailed footnotes, both as justification for the arguments made, and as a suggestion for further reading.

Integrative cross-disciplinary approaches may be seen as flimsy and superficial by those working on more focused and deeper approaches. Drawing on the ideas of the philosopher Isaiah Berlin, who in turn was inspired by the Greek poet Archilochus, we can draw a distinction between the foxes who skirt around working on many different issues trying to contribute a little to each, and the hedgehogs who burrow deeper and deeper, tunnelling down to learn more and more about a single issue. I'm afraid I've always been a fox, with the attendant disadvantages of this approach. I'm grateful to have colleagues with more profound knowledge, many of whom have mentored me; several will be highlighted during the course of the volume. A particular thank you to Anton van Niekerk, Andrea Palk, Christine Lochner, Damiaan Denys, Derek Bolton, George Ellis, Goodman Sibeko, Harold Kincaid, Karen Mare, Nastassja Koen, Sean Baumann, Thaddeus Metz, and Timothy Crowe for being willing to offer insightful critique and warm support.

As a teenager thinking about the big questions and hard problems, I was fortunate not only to have access to a good library, but also to thoughtful mentors: my two parents and my four much older siblings: all of whom seemed to have entirely different, but useful, advice about the world and our place in it. Each of them seemed to me to have something valuable to contribute to answering the big questions, with each proposing somewhat different solutions to my daily problems of living. Reconciling different views in the family about the value of modern medicine was a particularly informative challenge. While I like to think that the integrative approach taken here stands on its merits, it may well have roots in these early family practices and habits of thinking. Having my own family exposed me to new and fresh views on the question of how to live, and even more practice in valuing and synthesizing different ideas. My dedication of this volume to these two families is to express my gratitude and love; I hope they'll recognize some of their wise voices and views in it. I trust that my father, Solly, an enthusiastic Talmud scholar, would at least have approved of the approach taken to footnoting.

The table of contents of this volume divides the big questions and hard problems into seven key issues. The psychologist George Miller argued in a well-known paper, 'The Magical Number Seven, Plus or Minus Two: Some Limits on Our Capacity for Processing Information', that humans struggle to hold more than seven different ideas in short-term memory at any one time, so perhaps this is a reasonable number to begin with. We'll begin by asking what sort of thing the brain and the mind are, and how best to conceptualize the intersection of reason and emotion. Then we'll go on to ask what is happiness and how can we bring it about, how should we best conceptualize pain and suffering, and how do we best approach moral concepts such as good and evil? The final two questions are 'how do we know this is all true?' and 'what is the meaning of life, or what makes a life meaningful?' This set of questions encompasses, then, some of the very big fields within philosophy: metaphysics, epistemology, and ethics.

Along the way, however, the integrative positions proposed for addressing these big questions and hard problems will also allow us to address a range of related subsidiary questions that are also important and interesting. These include a number of questions at the intersection of philosophy and psychiatry, including ones about how best to conceptualize psychiatric disorders, about the intersection between evil and psychopathy, and about whether progress in psychotherapy is possible. They also include a range of questions about the sciences and humanities in general, including the difference between science and pseudoscience, whether scientific and moral progress is possible, and the nature of free will. I hope that you'll find that this volume provides a useful way forwards for thinking about life and its meaning in our current neurocentric age.

Chapter 1

Introduction

As a teenager trying to figure out how to cope with life, I was fortunate to be able to draw on the extensive resources of my librarian mother. The family's bookshelves were neatly and carefully arranged, and covered popular writing on philosophy, self-help, and pop-psychology books, as well as a range of science writing. The philosophy books were intriguing, but Dale Carnegie's volume on 'How to Win Friends and Influence People' and Eric Berne's volume on 'Games People Play' provided practical advice about day-to-day life. I was also completely smitten with Desmond Morris's description of humans as 'naked apes'; it seemed to me that thinking about humans in terms of biology provided an important foundation for answering the big questions, and addressing the hard problems of living.

Given this sort of reading, it is perhaps not surprising that I ended up wanting to work in psychiatry: a field that combines psychology and biology (indeed I like to use the word psychobiology, which emphasizes that these fields are inextricably linked, and that both are needed to fully understand animal and human behaviour). Still, as an adult the big questions persisted for me; not only did I have to figure out how to keep coping with the hard problems of living, but I was also responsible for helping others to do so. At this stage, however, these early resources seemed less useful. Work in philosophy was often highly technical: too abstract and narrow to be useful to me. Self-help and pop-psychology books, on the other hand, were often superficial, with little scientific evidence base for the advice offered.[1]

My career as a psychiatrist has been more surprising to me; I ended up working as a clinician–scientist in a wider range of areas, and using a broader variety of methods, than I would have anticipated. I was initially exposed to clinical issues and scientific questions in the area of obsessive–compulsive disorder, but later moved on to work on posttraumatic stress disorder, substance use disorders, and a number of other conditions. I was initially focused on psychopharmacology methods, but then moved onto work on animal models, on neurogenetics and neuroimaging, on psychotherapy, and on epidemiology. I have therefore

1. Donald Dewsbury has traced the history of the term **'psychobiology'** [1], noting that it has been employed by those aiming for integrated and nonreductionist approaches to behaviour. Adolf Meyer, a Swiss-born psychiatrist who worked at John Hopkins, was key in introducing the term into psychiatry, to refer to an integrated biopsychosocial approach. See later footnote on the term biopsychosocial.

Problems of Living. https://doi.org/10.1016/B978-0-323-90239-7.00004-3

been acutely aware of advances in the cognitive-affective sciences including neuroscience, and the argument that these would require a rethinking of key issues in both philosophy and psychiatry.[2]

In this chapter I want to outline in more detail some of the difficulties that philosophers and psychiatrists have faced in answering the big questions and hard problems, emphasizing some of the impasses between different approaches within their fields. Over the course of the volume, I will argue that in the past several years a number of resources have been developed in both philosophy and psychiatry which allow a more integrative way of approaching perennial debates in these fields, and which may be helpful in our attempts to answer the big questions and to address problems of living. But to begin with it's useful to explore some of the key struggles in these disciplines: this exploration helps provide us with a clear road map for moving forwards.

1.1 Perspectives of philosophy

Socrates went onto the streets of Athens and engaged its citizens in a conversation about what they really knew. Socrates's love of wisdom (philosophy) was not merely ethereal; he was engaged in a robust critique of Athenian society. The Socratic dialogue depicted by his student Plato remains an extraordinary powerful tool; rigorous debate with others may challenge implicit beliefs and may clarify one's own thinking, sometimes in transformative ways. The ancient dictum 'Know Thyself' has been a foundational pillar not only for philosophy, but also for a range of insight-oriented psychotherapies. Socrates went so far as to argue at his trial that the unexamined life is not worth living, a point that we'll come back to, though, as this may take things a tad too far for some of us.[3]

The ancient Greeks were not the only ones who were thinking through questions of how to live at a time when agriculture had made great strides, and when the start of city life posed a range of new issues. The pioneering

---

2. The terms 'cognitive science' or 'cognitive neuroscience' are much more commonly used than **'cognitive-affective science'**. For useful histories of cognitive science, see Howard Gardner (who has described how cognitive science brings together philosophy, neuroscience, cognitive psychology, linguistics, and anthropology) [2], Margaret Boden [3], and Michael Dawson (who has described early cognitive science, later connectionism, and the more recent focus on embodiment) [4]. For a useful compendium of cognitive neuroscience, see David Poeppel, George Mangun, and Michael Gazzaniga [5]. Later, we will emphasize how closely linked cognition and affect are, hence the use of the term 'cognitive-affective' here. The terms 'scientist' and 'neuroscience' are surprisingly new, and even more recently multiple neurodisciplines, including neurophilosophy and neuroethics, have emerged as part of the **'neuroscientific turn'** or 'neuroturn' [6].

3. We should be careful not to see Socratic or other philosophical arguments as emerging 'ex nihilo'; rather such thought is arguably best understood in the context of specific conceptual debates (e.g. Socrates versus the Sophists) as well as socio-political conflicts (for one, perhaps idiosyncratic, reading of Athenian politics, see Irving Stone [7]). Furthermore, our reading of key philosophers, such as Socrates, may change considerably over time [8, 9]. I make no apology for often citing **aphorisms** here; for more on this compact but powerful genre, see Andrew Hui [10] and for collections of aphorisms by philosophers, see Mark Vernon on 42 [11] and Daniel Klein [12].

philosopher–psychiatrist, Karl Jaspers, has referred to an 'Axial Age'; during the first century BC, the Greek, Hebrew, Indian, and Chinese civilizations were contributing their classical responses to the big questions. Sophisticated responses were also being provided in Africa, the Americas, and elsewhere, but unfortunately were not as well recorded. Plato's protégé, Aristotle, exemplifies the breadth and depth of Axial Age work; for Aristotle the sciences included both metaphysics (first philosophy or the theory of reality) and physics (a second philosophy devoted to the study of nature, later known as 'natural philosophy'). His thinking remains hugely influential in discussions of the big questions, and we'll keep drawing on his far-sighted discussions about the virtues and about human flourishing.[4]

It took a while for science to develop to the point where it was clear that much of what the ancient Greeks had written about the specifics of biology was incorrect. Ancient views, as well as many modern metaphors, locate the soul in the gut, the heart, and the brain. That it took so long for the brain to be seen as the seat of the mind was in part due to Galen, a Roman physician who influenced medical practice for centuries, despite his deeply flawed dissections. Thomas Willis, a 17th century physician working in Oxford, was a key father of modern neuroscience: he dissected out brains in extraordinary detail, showed similarities and differences in brain structure across species, and indicated how neurological diseases were linked to brain abnormalities.[5]

In thinking about the big questions, philosophers of the modern era were keenly aware of such advances in science. Baruch Spinoza, whose family had fled from Portugal to the Netherlands, was well versed in classical Greek and Hebrew works, and also a man who worked on a daily basis with the laws of optics (as a lens grinder). His answers to the big questions, as we'll see later, have a great deal in common with those of contemporary psychotherapists. By the time the Scottish Enlightenment author, David Hume, was working, the modern era was well underway. Like Aristotle and Spinoza, Hume was interested in the question of how best to live, and based his answers on a close examination of human nature in general, and on the workings of human reason and emotion in particular.[6]

Histories of philosophy often simplistically classify Spinoza as one of the continental rationalists (alongside René Descartes and Gottfried Leibniz), and

4. For more on the **'Axial Age'**, see Karen Armstrong as well as more sceptical authors [13–16]. City life not only raised new questions, but also provided some citizens with the time and means to address them.

5. Carl Zimmer has an excellent volume on this transition from **Galenic views** to modern neuroscience [17]. See also George Makari [18].

6. We'll keep returning to **Spinoza and Hume's** naturalistic and humanistic ideas throughout this volume. The literature on each of these philosophers is massive. On Spinoza, I've particularly enjoyed Antonio Damasio [19], Rebecca Goldstein [20], and Irvin Yalom [21]. A number of readable volumes explore Hume's work by investigating relationships with his contemporaries Samuel Johnson [22], Jean-Jacques Rousseau [23], and Adam Smith [24]. See also Russ Roberts's work, which although about Smith rather than Hume, is relevant to our focus here [25].

Hume as one of the British empiricists (alongside John Locke and George Berkeley). A contrast is then drawn between the rationalist claim that knowledge is gained independently of sense experience, and the empiricist claim that sense experience is the ultimate source of knowledge. Certainly, the German philosopher Immanuel Kant saw his system of metaphysics as bridging rationalism and empiricism. As he famously said "Concepts without percepts are empty, percepts without concepts are blind".[7]

As we'll see during the course of this volume, the views of these authors, as well as of key 18th- and 19th-century authors who responded to their works, including Arthur Schopenhauer and Friedrich Nietzsche, continue to shape contemporary debates in philosophy, including debates about the brain–mind, and about reason and emotion. In his attempt to reconcile rationalism and empiricism, Kant usefully proposed the notion of schematas, which are simultaneously in the person, but informed by the world; we'll say more about the value of schema constructs in the next chapter. Note, though, that Kant's view is rooted in idealism rather than realism, and so rather different from the approach we will take to schemas.[8]

Impressed by the ongoing success of the natural sciences, in the early 20th century several leading philosophers aimed to make philosophy more like physics and other sciences. The Vienna school of logical positivism, for example, argued for continuity across the sciences, which were all based on empirical observation and mathematical logic.[9] A key idea was that only propositions that could be verified by direct observation or logical proof were meaningful; a scientific philosophy would necessarily be based on such

---

7. This citation is from Kant's 'Critique of Pure Reason' [26], which has a chapter on 'schematism' (see next footnote). **Kant** claimed to have been awoken from his slumbers by Hume, and regarded his integration of rationalism and empiricism as his 'Copernican revolution'; however, this standard narrative seems to be an oversimplification [27].

8. In discussing how there must be some connection between the experienced idea and the experienced world, the schema is "properly, only the phenomenon, or sensible concept, of an object in agreement with the category". Thus Kant contrasts his **transcendental idealism** ('things-in-themselves' are unknowable), with Descartes' problematic idealism (the existence of matter can be doubted) and Berkeley's dogmatic idealism (matter does not exist), although not mentioning Berkeley by name. His transcendental idealism also differs from **transcendental realism**, which he regards as the 'common prejudice' that objects are 'things-in-themselves', independent of our sensibility. Work on Kantian schemata has recently been reviewed [28]. For a realist critique of Kantian idealism, see Roy Bhaskar [29], and for a defence of Kant's approach, see Robert Hanna [30].

9. Useful books on the **Vienna Circle** include work by Allan Janik and Stephen Toulmin [31] and Karl Sigmund [32]; these volumes cover not only key philosophical ideas, but also the politics and personalities which drove these ideas forwards. The leader of the Circle, Moritz Schlick, trained in physics, and was one of the first to consider the philosophical implications of Albert Einstein's work. We'll later return to the questions of the relationship between philosophy and science, and between the sciences.

propositions. A distinction between synthetic propositions (true by virtue of how their meaning relates to the world) and analytic propositions (true by virtue of their meaning) was a key foundation for subsequent Anglo-American analytic philosophy.[10]

Key arguments in logical positivism exemplify what I'll call **a classical position** in philosophy. This position emphasizes objectivity; it sees science as discovering the laws of nature, and it views meaning as involving verification of the natural world. These views lead in turn to particular views on more focused questions (e.g. in the philosophy of medicine and psychiatry) and on broader questions (e.g. about the meaning of life), which we will explore in more detail over the course of this volume. Table 1 provides a sketch of this position, including its historical roots in the work of particular ancient and renaissance philosophers. I'd emphasize that this is just a sketch; the cells represent heuristic abstractions, and no particular thinker may fit neatly within any. We'll return to the classical position and this table, adding more detail, as the volume proceeds.

It's relevant to remember that at the time the Vienna Circle was meeting, the political situation in Europe was becoming darker and darker. The hope that philosophy would be placed on a firm scientific footing, and that it would move beyond continental metaphysics and its associated values, was therefore linked also to a broader goal of the emancipation of humankind. These kinds of concerns became ever more pressing as the 1930s saw irrationalism and anti-Semitism dominate public discourse in Austria. The Vienna Circle was forced to disband, although many of its members were able to continue their work in the United States, England, and other English-speaking countries.

The philosophical programme of logical positivism also proved difficult to complete. It turns out that while some constructs, like the elements of the periodic table, can be defined by necessary and sufficient criteria, most language is not like that. Our definition of weeds, for example, varies from time to time, and place to place. In his early work, Ludwig Wittgenstein, one of the most important philosophers of the 20th century, aimed to formalize the relationship

10. Analytic philosophy or ordinary language philosophy has roots in the work of Bertrand Russell, Gottlob Frege, and the early Ludwig Wittgenstein. The **analytic/synthetic distinction** has roots in the work of Hume: his work distinguished between 'matters of fact' and 'relations of ideas'. Kant also discussed analytic and synthetic propositions, introducing a distinction between a priori propositions that do not rely on experience, and a posteriori propositions that are validated by experience. Key early figures in logical positivism and analytic philosophy, including Alfred Ayer and Rudolf Carnap, further contributed to discussion on these issues. The Harvard philosopher, Willard Quine, criticized the analytic-synthetic distinction in his paper on 'Two Dogmas of Empiricism' [33], and many important figures in modern philosophy have weighed into the debates, including Nelson Goodman, Wilfred Sellars, Peter Strawson, John Searle, Hilary Putnam, Jerry Fodor, Noam Chomsky, George Lakoff, and Mark Johnson [34].

TABLE 1 Classical, critical, and integrative philosophical positions.

	Classical position	Critical position	Integrative position[a]
Historical sources	Plato, Descartes, early Wittgenstein, logical positivism, analytic philosophy (cogito, ergo sum)	Vico, Herder, later Wittgenstein, Kuhn, continental philosophy (I exist, therefore I think)	Aristotle, Dewey, pragmatic realism, Bhaskar, critical realism, embodied realism[b] (ago, ergo cogito)
Philosophy of science	Science describes the world objectively, and deduces covering laws	Science depicts the world subjectively, constructing a narrative	Science is a social activity, and it discovers real structures, processes, and mechanisms[c]
Philosophy of language	Meaning requires verification	Meaning is based on validation	Language is embodied and embedded, but does refer to real phenomena[d]
Natural kinds	Natural kinds are essential and defined by necessary and sufficient criteria	Human kinds are constructed and defined by their use	Different kinds of kind involve different structures, processes, and mechanisms; many natural kinds are fuzzy
Naturalism and normativism	The sciences entail conceptual and explanatory monism (strict naturalism)	The sciences entail particular ways of evaluating and judging (normativism)	The sciences entail conceptual and explanatory pluralism (soft naturalism)
Answering the big questions	Propositions about the big questions often do not make sense	Multiple valid propositions in response to the big questions are possible	Careful reasoning-imagining about facts and values leads to better cognitive-affective maps

[a]*My distinctions between the **'classical' and 'critical'** draw on the work of a range of authors who have developed analogous distinctions in the literature. They include tough-minded and tender-minded (William James), platonism and nominalism (Willard Quine), Enlightenment and Counter-Enlightenment (Isaiah Berlin), objectivism and intellectualism (Maurice Merleau-Ponty), rationalism and romanticism (Ernest Gellner), metaphysicists and ironists (Richard Rorty), objectivism and subjectivism (Thomas Nagel), objectivism and relativism (Richard Bernstein, Mark Johnson), universalism and historicism (Joseph Margolis), legend and relativism (Philip Kitcher), essentialism and normativism (Kenneth Schaffner), empiricism and postempiricism (Derek Bolton), modernity and postmodernism (Hans Bertens), positivism and antirealism (Richard Miller), old deferentialism and new cynicism (Susan Haack), and affirmative and negative view (David Cooper) [53]. Such authors may focus on ontology, epistemology, and/or morality. Many authors hope to integrate naturalism and humanism, although this is perhaps an overly simplistic characterization in that within the critical position there is a humanist versus antihumanist debate [54, 55].*

[b]*Plato and René Descartes are listed as classicists, partly because of their emphasis on* ***mathematical truths*** *[56, 57] (Harry Frankfurt's volume also provides the social context to Descartes' work). In opposition to Descartes, Giambattista Vico's verum factum (****'truth is made'****) principle holds that truth is not verified through observation and mathematics, but rather that knowledge entails understanding the role of human construction. Johann Herder was one of the first to argue that language is key for framing our understanding of the world; it is the 'organ of thought'. Vico and Herder represent a Counter-Enlightenment movement that is redolent of romanticism [58]. Any attempt to put Wittgenstein in boxes of the sort tabulated here is bound to fail; see Alice Crary and Rupert Read's volume for essays that emphasize continuity across his work [51]. For more on Wittgenstein's naturalism, see Kevin Cahill and Thomas Raleigh [59]; for more on Wittgenstein's antiscientism, see Jonathan Beale and Ian Kidd [60]. Thomas Kuhn is well known for his work on scientific paradigms [61], although there is debate about the extent to which he is a relativist [62]. Aristotle and Dewey's views have much in common [63], and it is notable that each addressed the dualisms of their day [64]. For more on pragmatic realism, critical realism, and embodied realism, see later footnotes.*

[c]*Hume is often cited as arguing that causation is inferred from repeated observations of the conjunction of events, but this may well be an incorrect reading [65]. Carl Hempel further argued that explanation involved the deduction of laws that allow accurate prediction or postdiction; Hempel played a key role in transforming early 'logical positivism' into subsequent 'logical empiricism', and his view of scientific explanation, which others have termed the* ***'covering law model'****, was an influential contribution to the classical position. A particularly influential account of scientific explanation as more story-like was the work of Thomas Kuhn, cited in a previous footnote. The range of other critiques of the covering law model includes Nelson Goodman's work on the 'grue paradox' [66], Roy Bhaskar's work on explanatory power [67], and Nancy Cartwright's work [68]. For more on* ***scientific explanation****, see a number of volumes [69–71], as well as later footnotes on explanatory pluralism and on metaphors of causation.*

[d]*Within logical positivism there is a range of positions on the* ***verification principle****, for example, Alfred Ayer, who was sent by his Oxford University mentor Gilbert Ryle to visit the Vienna Circle, and who was key in bringing their ideas to English analytic philosophy, differentiated between strong and weak verificationism [72]. A hermeneutic tradition, in contrast, has focused on the importance of meaning as validation. Paul Ricoeur, for example, emphasized that understanding a text requires a continuous process of validation, and noted that "The* ***logic of validation*** *allows us to move between the two limits of dogmatism and skepticism" [73]. Richard Rorty has usefully contrasted meaning realism and linguistic idealism [74]. The notions of embodied and embedded cognition date back to a volume on the embodied mind by Francisco Varela, Evan Thompson, and Eleanor Rosch [75], and also build on the work of the pragmatic philosopher John Dewey and the continental philosopher Maurice Merleau-Ponty; see later footnote on '4E'. There is arguably also a link between Arthur Schopenhauer and constructs of the embodied mind [76].*

between language and the world.[11] But in his later work, he argued that in order to understand the meaning of words one had to use them in context. In the context of a 'language game', or a particular 'form of life', one made use of particular words and this use was their meaning, and one could appreciate the 'family resemblances' between different constructs.[12]

This shift moved Wittgenstein closer to what has been termed 'continental philosophy', in opposition to Anglo-American analytic philosophy. Continental philosophy spans a longer period of time than analytic philosophy, dating back to the 19th century German idealism which followed from the earlier work of Immanuel Kant. Still, it too needs to be understood against the backdrop of particular times and places. Certainly, for those working during and after World War II, it was difficult to put one's hope in a scientific vision. After all, the science of eugenics was one stimulus for Hitler's vision of an Aryan future. A different vision was needed; a view that sees ideas as closely linked to time and place, that questions power, and that emphasizes the importance of agency, is more appropriate in this context.[13]

---

11. Bertrand Russell, a polymath at Cambridge University, remarked that his student **Ludwig Wittgenstein** was "perhaps the most perfect example I have ever known of genius as traditionally conceived; passionate, profound, intense and dominating" [35]. Certainly, the two did not always see eye to eye, as exemplified in their early discussion about the proposition that "there is no rhinoceros in the room". According to Joseph McDonald, Wittgenstein's argument was that such a proposition is an everyday statement rather than a scientific proposition; Russell responded by appearing to search under the desks for a rhinoceros and concluded that "My German engineer, I think, is a fool" [36]. A notable debate also took place between Wittgenstein and another Austrian philosopher, Karl Popper; Popper had been invited by the Cambridge University Moral Sciences Club, to give a lecture on 'Are There Philosophical Problems?' Wittgenstein was the chair of the meeting, and took the position that there were no substantial philosophical problems, only linguistic puzzles. According to Popper, Wittgenstein began using a fire poker to emphasize his points, and challenged Popper to put forward a moral rule. Popper answered with "Not to threaten visiting lecturers with pokers", whereupon Wittgenstein threw the poker down and marched out the room [37].

12. Many have divided **Wittgenstein's career** into an early phase (more associated with logical positivism) and a later phase (more associated with positions in continental philosophy). The early Wittgenstein is the author of the 'Tractatus Logico-Philosophicus' [38], while the later Wittgenstein is the author of the 'Philosophical Investigations' [39]. The title of the 'Tractatus' is redolent of Spinoza's 'Tractatus Theologico-Politicus' and may have been suggested by George Moore [40]. Both volumes were written in the context of, and so partly in response to, considerable political turmoil [41]. The 'Philosophical Investigations' raises the **'rule-following paradox'**; as stated by Wittgenstein, "This was our paradox: no course of action could be determined by a rule, because any course of action can be made out to accord with the rule"; see Saul Kripke's discussion [42]. Along parallel lines, it is noteworthy that Russell and Alfred Whitehead's work on the principles of mathematical logic constituted a core component of analytic philosophy, but that Kurt Gödel later argued that any such formal system contains propositions which cannot be proven true or false from within the system, and so is necessarily incomplete [43–45].

13. This kind of point again (see previous footnote) raises the question of just how much philosophy itself is a social activity versus an activity that addresses the real structures, processes, and mechanisms of the world [46]. We return to this question later. For now, I can cite Karl Jaspers, who wrote "Since the basic questions of philosophy grow, as practical activity, from life, their form is at any given moment in keeping with the historical situation; but this situation is part of the continuity of tradition" [47].

Key stances taken by the latter Wittgenstein and continental philosophy exemplify what I'll call **a critical position** in philosophy. This position emphasizes subjectivity; it views knowledge as an interpretation or narrative, and it views meaning as a matter of interpersonal validation. Descartes' slogan of 'cogito, ergo sum' (I think therefore I am) exemplifies the focus of the classical position on objective rationality; in contrast, existential philosophers have argued 'I exist, therefore I think', so underscoring the importance of subjective experience. Indeed, the critical position leads to particular views on both more focused questions and on broader issues, which contrast markedly with those of the classical position. Table 1 provides a rough sketch of this position, including its historical roots in the work of particular philosophers; the aim again is to draw out some family resemblances, rather than to provide necessary and sufficient characteristics. Again, we'll return to the critical position and this table, adding more detail, as the volume proceeds.[14]

Wittgenstein hoped that philosophy would bring clarity and that its methods would be like therapies. Instead, it turned out that battles between analytic (or ordinary language) philosophy and more continental (and later postmodern) philosophy have continued to play out over decades and across many different places. Table 1 also provides a rough sketch of some key perennial and persistent debates in philosophy. Once again, these sharp foci and contrasts are intended as a heuristic device; they oversimplify thousands of years of complex philosophical debate. For interested readers, the footnotes provide more detailed justification for some key contrasts and for some key cells depicted in the table.[15]

For the average teenager looking for practical answers to everyday problems of living, neither the classical nor the critical position is particularly appealing; ordinary language philosophy tends to focus on very narrow questions in a very technical way, while continental philosophy can be extremely opaque in its writing style. Nevertheless, authors in these traditions have made crucially important contributions; we'll return to this point later on. For now, I'd want to say that the approach in this volume is to try draw on the best of both the classical and the critical perspectives, to try avoid errors that have been made within each of these perspectives, and to try develop an integrative way forwards.[16]

---

14. **Søren Kierkegaard** labelled Descartes' slogan a tautology, arguing that existence is presupposed in order for thinking to occur [48]. Along similar lines, for **Jean-Paul Sartre**, 'existence precedes essence' [49]. In his volume, Kierkegaard also proposes that 'subjectivity is truth' [50].

15. Much has been written on Wittgenstein's notion of philosophy as therapy [51]. For an early volume on persistent philosophical problems, see Mary Calkins' work [52].

16. Allan Bloom writes "Positivism and ordinary language analysis have long dominated, although they are on the decline and evidently being replaced by nothing. These are simply methods of a sort, and they repel students who come with the humanizing questions. Professors of these schools simply would not and could not talk about anything important, and they themselves do not represent a philosophic life for the students" [77]. For similar sentiments about **fashionable nonsense**, see the work of Alexander Nehamas [78], Susan Haack [79], Robert Solomon [80], and Anthony Kronman [81]. For one infamous critique of postmodern writing, see Alan Sokal and Jean Bricmont [82].

In the 21st century, at least in some parts of the world, philosophy seems to have returned to the Socratic approach: engaging with people, and pushing them to think about their assumptions. Thus, for example, philosophy clubs, where debates about contemporary issues are held, have been formed around the world. These don't get mired down in analytic-synthetic distinctions, nor do they revert to postmodern critique. Instead, they tackle the big questions of life as well as problems of living, drawing on a range of ancient and modern resources to answer them. Similarly, in developing an integrative view we will draw on the resources of prescient philosophers and of contemporary science, including cognitive-affective science.[17]

The Columbia University philosopher John Dewey addressed the big questions at a time when there had been particularly important advances in cognitive-affective science, including work in neurophysiology and Charles Darwin's research on the evolution of emotions. While European authors were taking the 'linguistic turn' towards analytic philosophy, in America Dewey was exposed to the 'pragmatic turn' of Charles Pierce and William James. In Leipzig, the physician–psychologist–philosopher Wilhelm Wundt was pioneering experimental psychology, but at Harvard, the similarly trained James had taken a much more practical focus to everyday psychology. To help us develop an integrative approach Dewey is a particularly useful author, as he attempted to resolve a range of conceptual dualisms (e.g. between body and mind, reason and emotion, nature and culture), drawing on the work of ancient philosophy, especially Aristotle, as well as modern science. We'll rely on a number of his ideas; let me mention four to begin with.[18]

First, like the ancients, Dewey emphasized that philosophy should be focused not so much on abstruse technical issues, but on the question of how to live. This question was key for Aristotle, Spinoza, Hume, and many others. Epicurus put it well: "Vain is the word of a philosopher which does not heal any suffering of man. For just as there is no profit in medicine if it does not expel the diseases of the body, so there is no profit in philosophy either, if it does not expel

17. I have enjoyed the popular work of Alain de Botton [83], Julian Baggini [84], Jules Evans [85], and Massimo Pigliucci [86] amongst others.

18. The phrase '**linguistic turn**' was popularized by Richard Rorty in his book of the same name [87], while the phrase '**pragmatic turn**' is employed (once) in his later 'Philosophy and the Mirror of Nature' [88], and is used in the title of a number of books, including work by William Egginton and Mike Sandbothe [89] and Richard Bernstein [90]. For more on Dewey's contributions in the context of evolutionary and psychological theory, see W. Teed Rockwell [91], Jerome Popp [92], and Svend Brinkmann [93]. For more on the contrast between **Wundt and James**, see Richard Wiseman [94]. Kurt Danziger, who played an influential role at the University of Cape Town, has emphasized that Wundt's approach eschewed scientism and psychologism; he saw psychology as an empirical science that crossed the natural sciences and the humanities, and that employed both physical and psychic causality [95]. Indeed, Wundt had a sophisticated approach to the mind–body problem, spoke of 'physiological psychology', and used the term 'critical realism' [96]. For more on commonalities between **Aristotle and Dewey**, see Joseph Chambliss [63]; it is notable that each addressed the dualisms of their day [64, 97].

the suffering of the mind". And as Dewey wrote, "Philosophy recovers itself when it ceases to be a device for dealing with the problems of philosophers and becomes a method, cultivated by philosophers, for dealing with the problems of men".[19] Along these lines, there's an argument that American pragmatism has the potential to increase engagement between analytic and continental philosophy, with a clearer focus on addressing the big questions and hard problems.[20]

Second, Dewey emphasized how organisms operate in the world by employing their sensorimotor systems. Once again, Aristotle is prescient; in his 'De Anima' he pointed out that the animate and the inanimate differ in two regards, sensation and motion. Whereas the linguistic turn had led to a focus on 'mental representations', Dewey's focus on the organism's direct engagement with the world can be thought of as a turn to the body. This 'corporeal turn' was an important precedent for the emphasis in current cognitive-affective science on cognition as embodied (in sensorimotor systems) and embedded (in particular environments).[21] Descartes' 'cogito, ergo sum' (I think therefore I am) becomes 'ago, ergo cogito' (I act therefore I think). In attempting to resolve dualisms, Dewey coined the term 'body–mind' and he used the word 'psychophysiology' (rather than 'psychobiology'); indeed the **pragmatic realism** that he pioneered provides us with an important integrative approach.[22]

Third, while Dewey saw continuities across scientific and everyday knowledge, he also noted that natural sciences such as physics and human sciences

---

19. For more on the notion of **philosophy as therapy**, pointing towards a way of living, see Martha Nussbaum [98], Richard Sorabji [99], Clare Carlisle and Jonardon Ganeri [100], James Wiley [101], and Matthew Walker [102], as well as the later footnote on philosophical counselling. More recently, during World War II, Wittgenstein asked, "What is the use of studying philosophy if all that it does for you is to enable you to talk with some plausibility about some abstruse questions of logic, etc., and if it does not improve your thinking about the important questions of everyday life?" [103].
20. For more on **pragmatism as integrating** analytic and continental philosophy, see William Egginton and Mike Sandbothe [104]. Note also the view that both the linguistic turn and the pragmatic turn have emphasized the human origins of and constraints on knowledge, and given key interactions between philosophers in Cambridge, Massachusetts and in Cambridge, England, a range of the latter (e.g. Frank Ramsey, Wittgenstein, Georg von Wright, Elizabeth Anscombe, Bernard Williams, Simon Blackburn) can even be considered as representing 'Cambridge pragmatism' [105–107]. For Huw Price, an Australian-born philosopher and erstwhile Bertrand Russell Chair of Philosophy at Cambridge University, such work begins with Hume, and allows a 'naturalism without representationalism' [108]. See also later footnote on Price's 'subject naturalism'.
21. For more on Dewey's notion of the sensorimotor system, see his paper on the reflex arc concept in psychology [109]. In more recent times, Jean Piaget has emphasized how sensorimotor reflexes form the basis for later more abstract thought [110, 111]. For a contemporary update on the role of the sensorimotor system in conceptual knowledge, see Gallese and Lakoff [112]. For more on the **'corporeal turn'**, see volumes with this phrase in their title [113, 114].
22. The pithy Latin here is from Leslie Marsh [115]. For Dewey's use of the term 'body–mind', see his 'Experience and Nature' [116]; and for his use of 'psychophysiology', see his 'Soul and Body' [117]. For more on how Dewey's **pragmatic realist** view of knowledge as emerging from an interaction between organism and environment foreshadows much subsequent work on embodied cognition, see Mark Johnson [118]. For more on pragmatic realism, see Dewey [116], Hilary Putnam [119], Susan Haack [120], Willem DeVries [121], Kenneth Westphal [122], and H. Chang [123].

such as psychology studied different topics and required different approaches. Science involves human activity and is both theory-bound (i.e. observations are made using particular conceptual frameworks) and value-laden (in that particular activities are prioritized as more important). Given that the human scientist is also a subject of psychology, in the human sciences particular attention must be reflexively paid to how the theories and values of the scientist influence the science. Yet again Aristotle is prescient, remarking that "it is the mark of an educated man to look for precision in each class of things just so far as the nature of the subject admits; it is evidently equally foolish to accept probable reasoning from a mathematician and to demand from a rhetorician demonstrative proofs".[23]

Fourth, Dewey was a fallibilist, who saw philosophy and science as characterized by epistemic virtues, such as openness to criticism and toleration of alternative views. Socrates famously asserted, "I know one thing, that I know nothing", and Confucius echoed that, "Real knowledge is to know the extent of one's own ignorance". A view of philosophy and science as a sketch that is open to expansion and revision was put forward by Aristotle, and many have subsequently emphasized the importance of epistemic humility.[24] Hume is famous for his skepticism; he insightfully noted that "The greater part of mankind are naturally apt to be affirmative and dogmatical in their opinions; and while they see objects only on one side, and have no idea of any counter-poising argument, they throw themselves precipitately into the principles, to which they are inclined; nor have they any indulgence for those who entertain opposite sentiments". The term fallibilism was, however, coined by Charles Pierce, and its importance was emphasized by other early American pragmatists.[25]

---

23. For more on Dewey's view of the reflexive nature of psychology, see Brinkmann [124]. Aristotle's words are from his 'Nicomachean Ethics' [125]. For a discussion of how Aristotle was the first to focus on **'scientia'**, knowledge based on both science and humanities, see Massimo Pigliucci [86].

24. While the notion of **epistemic virtues** arguably begins with Aristotle's 'Nicomachean Ethics' [126], it has been expanded by many, including Dewey (as noted here), Thomas Kuhn [61], Linda Zagzebski [127], and Ernest Sosa [128]. For more on epistemic virtues and vices in general, as well as their application to science, see Heather Battaly [129], Quassim Cassam [130], and Naomi Oreskes [131]. For more on explanatory virtues in particular, see Kevin McCain and Ted Poston [132].

25. Hume's words on **epistemic humility** are from his 'An Enquiry Regarding Human Understanding' [133]. Nietzsche put it more succinctly; "Convictions are more dangerous enemies of truth than lies" [134], as did William Yeats (writing "The best lack all conviction, while the worst/ Are full of passionate intensity" [135]) and Bertrand Russell (writing "The fundamental cause of the trouble is that in the modern world the stupid are cocksure while the intelligent are full of doubt" [136]). For more on epistemic humility, see Richard Popkin [137], Leah Cohen [138], and David Cooper [55]. For more on the origins of the term **fallibilism**, see Joseph Margolis [139] and Ben Kotzee [140]. For more on **ignorance and science**, see Peter Medawar [141], Richard Feynman [142], Stuart Firestein [143], Mario Livio [144], and Rik Peels and Martijn Blaauw [145]. We'll return to the question of fallibility when addressing moral fallibility, and when discussing the demarcation problem.

In the century since Dewey developed his pragmatic realism, various other approaches to realism have emerged, offering an integrative alternative to the classical and critical positions. As a student at the University of Cape Town, I frequented a small local bookstore that sold a range of progressive books; its shelves stocked a wide range of critiques of apartheid, of capitalism, and of medicine and psychiatry. These critiques captured my imagination, but also made me query my choice of profession. On one of many visits to this bookstore, I come across the work of the British philosopher, Roy Bhaskar, the founder of **critical realism**. Writing at a time when the natural sciences had made considerable progress, but were also the subject of considerable criticism, Bhaskar queried whether the human sciences could be both powerful and critical. I found his answers enormously persuasive; he was enthusiastic about explanatory accounts in the natural and human sciences, as well as about the critical light that they shed on social structures and activities, including science itself. His work encouraged me to continue my medical studies, while at the same time being open to their criticism.[26]

Bhaskar's integrative approach attempts to go beyond both the classical and the critical positions, given that "For the positivist, science is outside society; for the hermeneuticist, society is outside science". He argues articulately that while both natural and human sciences are social activities, each also delineates the structures, processes, and mechanisms of the world, and that each requires different methodologies. For Bhaskar, such a view of the natural and human sciences avoids reductionism (the view that their subject matters are ultimately the same) and scientism (the view that they must all use the same method). Indeed it's remarkable how many resonances there are between Bhaskar and Dewey; Bhaskar's naturalism, his fallibilism, and his emphasis on philosophy as a means to change the world are all consistent with Dewey's thinking and writing. Importantly, Bhaskar also emphasizes that science advances by discovering real entities and real capacities rather than by describing laws. This view resonates with Aristotelian ideas about the properties and capacities of entities, and there has since been interest in philosophy of science in neo-Aristotelian approaches as well as a good deal of work on scientific realism.[27]

---

26. For more on Bhaskar's **critical realism**, see his work on the possibility of naturalism [29]. The term 'critical realism' has, however, been used in connection with a number of different authors and positions [96]. Just as scientism is an oversimplification [146–148], so is culturalism [149].

27. For more on Aristotle's prescience regarding **scientific realism** (a term encompassing practical, critical, and embodied realism), see the work of Nancy Cartwright [150], Ruth Groff and John Greco [151], Brian Ellis [152], and William Simpson, Robert Koons, and Nicolas Teh [153]. For a discussion of the contrast between Putnam's pragmatic realism and Bhaskar's critical realism, see Ruth Groff [154]. Many others have advanced scientific realism, arguably including Spinoza (who has been read as contributing to critical realism [155]) and Hume (his realism is the subject of ongoing study [156]). For a useful informal account of the history of scientific realism, including acknowledgement of contributions from the southern hemisphere (e.g. Jack Smart [157]), see Gironi and Psillos [158].

By the time of my psychiatric residency, I was more likely to frequent bookstores that specialized in contemporary cognitive-affective science. Advances in the cognitive-affective sciences have engaged a broad range of philosophers, but the work of George Lakoff and Mark Johnson has been particularly sophisticated and significant. These authors have emphasized work showing that we employ basic sensorimotor and abstract cognitive-affective maps (or metaphors) to make sense of our world and have rigorously explored the conceptual consequences of these findings. As mentioned earlier, this **embodied realism** can be seen as building on Dewey's foundational work on organisms' sensorimotor engagement with the world. Ongoing work on embodied and embedded cognition in the cognitive-affective sciences aims at 'putting brain, body, and world together'.[28]

Lakoff and Johnson's work has particularly important implications for our understanding of human constructs. A focus on embodied and embedded cognition, including metaphoric maps, suggests the impossibility of finding necessary and sufficient criteria for many of our categories, and supports the view that meaning is grounded in our day-to-day use of language. At the same time, on the basis of careful argument, employing reason and imagination, we can choose one metaphoric framework over another; thus this approach does not hold that 'anything goes'. Indeed, work that sharpens our knowledge, understanding, and use of embodied metaphors may help bridge the gap between the classical and the critical positions, avoiding the reductionism or scientism of the former, while at the same time seeing science as rigorous and progressive.[29]

Taken together, the work of Dewey and later pragmatic realism, of Bhaskar and subsequent critical realism, and of Lakoff, Johnson, and others on embodied realism provides the foundations for an integrative position in philosophy (see again Table 1, for a sketch). The **integrative realism** proposed here values both objectivity and subjectivity; it sees science as a social activity that involves human engagement with the physical and human worlds, with this engagement

28. For more on **embodied realism**, see George Lakoff and Mark Johnson [159]. Lakoff and Johnson argue that the cognitive-affective sciences do not support the underlying assumptions of analytic philosophy; human thought is not so much based on the use of precise propositions, but rather often involves the use of metaphoric language. Even our concepts of causation rely on metaphoric thinking, a point to which we will return later. This perspective is consistent with Dewey's pragmatic realist view and Bhaskar's critical realist view that while science is a social activity, it does discover real structures, processes, and mechanisms, hence the phrase 'embodied realism'. This specific phrase is from Andy Clark and David Chalmers [160, 161]. See also later footnote on embodied, embedded, enactive, and extended cognition.

29. The antirealist phrase '**anything goes**' is used by Paul Feyerabend [162, 163]. For a discussion of overlap between Bhaskar's critical realism and Lakoff and Johnson's embodied realism, see T. Murray [164].

leading to more useful or sharper metaphors, that provide us with greater knowledge and improved understanding of the nature of underlying structures, processes, and mechanisms. These strands of realism also align in underscoring the virtue of 'epistemic humility', emphasizing the fallibility of science; at any particular stage, science offers only a sketch that facilitates further detail or revision.

A number of objections might immediately be made at this point. First, these are quite different positions within realism that cannot simply be lumped together. Second, there are other potentially useful positions within realism that have not been explored here. Third, integration per se is not sufficient justification for moving in a particular direction; what are the specific strengths of a position that draws on these strands of realism?

I would counter that these positions align in key ways, incorporating key strengths of the classical tradition (they see science as rigorous and progressive) and of the critical tradition (they see science as theory-bound and value-laden), while avoiding key weaknesses (the reductionism or scientism of the classical view, and the relativism or skepticism of the critical view) (as outlined in Table 1). While this is a broad strokes alignment, it has the strength of drawing together different areas within philosophy (for now we have focused on philosophy of science and language, and later we will include additional areas such as ethics) as while as various complementary approaches within each of these areas.[30]

While a classical position tends to adopt a 'strict naturalism' with conceptual and explanatory monism (e.g. chemical entities in the Periodic Table can be grouped on the basis of their electronic configuration, and atomic number accounts for distinctions between entities), the integrative approach here aligns with a '**soft naturalism**' that welcomes **conceptual and explanatory pluralism**.[31] Soft naturalism sees reality as 'dappled' or 'promiscuous'—the philosopher Philip Kitcher therefore refers to 'pluralistic realism', while the physicist Sean Carroll refers to 'poetic naturalism' to emphasize the many ways

30. This sort of alignment may itself be considered an important epistemic virtue; see Susan Haack's work on the importance of **reintegrating psychiatry** [165].

31. Naturalism may refer to an ontological commitment (i.e. one that excludes 'supernatural' entities) or to a methodological commitment (i.e. one that emphasizes scientific explanations) [166, 167]. For more on **strict naturalism** (e.g. the work of Willard Quine and Wilfrid Sellars, which emphasizes the ontology and methodology of the physical or natural sciences) versus **soft or liberal naturalism** (e.g. the work of Hilary Putnam, which welcomes conceptual and explanatory pluralism), see Peter Strawson [168], Owen Flanagan [169], Mario De Caro and David Macarthur [170], and D. Christias [171]. Soft naturalism arguably dates back to Aristotle, and includes Hume [168, 172, 173]. Strict and soft naturalism also differ in their view of the relationship between philosophy and science, an issue to which we will return.

there are of talking about the world, some of which deserve to be called real.[32] Furthermore, soft naturalism is not only concerned with mechanism, but also with meaning; it is interested in 'erklären' (explanation, or the space of causes) as well as in 'verstehen' (understanding, or the space of reasons). The philosopher Huw Price uses the phrase 'subject naturalism' to emphasize philosophy's interest in what science tells us about ourselves and our concerns.[33]

Both integrative realism and soft naturalism, then, are intent on emphasizing the complexity and fuzziness of the world, avoiding reductionism and scientism, and bridging the erklären–verstehen divide.[34] The South African-born philosopher, John McDowell, for example, criticizes what he terms 'bald naturalism' or a 'naturalism of disenchanted nature' that ignores the space of reasons, and draws on Aristotle to propose a 'relaxed naturalism' that keeps nature

---

32. John Dupré has argued that there are many legitimate ways of dividing the worlds into kinds, consistent with his notion of 'promiscuous realism' [174], and Nancy Cartwright has emphasized that we live in a "dappled world, a world rich in different things" [175]. Philip Kitcher, the John Dewey Professor of Philosophy at Columbia University, has emphasized the prescience and importance of Dewey's contributions in particular, and has proposed a **'pluralistic realism'** [176]. Along similar lines, others influenced by pragmatism include Susan Haack, who reminds her readers of William James's phrase 'pluralistic Universe' [120], and Israel Scheffler, who refers to 'plurrealism' [177]. Sean Carroll's phrase is from his 'The Big Picture: On the Origins of Life, Meaning, and the Universe Itself' [178]. For more on **graded categories**, see Eleanor Rosch [179] and Lakoff and Johnson [159]. There is a related philosophical literature on constructs such as indeterminacy, vagueness, and essential vs fuzzy natural kinds (see later footnote). For more on **indeterminacy and vagueness**, see Bertrand Russell [180], Timothy Williamson [181], Rosanna Keefe and Peter Smith [182], Roy Sorenson [183], Jose Ciprut [184], Israel Scheffler [185], and Richard Dietz [186].

33. The contrast between **'erklären'** and **'verstehen'** emerged during 19th century debates in philosophy, sociology, and other fields, about whether the natural sciences and the social sciences required different methods. Positivists such as Auguste Comte and John Stuart Mill argued that the natural and social sciences both discovered laws, while antipositivists such as Max Weber and Wilhelm Dilthey argued that natural and social sciences required different methods. A range of influential constructs emerged from these debates, for example, Wilhelm Wandelbind's contrast between the nomothetic and ideographic, which in turn drew on Immanuel Kant's division between the tendency to generalize and the tendency to specify, and which was subsequently used by Gordon Allport, Michael Schwartz and Osborne Wiggins, and others. More recent authors (e.g. Wilfrid Sellars, John McDowell, Robert Brandom, also known as the Pittsburgh School), working on what Sellars called 'Kantian Themes', and rejecting what he termed 'the myth of the given', have referred to the space of causes versus a norm-governed space of reasons [187]. For more on 'subject naturalism', see the work of Huw Price [188]. In similar vein, David Macarthur argues that "we do better to humanize nature than to naturalize the human" [189].

34. Pragmatic, critical, and embodied realism can all be read as attempting to **bridge the erklären–verstehen divide**. See, for example, the work of David Leary on William James [190], and the work of Andrew Sayer which argues that for critical realists, the human sciences are "neither nomothetic (that is, law-seeking) nor idiographic (concerned with documenting the unique)" [191]. Wilfred Sellars, who can be considered a pragmatic realist [192], has attempted to reconcile what he terms the 'scientific image' and the 'manifest image' (through which humans encounter themselves) [193]. See also the work Stellenbosch University philosopher Anton van Niekerk [194], as well as writing by Owen Flanagan [195] and Dominic Murphy [196] that draws on Sellars's distinction. See also later footnotes on explanatory pluralism and on 'explanation-aided understanding'.

'partially enchanted'.[35] A classical position may focus on essential natural kinds (e.g. chemical elements), while a critical position may emphasize constructed conventional kinds (e.g. edible species). Consistent with conceptual and explanatory pluralism, an integrative position emphasizes that the natural and human sciences are interested in multiple kinds of kinds, and in the structures, processes, and mechanisms that underlie reality.[36]

Consider, for example, the notion of 'invasive species'. There are multiple different kinds of invasive species and they can be invasive in many different ways. To fully understand the construct of invasive species, it may be useful to employ a range of concepts and explanations, ranging from those focused on molecular mechanisms through to those focused on societal factors. Scientific work on such species carefully considers both general principles (e.g. about the evolution of plants) as well as relevant particulars (e.g. about a species in a specific place and time). Invasive species are neither essential natural kinds, nor entirely conventional kinds; rather we can conceptualize them as fuzzy natural kinds.[37]

Consider another key issue in philosophy, the fact-value dichotomy. It has been argued by some that one cannot move from is to ought, from descriptions (about what is) to prescriptions (about what ought to be). The logical positivists of Vienna set a firm division between positive statements of fact based on observation and reason, and normative statements of value. But when it comes to everyday concepts, such as our notions of well-being and disease, facts and values seem to be intertwined; they are 'fraught with ought'.[38] An exploration

---

35. See McDowell's volume on 'Mind and World' [197] and previous footnote on the Pittsburgh school. For more on **pragmatic naturalism**, see the work of M. Williams [198], Peter Manicas [199], Phillip Kitcher [176], P.D. Magnus [200], M. De Caro [201], and Paul Giladi [202]. See also S. Morris Eames [203], who covers the writings of Pierce, James, Dewey, and George Herbert Mead.

36. The term '**natural kind**' arguably has roots in the work of John Stuart Mill, and was revived by his godson, Bertrand Russell [204, 205]. Note that not all would agree that chemical entities are natural kinds, see Paul Churchland [206]. An integrative realism may emphasize that there are no bright line distinctions between different kinds of kinds; see P.D. Magnus [200]. For more on natural kinds in general, see the work of Nigel Sabbarton-Leary and Helen Beebee [207], Joseph Campbell, Michael O'Rourke, and Matthew Slater [208], P.D. Magnus [200], and Catherine Kendig [209]. For more on natural kinds in psychology, including a discussion of Ian Hacking's concept of 'human kinds', see Svend Brinkmann [210]. See later footnotes on essentialism, as well as on whether emotions, biological categories, and mental disorders are essential natural kinds.

37. The importance of considering both scientific generalizations and particulars is emphasized by Susan Haack [120], who in turn draws on Mary Hesse [211]; a position that we might term **scientific particularism**. Later on we will discuss moral particularism. The notion of fuzzy natural kinds is consistent with views that natural kinds can vary along continua (see Rachel Cooper [212]), and that kinds may combine naturalistic and conventional features in complex ways (see Richard Boyd [213]). Note that there is a large literature on the philosophy of species, and on monist vs pluralist views of species [214, 215].

38. The phrase '**fraught with ought**' comes from Wilfrid Sellars [216], and indeed Sellars's work has been described as naturalism with a normative turn [217]. For another early argument along these lines, see John Searle [218]. In philosophy of language, there is ongoing debate between attempts to define representation with a naturalist framework (i.e. **naturalism**), and the view that all meaning is normative (i.e. **normativism**), a view that can be traced to work by Wittgenstein and Saul Kripke [219]. Hilary Putnam, an important contributor to this debate, refers to the 'horror of the normative' that is apparent in (strict) naturalism [220].

of metaphors of disease helps us to understand both naturalistic and normative aspects of disease, another point to which we will return. If the fact-value distinction doesn't entirely hold, then it may well be the case that scientific advances inform value judgements, and that scientific activity reflects specific values.[39] Indeed, over the course of this volume we'll argue that like other human practices, science entails a range of epistemic virtues as well as practical judgement or wisdom (what Aristotle termed 'phronesis').[40]

Lakoff and Johnson's contribution to an integrative position draws closely on the cognitive-affective sciences, one component of which is neuroscience. Drawing on neuroscience in particular as a resource for philosophy is a key idea in two newly emergent fields relevant to this volume: neurophilosophy and neuroethics. Neurophilosophy has argued that just as philosophical questions about metaphysics have been resolved by sciences such as cosmology, so philosophical questions about the mind will increasingly be resolved by the neurosciences. Neuroethics is concerned with both the ethics of neuroscience research and practice, as well as the neuroscience of ethical decision-making; it too therefore highlights the importance of using neuroscience to understand philosophy. Psychiatry, also, has become increasingly focused on neuroscience, and we'll turn to this field next.[41]

---

39. Hume is often cited as a key source for the **fact-value dichotomy**, but there is considerable controversy about this point [221]. For some logical positivists, just as metaphysical doctrines are typically meaningless, so ethical statements merely reflect feelings; see, for example, see Ayer's influential emotivist position [222]. Nevertheless, there are other positions within logical positivism; Moritz Schlick, the leader of the Vienna Circle, for example, regarded ethics as a science focused on values, viewed moral behaviour as having its origin in pleasure and pain, and concluded that man is noble because he enjoys moral behaviour [223], and his student, Herbert Feigl, proposed a scientific humanism [224]. Related to the fact-value distinction is the argument of moral nonnaturalism that the description of the natural properties of X cannot be used to reach an evaluation of X as good (this is the **'naturalistic fallacy'** or 'open question argument' of George Moore, who worked alongside Russell and Wittgenstein at Cambridge University, and who was another key architect of analytic philosophy, particularly its approach to ethics). For pragmatic realism, critical realism, and embodied realism, facts and values are entangled [29, 159, 225, 226]. For recent discussions of the fact-value dichotomy, see Harold Kincaid, John Dupré, and Alison Wylie [227], Louis Charland and Peter Zachar [228], Svend Brinkmann [210], P. Gorski [229], S. Elqayam [230], Anna Alexandrova [231], and Mark Okrent [232]. We'll return to the fact-value distinction when discussing the nature of disease, as well as to questions of the relationship of between science and philosophy (including ethics) and of the values of science.

40. For more on the relationship between **scientific activity and phronesis**, see Bent Flyvbjerg, Todd Landman, and Sanford Schram [233], and Elizabeth Kinesella and Allan Pitman [234]. See also later footnotes on medicine as a craft, epistemic virtues in medicine, and medical phronesis.

41. Paul and Patricia Churchland have made particularly important contributions to **neurophilosophy** [235–238]. The idea that **neuroethics** can be subdivided into these two areas was put forward by Adina Roskies [239].

1.2 Perspectives of psychiatry

With philosophers in the Anglo-Saxon analytic tradition often unable to provide much in the way of advice to those looking for answers to the big questions of life, other professionals and practitioners have filled the gap during the last century. Theological answers to the big questions have played an immensely influential role during the course of history, and in the health arena a range of indigenous healers, including the sangomas and inyangas of South Africa, have long provided advice about how to cope with life. With the growing influence of modern science and medicine, and a wish for naturalistic answers and advice, the stage was set in the 20th century for psychiatry to also get into the business of addressing the big questions, albeit by focusing on what it termed 'problems in living'.[42]

As a psychiatrist, I'm undoubtedly a bit biased, but there are perhaps one or two good reasons for psychiatry to be involved in answering the big questions. First, as a branch of medicine, psychiatry is based on science, including cognitive-affective sciences such as neuroscience, where there have been extraordinary advances. Second, clinicians are experienced in addressing the suffering of their patients, and in helping to provide practical real-world solutions to the problems of living; in that sense medicine is a craft. That said, what was perhaps most important in propelling psychiatry into the arena of big questions was the emergence of some truly influential figures whose work captured the world's imagination. Of these, the Viennese physician Sigmund Freud was the most important.[43]

Shifting paradigms in psychiatry[44]

Freud was deeply concerned with the big problems, including "problems of self-knowledge, or rather of lack of self-knowledge, of the nature of desire,

---

42. The phrase **'problems in living'** seems to have been coined by Herbert Stack Sullivan, and popularized by Thomas Szasz, but it has been used by a range of other authors [240].
43. The conceptualization of medicine in terms of science is to some extent a new one; the ancient Greeks would have classified **medicine as a craft** or techne [241]. There is also an interesting literature on whether medicine is more of a science or an art; William Osler viewed **medicine as an art informed by science**, and so advocated that physicians study both biology and the humanities (see S. Nassir Ghaemi who argues, "For medicine, in Osler's view, was an art based on a science; not just an art, and not just a science. Without science, medicine would be empty; without art, it would be irrelevant" [242]). We'll return to these sorts of issue when we discuss in more detail how judgements of disease involve both facts and values, how medicine entails practical judgement or wisdom, and the relationship between the sciences and humanities.
44. The use of this term does not mean an endorsement of Kuhn's views (see earlier footnote) or of related frameworks (such as Michael Foucault's 'episteme').

and of the relationship of both to our actions".[45] He put forward a novel theoretical framework that provided a sophisticated description of the brain–mind, that accounted for the emergence of brain–mind disorders, and that provided a solution for patients—psychoanalysis. While some of his ideas remain influential, Freud's reputation has taken rather a battering during the 'Freud Wars'. There is growing evidence, for example, that some of his case reports were entirely concocted, that the success of others was vastly exaggerated, and that some of his singular focus on sexuality was due to cocaine addiction.[46]

Perhaps the most crucial criticism of psychoanalysis is that it simply is not a science. We'll discuss details of the psychoanalytic model, and whether it is scientific, in more detail later on. However, for now we can say that the weight of the evidence suggests that psychoanalysis is rather different from the sciences: its adherents seem to give more weight to authority than to evidence, many of its constructs remain influential but poorly validated, and the overall evidence base for psychoanalytic treatments is relatively thin. Taken together,

---

45. This citation is from the philosopher Alasdair MacIntyre [243]. A large body of literature at the intersection of **psychoanalysis and philosophy** has developed [244, 245]. Contributors to this literature have taken a range of stances, including those focused on how Freud integrated the sciences of his time (e.g. Frank Sulloway, David Livingstone Smith, George Makari), those who have seen psychoanalysis as a hermeneutic practice (e.g. Hans-Georg Gadamer, George Klein, Paul Ricoeur, and Roy Schafer), and those have put forward the related views that psychoanalysis is based on problematic evidence (e.g. Adolf Grünbaum), is a pseudoscience (e.g. Karl Popper, Frank Cioffi, Hans Eysenck), or is a mythology (e.g. Ludwig Wittgenstein, Ernest Gellner, Karl Jaspers, Roger Scruton). A range of other positions has been put forward, including that psychoanalysis can be considered an extension of folk psychology (e.g. Thomas Nagel, Richard Wollheim, Marcia Cavell), that psychoanalysis is important for moral inquiry (e.g. Alasdair MacIntyre, Alfred Tauber), that psychoanalysis is key for a critique of contemporary society (e.g. Herbert Marcuse, Norman Brown, Eric Fromm), or that psychoanalysis is a precursor of cognitive science (e.g. Matthew Erdelyi, Patricia Kitcher, Jerome Wakefield). There is the view that only a compartmentalized model of the mind can explain the irrational (e.g. Donald Davidson), and a view that psychoanalysis provides a window into understanding modernity (e.g. Philip Rieff). Note also the argument that several of Freud's ideas were preceded by a range of philosophers including Schopenhauer and Nietzsche [246, 247]. Many of these viewpoints have been put forward in book length contributions, but some key contributions are briefer (e.g. see Donald Davidson's 'Paradoxes of Irrationality' [248]). Wittgenstein's ambivalence towards psychoanalysis is of particular interest: on the one hand he admired it for developing a way of speaking that was potentially able to clarify psychological issues, but on the other hand he criticized psychoanalysis as unscientific [249]. Jasper's critique of psychoanalysis is also noteworthy; his views changed over time, and included a range of concerns, including what he saw as totalitarian characteristics of the psychoanalytic movement [250].

46. Frederick Crews provides a particularly devastating critique [251]. For more on the **Freud Wars**, see John Forrester [252], Stephen Wilson and Oscar Zarate [253], Todd Dufresne [254], and Mikkel Borch-Jacobsen and Sonu Shamdasani [255]. Paul Roazen writes that "the history of psychoanalysis consists of a continuous conversation with Fraud" [256]; although this sentence has a typographic error (Roazen is incorrectly citing Joseph Schwartz [257]), it may well be a prescient and perspicacious assertion.

although psychoanalysis has provided a number of productive concepts for clinicians to explore, it is not altogether surprising that critics have insisted that it is a pseudoscience, and that it has lost some of its power and panache when it comes to addressing the big questions.

Psychoanalysis was a predominant influence in American psychiatry for many years. During the second half of the 20th century, however, there was a seismic shift towards a more biological psychiatry. This was due in part to the introduction of a range of effective medications and new neuroscientific methods. Whereas chairs of American Departments of Psychiatry were once mostly psychoanalysts, many now are clinician–scientists with strong research programmes in neuroscience. The National Institute of Mental Health (NIMH), the world's largest funder of research in psychiatric science, has increasingly conceptualized mental disorders as brain disorders.[47]

My own training in psychiatry took place at a time when this shift was well underway, with some psychiatric residencies very much weighted towards biological psychiatry. I applied to residencies that offered training in both psychoanalytic and biological psychiatry and was delighted to be accepted by Columbia University. It turned out that my treatment of any particular patient during the residency was significantly influenced by who my supervisor was. Thus, for example, as part of my psychoanalytic training, I did psychoanalytically informed psychotherapy with a young man with obsessive–compulsive disorder (OCD) three times a week for several years. Each week I met with my supervising psychoanalyst, and we reviewed my verbatim notes of the sessions. While I learned a great deal about psychoanalytic theory, my patient's symptoms didn't shift at all. This experience was one of several during my residency that did not inspire particular confidence in psychoanalytic approaches.

Although supervision in psychotherapy and pharmacotherapy treatment took place in the same Columbia University building, they were often quite different experiences. Roger MacKinnon was our main psychoanalytic teacher: he was a tall and imposing man who had been in the navy, and who exuded authority. In the pocket of his starched white shirt was a blue pen for recording sessions, and a red pen for recording patients' dreams; exemplifying his (and, of course, Freud's) view of dreams as the royal road to the unconscious, and ensuring that supervision of my OCD patient focused a great deal on dream material. Discussions with Dr. MacKinnon would often involve how best to really understand patients; he had a remarkable knack for getting to the heart of a person's issues and coming up with patient-centred solutions. To the residents he would

47. For a review of the **transitions in American psychiatry** during the 20th century, see the work of M. Wilson [258]. The split between psychoanalytic and biological psychiatry is detailed by Tanya Luhrmann [259]. Also interesting is the work of South African-born Norman Rosenthal [260]; this describes his own time at Columbia University, during the transition from psychoanalytic to biological psychiatry.

freely dispense with excellent advice on how to be a good psychiatrist, and how to tackle life more generally.[48]

In contrast, Donald Klein, a psychopharmacologist, was director of psychiatric research at Columbia, and rumour had it that he had dropped out of psychoanalysis, a sign of tremendous antiestablishment rebellion at that time. My first interaction with him revolved around his decision to put one of my patients on quite a dangerous combination of medication: he seemed rather a risk-taker. Later interactions involved him suggesting new statistical analyses of data (he was extraordinarily expert in this area), and new research to test one or other creative but integrative scientific idea (e.g. his false suffocation theory of panic disorder).[49] In some ways, MacKinnon represents the best of a psychiatric tradition that is focused on understanding ('verstehen'), while Klein represents the best of a contrasting tradition focused on explanation ('erklären') (Table 2). That said, given the extraordinary skill with which both MacKinnon and Klein interviewed patients, it seems to me that each was using an integrative framework that brought together mechanism and meaning.

Klein inspired a generation of mentees who were passionate about the new biological psychiatry, and committed to advancing psychopharmacology in particular. Towards the end of my residency I heard an inspiring talk from one of them, Eric Hollander, about advances in the neurobiology of OCD and about how medication worked in this disorder. NIMH psychiatrists Thomas Insel and Judith Rapoport had contributed to work showing that a particular neurotransmitter system—the serotonin system—played a key role in OCD. A serotonergic agent that revved up the system made symptoms worse, and medications that blocked the serotonin system made symptoms better. I was hooked and began a postdoctoral fellowship in psychopharmacology under Eric's thoughtful guidance.

An emphasis on the biological aspects of psychiatry has had a number of important successes: First, knowledge of the biology of mental disorders has helped dispel potentially harmful and stigmatizing ideas about their aetiology. For example, the idea that mothers are 'schizophrenogenic'—that abnormal mother–child interactions are the cause of psychosis—makes little sense in this framework, which essentially blames the victim. Second, knowledge of the psychobiology of mental disorders encouraged research on a range of psychiatric medications. Evidence-based pharmacotherapy has helped many individuals and paved the way for deinstitutionalization of psychiatric asylums: with

---

48. **Roger MacKinnon's** textbook, recently co-edited with Robert Michels and Peter Buckley, on 'The Psychiatric Interview in Clinical Practice' [261] remains a go-to source of advice for me when thinking about difficult clinical issues.

49. **Donald Klein** posited that we have a suffocation alarm that is triggered by low levels of carbon dioxide (CO_2), and that in panic disorder this alarm has an abnormally low threshold [262]. Although this hypothesis lies at the borderline of academic respectability, Klein marshalled an extraordinary amount of evidence for it, and defended it with considerable skill [262].

TABLE 2 Classical, critical, and integrative positions on psychiatric classification and psychiatric science.

	Classical position	Critical position	Integrative position
Historical roots	Kraepelin, Freud	Sartre, Foucault	Wundt, Jaspers[a]
Psychiatric practice	Focus should be on improving our knowledge of disorders (erklären)	Focus should be on improving our understanding of people (verstehen)	Both mechanism and meaning are key; science is theory-bound and value-laden[b]
Psychiatric science	Psychiatry is a science with laws (of psychoanalysis, of neurobiology, of learning) that allow generalization	A focus on brain–mind laws leads to simplistic models, with reductionism and scientism, ignoring contexts	Structures, processes, and mechanisms underlying mental illness can be discovered
Psychiatric classification	Classifications are science based: operationalize disorders using necessary and sufficient criteria	Classifications are socially constructed: they differ from time to time and place to place	Classification is a scientific and a social process: our classifications can improve iteratively over time[c]
Psychiatric epidemiology	Operationalization has demonstrated that psychiatric disorders are prevalent	Operationalization is a flawed approach, surveys enable medicalization	Psychobiological structures underlie mental disorders, thresholds reflect facts and values

[a]***Emil Kraepelin*** *was a German psychiatrist who made an important contribution by carefully describing and differentiating mental disorders; for example, his distinction between schizophrenia and bipolar disorder remains an important one. Freud saw himself as a scientist discovering the laws of the unconscious; some members of the Vienna Circle were in psychoanalysis, and Freud was a member of the Society for Positivist Philosophy [246].* ***Jean-Paul Sartre*** *criticized psychoanalysis, arguing that drive theory was contradictory, and that an existential approach based on human freedom was needed. Michel Foucault saw psychiatric practice as a means of controlling deviance and exerting power. See previous footnote on Wundt's integrative position.* ***Karl Jaspers*** *is a pioneering figure in psychiatry; he made a number of important contributions, including emphasizing that good psychiatric practice entails aspects of both erklären and verstehen [263, 264].*

[b]*A view of* ***science as 'theory-bound'*** *has various roots, including the work of Dewey [265], Norwood Hansen [266], and Thomas Kuhn [61]. A view of* ***science as 'value-laden'*** *emerges from an emphasis on science as a particular kind of social activity, and from the dissolution of the fact-value distinction (see earlier footnote). Table 2 is focused on philosophy of psychiatry, but there are overlaps with work on the philosophy of psychology and cognitive science. The 'interpretive turn' in philosophy of psychology, for example, resonates with a focus on verstehen [267–269]. The work of Barbara Held resonates with the approach taken here, insofar as she is focused on 'middle ground theory' that goes beyond erklären–verstehen and related divides [269].*

[c]***Carl Hempel****, a philosopher who was key in moving 'logical positivism' towards 'logical empiricism', gave an address at the American Psychopathological Association which argued the relevance of operationalizing mental disorders. However, it is not clear that this had any influence [270].*

appropriate interventions, many people do not require long-term hospitalization and can be successfully treated in their communities.[50]

Biological psychiatry has provided us with truly lovely images of the brain and with incredibly detailed knowledge of how genes contribute to mental illness. While this sort of knowledge may well be a useful resource for answering the big questions, it also has key limitations. First, despite tremendous advances, our actual understanding of how the brain–mind works, and how it goes awry, is terribly limited. Current biological theories of mental disorder entail a good deal of neuromythology.[51] Second, and relatedly, biological explanations always run the risk of reductionism and scientism. When addressing the big questions we wish to avoid the twin dangers of being mindless (a risk with biological psychiatry) and brainless (a risk with psychoanalysis).[52]

At the same time as the shift towards biological psychiatry occurred, and contributing to the loss of confidence in psychoanalysis in mainstream psychiatry, cognitive-behavioural psychotherapies emerged. Behavioural therapy has roots in the work of the American psychologists John Watson and Burrhus Skinner, and the Russian physiologist Ivan Pavlov. A key clinical innovator, though, was the South African Joseph Wolpe, who completed his medical studies at the University of the Witwatersrand during World War II. Like many graduates he volunteered to assist the Allies, joining the Cape Corps as a medical officer. My own father, born just 2 years before Wolpe, on completing his accounting degree, served as an infantryman in Egypt and Italy, where he experienced huge distress. Wolpe was assigned to work locally with similar soldiers, diagnosed with PTSD. This condition was conceptualized in psychoanalytic terms as 'war neurosis', and narcotherapy was used to help express the 'repressed' trauma. Unfortunately this didn't work well, and Wolpe begin to explore other ways of thinking about anxiety and related disorders.

---

50. The term **'schizophrenogenic'** was coined by the psychiatrist Frieda Fromm-Reichmann. Gregory Bateson was one of those who played a key role in developing the maternal causation theory of schizophrenia, arguing that mothers placed their children in a 'double-bind'. With deinstitutionalization, families of patients with schizophrenia were increasingly responsible for their care: they teamed up with biological psychiatrists to argue that schizophrenia is a brain disease. This is not to exclude the role of a range of factors involved in this condition; there is evidence that early adversity does increase risk for schizophrenia, and the idea that interventions to decrease expressed emotion in families may be helpful has led to productive research. It is also not to support deinstitutionalization in the absence of good community care—a terrible idea.

51. Karl Jaspers early on used the phrase 'brain mythology' [47]. Early uses of the term **'neuromythology'** are found in William Landau [271] and Raymond Tallis [272].

52. See earlier footnote for Roy Bhaskar's argument that it is possible for a naturalistic perspective to hold that both natural and social sciences are sciences, without giving way to **reductionism** (the view that their subject matters are ultimately the same) or **scientism** (the view that they must all use the same method) [29]. For more on different kinds of reductionism see, for example, Ingo Brigandt and Alan Love [273]. With regards to psychiatry, this differentiation between brainlessness and mindlessness is from Z. Lipowski [274]. For more on scientism, see earlier footnote on scientism and culturalism. For more on **neuroreductionism** and **neuroscientism**, see Dai Rees and Steven Rose [275], D.S. Weisberg and colleagues [276], Sally Satel [277], and Raymond Tallis [278].

Wolpe was referred by James Taylor, a University of Cape Town (UCT) psychologist, to the work of Clark Hull on 'Principles of Behavior'. This impacted him deeply and he decided to embark on a dissertation on feline anxiety. He induced fear of neutral stimuli in cats (by pairing these stimuli with a shock) and then desensitized them (by pairing these stimuli with food). Crucially, however, unlike many experimental psychologists and physiologists working in the behavioural tradition, he then took this work into the clinic. He taught patients with phobias relaxation techniques and then encouraged them to confront their fears, building up gradually along a hierarchy from smaller fears to larger ones. This process of desensitization turned out to be remarkably effective.

Wolpe's work falls into a number of proud African traditions. The South African poet, Eugene Marais, was the first to carefully observe nonhuman primates in the field. His work would not be considered scientific by today's standards and was published only after his death, but as Robert Ardrey pointed out in his foreword to 'The Soul of the Ape', Marais' observations on baboons laid out the field of primate studies long before it became a professional field.[53] Marais' work on termites was plagiarized by the Nobelist Maurice Maeterlinck, and he later suicided. The best known exemplar of a different, clinician–scientist, tradition in Africa is Christiaan Barnard, the University of Cape Town and Groote Schuur Hospital heart surgeon, who honed his heart transplantation skills in baboons, before completing the world's first human heart transplant.[54]

Wolpe described this desensitization process in his 1958 volume on 'Psychotherapy by Reciprocal Inhibition'.[55] His volume was severely criticized

53. This publication was roundly criticized by South African-born Solomon Zuckerman, who had emigrated to the United Kingdom, and was an early leader in primatology studies, undertaken at Regent's Park Zoo. Alan Morris provides the fascinating context to this rather unbecoming attack [279]. Interestingly, Zuckerman was also a critic of Australian-born **Raymond Dart's** pioneering work in South Africa on Australopithecus, a crucial step in demonstrating the origins of humankind in Africa [280]. Dart was interested in a range of areas, including the neuroscience of the neocortex as well as the origins of human aggression, which in turn influenced Robert Ardrey [281]. The notion that *Homo sapiens* emerged only recently in Africa, before dispersing to other continents (the 'Out of Africa' hypothesis), has since been supported by behavioural evidence (e.g. from Christopher Henshilwood's findings in the Blombos cave of the Western Cape), and confirmed by molecular evidence (e.g. New Zealand-born Allan Wilson's work on the molecular clock and on 'Mitochondrial Eve') [282].

54. Although narcotherapy might have failed for 'war neurosis', it turns out that there a good deal of current interest in using **pharmacotherapy to enhance cognitive-behavioural** interventions. Early on, Meldmand and Hatch treated a patient with aeroplane phobia using a combination of pharmacotherapy and desensitization [283]. More recently, Kerry Ressler and Barbara Rothbaum hypothesized on the basis of laboratory findings about the molecular basis of learning, that a medication targeting a specific neurotransmitter system (the glutamatergic system) would augment behavioural therapy: an elegant and pioneering step forwards for translational neuroscience (neuroscience that moves from the bench to the bedside).

55. Wolpe worked on this manuscript while a Fellow at the Center for Advanced Studies in Behavioral Sciences in Stanford. Other Fellows at the time included Karl Popper and **Ernest Hilgard**, who not only coined the term 'classical conditioning', but who later on did pioneering work on a non-Freudian approach to the unconscious [284].

by the psychiatric community; their psychoanalytic model predicted that unless patients resolved their underlying conflicts, they would relapse, substituting one symptom for another. However, Wolpe subjected his ideas to rigorous research. Early on, his doctoral student Arnold Lazarus compared the impact of psychoanalytic interpretation with systematic desensitization on a variety of self-report and direct observation measures, and found the latter more efficacious. In this early trial, experimenter bias may have been a confounding factor, but behavioural therapy (a term that appeared for the first time in print in a case report by Lazarus in the South African Medical Journal),[56] went on to be studied in dozens of rigorous trials, becoming the first evidence-based psychotherapy.[57]

Like any good clinician–scientist, Wolpe carefully acknowledged others' work. Jules Masserman had done partly similar work with cats (but had described his findings in terms of 'neurosis' and overcoming 'conflict'); Pavlov had earlier described reciprocal inhibition as counter-conditioning; and Mary Jones, a student of Watson, had predated his human work on desensitization.[58] Watson himself had suggested that fear could be conditioned in infants. First, he showed that a research subject, known by the moniker Little Albert, was not afraid of a laboratory rat, but that he cried when Watson made a loud noise. Then, by showing Little Albert a rat and making a loud noise, the infant seemed scared of the rat, as well as of a number of other objects. A few years later, Mary Jones worked with a toddler, Peter, who was afraid of rabbits; showing that she could diminish his fear by exposing him to the rabbit, while at the same time calming him with candy. This procedure is similar to a variant of desensitization termed in vivo desensitization.[59]

Taken together, Wolpe's approach to scientific inquiry differed from that of key figures in the psychoanalytic movement in a number of crucial

---

56. See the work of R. Poppen and A. Lazarus [285] for more about Wolpe and the early history of **behavioural therapy**, a term also used early on by Burrhus Skinner and Hans Eysenck.

57. The idea of **evidence-based medicine** has become increasingly important. However, there have been criticisms of its application to psychiatry and psychology. We'll return to this point in the next chapter.

58. **Jules Masserman** was a psychiatrist (he studied under Adolf Meyer) and psychoanalyst who worked with animals, but who was critical of behaviourism. In a fascinating article, Alison Winter argues that although Masserman achieved significant professional success, his attempt to bend psychoanalytic constructs to animal behaviour was unconvincing [286]. **John Bowlby**, another psychiatrist-psychoanalyst, had a more sophisticated understanding of animal behaviour; his attachment theory drew on and contributed to ethology [287]. Watson (a colleague of Adolf Meyer) was well aware of Freud's work, including his Oedipal-focused conceptualization of 'Little Hans', who had a phobia of horses. He and Rayner wrote, "The Freudians twenty years from now, unless their hypotheses change, when they come to analyze Albert's fear of a seal skin coat – assuming that he comes to analysis at that age – will probably tease from him the recital of a dream which upon their analysis will show that Albert at three years of age attempted to play with the pubic hair of the mother and was scolded violently for it" [288]. For John Bowlby's conceptualization of Little Hans, see the work of J. Wakefield [289].

59. A detailed examination of the original reports of both **little Albert and Peter** suggests that some features of the work have been 'cherry-picked' or altered; Watson did not create an enduring phobia to rat-like objects [290, 291] and Jones's approach involved interventions other than mere desensitization [292].

respects. Whereas Freud emphasized the originality of psychoanalysis, was confident in his own clinical findings, and dismissive of laboratory research,[60] Wolpe saw himself as contributing to an empirical tradition, where rigorous testing was required, so that more comprehensive laws and more efficacious interventions could be found.[61] The Association for the Advancement of Behavioral Therapy that he was instrumental in founding continues to focus on the importance of rigorous research to this day (as the Association for the Advancement of Behavioral and Cognitive Therapies).[62] There is a now a massive evidence base for cognitive-behavioural psychotherapies, and they are receiving ongoing study across the world in the growing discipline of global mental health.[63]

Wolpe framed his behavioural therapy in terms of the laws of learning (Table 2), but more recent cognitive-behavioural therapy (CBT) employs cognitive models of the mind, influenced in good part by the advent of the computer, and the so-called cognitive revolution.[64] We'll discuss these models in more detail in the next chapter, where we focus on how best to conceptualize the brain–mind. For now we can simply note that Wolpe and members of his Association were primarily interested in treating psychiatric disorders: answering the big questions was not in their immediate sights. Indeed, one of the criticisms of the behavioural tradition has been that it fosters reductionism and scientism, that its view of humans is 'thin' rather than 'thick'.[65]

---

60. Hans Eysenck cites **Freud's response to Rosenzweig**, who had attempted to study repression experimentally: "I cannot put much value on these confirmations because the wealth of reliable observations on which these assertions rest make them independent of experimental verification" [293]. The idea that Copernicus demonstrated the sun not the earth was at the centre of the universe, Darwin proved that there was nothing particularly special about the descent of man, and Freud found that the unconscious rather than the ego ruled behaviour is a pithy one, but its appeal is diminished by Freud himself having put it forward [255, 294]. In fact, the idea of the unconscious has a long history [246, 247].

61. It is interesting to consider **Joseph Wolpe**'s view of the relationship of behavioural therapy to other sciences. His earliest thinking drew on neuroanatomy (noting the existence of reciprocal neural pathways), and some of his first writings reflected on the neural basis of conditioning (suggesting networks of inhibitory and excitatory synapses) [284] and he would therefore presumably be heartened by ongoing advances in understanding the neurobiology of fear conditioning and extinction, and the translational of such findings to the clinic. However, he was not enthused by cognitive-behavioural approaches, and he and Lazarus fell out over this issue [295, 296]. Similarly, in a debate about the relationship of behaviourism and humanism, it seems that Wolpe saw behaviourism as a comprehensive framework that did not need to integrate with other perspectives [297].

62. The spread of a scientific theory requires a good deal of social activity. **South African students and colleagues of Wolpe**, including Arnold Lazarus, Stanley Rachman, and Isaac Marks, played influential roles in moving behavioural therapy forwards.

63. See my volume with Judith Bass and Stefan Hofmann [298].

64. See previous footnote for histories of cognitive science.

65. The contrast between **'thin' vs 'thick'** descriptions was introduced by the analytic philosopher Gilbert Ryle to differentiate descriptions of behaviour that were superficial versus those that fully addressed context. This contrast was later extended by the anthropologist Clifford Geertz, to explain his approach to ethnography. The notion was later also used by Bernard Williams for terms that entail both descriptive and evaluative elements, and so is relevant to the fact-value distinction [299]. As we've already noted, and later will return to, Aristotle sees science and ethics in terms of a sketch that moves from thin to thick constructs (in his 'Nicomachean Ethics', cited by Martha Nussbaum [300]).

We should be careful not to stereotype psychoanalysts as intellectuals who develop deep relationships with patients as they together resolve the problems of living, in contrast to cognitive-behavioural therapists who are simple technicians applying rote exercises to change behaviour. A formative experience for me as a resident in psychiatry was watching Roger Mackinnon and Jeffrey Young do psychotherapy. Dr MacKinnon was a doyen of psychoanalysis, while Jeff was a young Turk—he'd been trained by a father of cognitive-behaviour therapy, Aaron Beck, in Philadelphia, and was now striking out on his own in New York. Both were outstanding therapists: they were able to understand patients and to build rapport, and perhaps most importantly, to challenge patients to change while maintaining this rapport. While conceptualizing things differently, they intervened similarly.[66]

On the other hand, in specific cases, there may be marked differences between psychoanalytic and behavioural interventions. A psychoanalytic model arguing that OCD involves excessive 'anal drives' and requires insight-oriented psychotherapy has been replaced by a neuropsychiatric model of OCD which provides a rationale for medication as well as quite different forms of psychotherapy.[67] The work at NIMH on the role of the serotonergic system in OCD and the value of serotonergic medications in treating symptoms in OCD played a key role in bringing this shift about. Whereas psychoanalysts have often been loath to use psychiatric medications, the evidence base for these agents grew substantially at this time, and in retrospect my supervising psychoanalyst should have been quicker to insist that we use medications to treat my patient with OCD.

Even before the work on serotonin in OCD began, the South African-born psychiatrist Isaac Marks had begun to extend the principles of behaviour therapy to OCD. Desensitization in the form of exposure therapy turned out to be an effective treatment. While I was a postdoctoral fellow, work by the psychiatrist Lewis Baxter showed that OCD was characterized by altered activity in specific parts of the brain, and that both pharmacotherapy and psychotherapy normalized brain circuitry. This was amazing for me; it provided a way of thinking about psychiatric symptoms as involving specific neurocircuitry, as well as a way of integrating pharmacotherapy and psychotherapy. Tom Insel, who played such a key role in early OCD research, later went on to become Director of

66. Later we will discuss the Dodo Bird Verdict that "Everybody has won and all must have prizes" (from Lewis Carroll's "Alice's Adventures in Wonderland" [301]), which was used by Saul Rosenzweig to suggest that all psychotherapies are equally effective. See also later discussion on progress in psychotherapy and later footnotes on psychotherapy integration.

67. During my postdoctoral fellowship in psychopharmacology, together with the psychoanalyst **Michael Stone** I edited a volume in the 'Essential Papers' series, which are collections of psychoanalytic papers [302]. Our volume was on obsessive–compulsive disorder, and tracks how different conceptual approaches to this disorder have emerged over time. Since the time I was a resident, Mike has generously encouraged my career.

NIMH, where he developed a vision of psychiatric disorders as disorders of neurocircuitry.[68]

DSM-III: Diagnostic criteria for psychiatric disorders

Together with the shift in American psychiatry from psychoanalysis to pharmacotherapy and cognitive-behavioural therapy, one additional change is key for understanding the progress of modern psychiatry, as well as its limitations. Down the hallway from where I saw my patient with OCD at Columbia University, Robert Spitzer was continuing his work on psychiatric nosology. In 1980 he had published the 3rd edition of the Diagnostic and Statistical Manual Disorders (DSM-III), a landmark in 20th century psychiatry. DSM-III provided diagnostic criteria that rigorously defined, or 'operationalized', a wide range of mental disorders. These diagnostic criteria were rapidly and widely adopted by clinicians and researchers. Thus, for example, DSM-III facilitated rigorous research on the epidemiology of psychiatric disorders, so leading to the current evidence base showing that mental disorders are widespread, and at least as disabling as other medical conditions.

Conceptually, DSM-III aimed to be theoretically neutral; for example, it replaced the term 'anxiety neurosis' with 'anxiety disorder'.[69] This sort of change didn't go down well with psychoanalysts, and indeed Spitzer was influenced by Don Klein, the senior biological psychiatrist at Columbia University: one reason for separating out different disorders in DSM-III seemed to be that they responded to different medications. This was Klein's concept of 'pharmacological dissection'; he had shown, for example, that panic disorder responds to tricyclic antidepressants but is worsened by antipsychotic agents, and NIMH colleagues had shown that OCD responded to antidepressants that acted on the serotonergic system, but not to those that acted on other neurotransmitter systems.

Practically, Spitzer's volume was a game-changer for psychiatry in good part because it allowed reliable clinical diagnosis.[70] Previously, for example, UK and USA psychiatrists had had different diagnostic approaches to schizophrenia. With clinicians now on the same wavelength, they could more meaningfully discuss treatment approaches. In addition, DSM-III allowed biological psychiatrists to reliably diagnose psychiatric conditions prior to embarking on their research.

---

68. Although **Tom Insel** clearly saw himself as primarily a neuroscientist, he invited **Isaac Marks** to visit NIMH, where Marks demonstrated how to do exposure therapy. I've followed the work of both men avidly over the years, and I am fortunate to have had the opportunity to co-author papers with both (although I don't agree with all of their views).

69. Although psychoanalysts used the term **'neurosis'** widely, it is notable that the origins of the word lie in biological theory about the neuronal basis of symptoms, and that some behavioural therapists, such as Joseph Wolpe, have also relied on this terminology.

70. There is evidence that **reliability in psychiatry** is as high as in any medical specialty. (In addition, we'll later note that effect sizes of psychiatric treatments are as high as those of other medical treatments.)

Whereas the earlier DSM-II was based on the 8th edition of the World Health Organization's International Classification of Diseases (ICD), DSM-III had significant influence on ICD-9. Together DSM-III and ICD-9, and their subsequent editions, helped establish uniformity of psychiatric diagnosis across the globe.

Both Mackinnon and Spitzer were key figures in the lives of Columbia University residents in psychiatry. Spitzer had completed a text revision of DSM-III, known as DSM-III-TR, in 1987, and perhaps had more time on his hands. He turned up in our emergency room, helping to see patients with the residents, and getting his own hands on experience of whether DSM was useful. Whereas I could barely stammer through a conversation with Mackinnon, Spitzer's down to earth approach made him easy to talk to. I wound up assisting him with some research during the day and playing squash with him after hours. A similar shift occurred in my fellow residents. At the start of our residency Mackinnon was our hero—a master of the sophisticated world of psychodynamics, while Spitzer was merely a second rate purveyor of simplistic books and measures. By the end of residency we felt we had learned as much, if not more, from Spitzer's logical and practical approach.

DSM-III and its subsequent revisions have, however, received enormous criticism, both from within psychiatry, and from without. First, psychodynamic psychiatrists have argued that the use of standardized checklists of symptoms deflects attention away from the patient as a unique individual. They may also argue that the classification ignores the unconscious conflicts which underlie symptoms, and so which are the more appropriate target for diagnosis and assessment. Second, biological psychiatrists have argued that psychiatric assessments need to pay more attention to neuroscientific findings. Spitzer's earlier work on diagnostic criteria, known as the Research Diagnostic Criteria, had informed the approach taken in DSM-III.[71] As Director of NIMH, Tom Insel led work on an entirely different framework—the Research Domain Criteria (RDoC)—which focuses on dimensions that are drawn from neuroscience research and that are relevant to the domains of cognition, emotion, and behaviour. Thus a patient who meets DSM-III criteria for major depressive disorder (a categorical diagnosis) may be described by RDoC as having, say, diminished positive affect and impaired working memory (on dimensions that cut across different psychiatric diagnoses). Third, critics of psychiatry have argued that psychiatric diagnoses are social constructions (along the lines of a concept like 'weeds') and do not have scientific validity.[72]

---

71. **Kenneth Kendler**, a philosopher-psychiatrist who has done seminal work in both psychiatry and the philosophy of psychiatry, has contributed a series of conceptual and historical articles that shed light on the issues discussed here [303]. I'm fortunate to have had Ken's mentorship on a number of projects, beginning during my psychiatry residency.

72. Tom Insel published a blog criticizing the diagnostic approach of DSM-5, just weeks before its public release, creating some embarrassment for the American Psychiatric Association (the publisher of DSM). To limit the damage, Jeffrey Lieberman, President-Elect of the American Psychiatric Association and Insel then released a joint communiqué stating that DSM-5 represents the best information currently available for psychiatric diagnosis, that RDoC is intended to lay the groundwork for future diagnostic systems, and that **DSM-5 and RDoC** represent complementary rather than competing frameworks.

To fully address these criticisms, we need a more detailed concept of the nature of disease: a topic that we'll tackle later. However, some immediate rejoinders are possible (Table 2). First, the RDoC framework is a complex one, based on science that is still very much in its infancy, and currently too unwieldy for use in day-to-day clinical practice.[73] Second, a clinician can and should both employ objective measures, and at the same time understand the patient as an individual. Third, while all science is a social activity that is theory-bound and value-laden, science is nevertheless able to outline the nature of structures, processes, and mechanisms that exist in the world, and that underlie both medical and psychiatric disorders.

A paper in the journal 'Science', 'On being sane in insane places' is sometimes cited as pointing to how problematic psychiatric diagnosis is. David Rosenhan, a professor of psychology, admitted himself and a number of other volunteers to a psychiatric hospital, with each person complaining of psychotic symptoms. Once admitted, the volunteers indicated that symptoms had resolved. In each case it took some time for the person to be discharged and in each case they were advised to stay on medication, suggesting that psychiatric diagnosis was a deeply flawed procedure. Spitzer and others argued persuasively, however, that this study said little about the reality and reliability of psychiatric diagnosis; if the same volunteers had presented complaining of vomiting blood, for example, they would similarly have received standard management regimes for haematemesis.[74]

Taken together, DSM-III and later editions have been key in driving psychiatric research and practice forwards. That said, work on psychiatric classification remains limited in important ways. First, there is the threshold problem: there is debate about whether specific conditions are really disorders rather than normal responses to the problems of living, and for any particular diagnostic entity there is debate about where to draw the line between normality and pathology. Second, there is the reification problem: there are a range of possible ways of clustering disorders and their symptoms, and current solutions are just some of many.[75] Third, there is the causality problem: our classification of psychiatric disorders (also called our 'nosology') is not closely based on causal mechanisms. Fourth, and relatedly, we have competing theoretical models of the aetiology of psychiatric disorders, and we have only partial knowledge of the underlying psychobiological mechanisms.

The rigorous community surveys based on DSM-III and ICD-10 diagnostic criteria for psychiatric disorders provide useful insights into some of these issues. Ronald Kessler's World Mental Health Survey consortium has now been undertaken in more than 30 countries with more than 100,000 participants.

---

73. We will discuss RDoC in more detail later on.
74. **David Rosenhan** [304] **and Robert Spitzer** [305] crossed swords, and Susannah Cahalan [306] suggests in her volume on this experiment that Rosenhan may have falsified some of his data.
75. Alfred Whitehead usefully referred to the **'Fallacy of Misplaced Concreteness'** [307]. Useful articles on **reification in psychiatric diagnosis** include those by Steve Hyman (another past Director of NIMH, and psychiatrist–neuroscientist) [308] and Kenneth Kendler [309].

This work has clearly demonstrated the high prevalence of psychiatric disorders and their associated burden of disease, together with a 'treatment gap' (only a small proportion of those with psychiatric disorders receive appropriate treatment). In addition, data from these surveys have been incorporated into public health studies, such as the influential Global Burden of Disease collaboration; crucially Christopher Murray and colleagues found that psychiatric disorders are amongst the most disabling of all conditions.[76] Such work has played a key role in evidence-based advocacy for better mental health services.[77]

Statements about the very high prevalence of disorders may, however, help fuel a 'credibility gap';[78] sadness and anxiety are normal emotions, and a high prevalence suggests overdiagnosis and overmedicalization. On the other hand, it's notable that there are some medical conditions, such as gum gingivitis, which are highly prevalent. In the presence of sugary diets plus insufficient dental hygiene, severe caries may result. Similarly, while sadness and anxiety are typically normal responses to the problems of living, under certain circumstances they may evolve into more severe psychiatric conditions, including depression and anxiety disorders. Kessler and his colleagues have pointed out that there is a graded relationship between symptom severity, burden of disease, and clinical outcomes; thus thresholds for diagnosis need to be set in such a way that 'minor' conditions are not simply ignored. Thresholds for diagnosis may also be impacted by the availability and cost of interventions. Thresholds for the diagnosis of hypercholesterolaemia, for example, were once set high; they were lowered when statins were introduced; and they were lowered again when statins were available as cheaper generics.[79]

These examples demonstrate clearly that medicine and psychiatry cannot set thresholds by 'carving nature at her joints', as Plato put it. While nature may have some joints, we've noted that more recent authors have emphasized the complexity and fuzziness of reality, as well as the indeterminacy of our categories and the vagueness of our boundaries, using words like 'dappled', 'promiscuous', and 'plural'. Certainly, as both biological psychiatrists and global mental health practitioners have emphasized, there is a spectrum between well-being and illness. Our cut-points reflect clinical reasoning at a particular time and place. Psychiatry is a theory-bound and value-laden social activity, and our clinical decision-making boundaries reflect the facts and values of any one particular time and place. That said, as more facts come to light, and as we sharpen

---

76. I count myself extremely fortunate that **Ronald Kessler** is one of my mentors. A recent summary of some of Ron's work on the World Mental Health Survey's is the edited volume by Kate Scott, Peter de Jonge, Ron, and myself [310]. For a book about Christopher Murray and the contribution of the Global Burden of Disease study, see Jeremy Smith [311].

77. The phrase **'no health without mental health'** has been a rallying cry of advocates for increased expenditure on mental health services [312].

78. See V. Patel [313].

79. See R.C. Kessler and colleagues [310].

our thinking about the relevant values, we are able to make better decisions about thresholds between more severe conditions, milder conditions, and normal responses to problems of living.[80]

Data from the World Mental Health Survey collaboration have also been useful in studying issues around the reification of mental disorders, of seeing our particular diagnostic sets as somehow given from on high, rather than as merely man- and women-made hypotheses. Thus, for example, different sets of diagnostic criteria for specific conditions can be compared, such as those of the DSM and those of the World Health Organization's International Classification of Diseases. Under Kessler's mentorship, I and other colleagues compared slightly different criteria sets for (or operationalizations of) PTSD; each identified a slightly different group of people, but each group had a similar level of impairment, indicating a similar need for treatment.[81] During the 11th revision of the International Classification of Diseases, studies of clinicians were undertaken to determine which criteria sets were viewed as more clinically useful. This sort of work indicates that despite the ever-present threat of reification, scientific progress in nosology is possible.[82]

What about the fact that psychiatric classification systems are not closely based on an understanding of the causal mechanisms underlying psychiatric disorders? I would note that in the rest of medicine, too, not all diagnostic entities link directly to specific causes. Take the cardiological entity of 'heart failure'. This is a very useful entity for clinicians, as it can be quickly assessed and as it can guide treatment. However, the term 'heart failure' says nothing about the specific cause of the illness; it may reflect any number of pathological processes. This is not to deny that psychiatry still has very little causal knowledge, with multiple competing perspectives on which causal mechanisms are most relevant.[83]

---

80. This phrase from Plato's 'Phaedo' is consistent with his **essentialism**. For contrasts between and critique of essentialist ideas in Socrates, Plato, and Aristotle, see Roy Bhaskar's discussion of the 'Platonic/Aristotelian fault-line' [314] and Lakoff and Johnson [159]. Indeed, there is significant dispute about the essentialism of these thinkers [315]. Similarly, while Aristotle referred to essences, he also frequently emphasized that entities could not be understood univocally, but only multivocally [316]. A range of subsequent philosophers have contributed to these issues [152, 317]. The term 'essentialism' was coined by Popper, with a pejorative connotation [318], but in keeping with a multivocal perspective it has also been seen as compatible with 'thick vague' constructs, with a positive connotation [319]. See previous footnotes on 'pluralistic realism' and different kinds of kind. We will later return more specifically to the questions of whether emotions, biological categories, and mental disorders are essential natural kinds.

81. See D.J. Stein and colleagues [320]. Thus Ron, I, and other colleagues have argued that there is a useful bidirectional relationship between **epidemiology and nosology** [321].

82. For example, Geoffrey Reed, myself, and other colleagues worked on an empirical study comparing different diagnostic guidelines for obsessive–compulsive and related disorders. Still, whether OCD belongs in the category of anxiety disorders or the category of obsessive–compulsive and related disorders is a more difficult choice [322]. We'll return to this issue later on.

83. Books on my parent's shelves included the work of Eugene Marais, Robert Ardrey, and others speculating about the evolutionary origins of human behaviour. As a young psychiatrist, I was delighted to come across the work of **Randolph Nesse**, a father of both evolutionary medicine and evolutionary psychiatry. I have been fortunate that Randy later mentored me on a number of projects, including an article on nosology from the perspective of evolutionary psychiatry, which includes the point made here [323].

Concerns about the credibility gap for mental disorders, inappropriate thresholding of these conditions, and their reification, together with shifts between explanatory models which emphasize entirely different causal mechanisms, have at times led to deep divisions both inside and outside the field of psychiatry. Such divisions raise the question of whether the field can provide an integrative framework for understanding and addressing mental disorders, and for contributing to discussions about the big questions. Over the course of the volume we will, however, also note many conceptual and empirical advances in psychiatry, and my hope is that the sketch of such an integrative framework will emerge from this discussion. For now I want to make a few assertions that draw on the integrative position in philosophy outlined earlier in this chapter, and that help lay some foundations for what follows (see Table 3).

First, psychiatry necessarily involves an engagement with psychobiological reality. That is, psychiatry is necessarily a cross-disciplinary field, addressing a range of phenomena from brain circuits all the way to social structures, and considering both natural and practical kinds. Building on the work of Adolf Meyer in psychiatry, George Engel put forward the biopsychosocial model in medicine, making the important argument that every patient needs to be considered in a comprehensive and nonreductionistic way. Although the field of biological psychiatry typically focuses on the neurobiology of mental disorders,

TABLE 3 Integrative positions in philosophy and psychiatry.

	Integrative position in philosophy	Towards an integrative psychiatry
Conceptual pluralism	Nature is dappled and promiscuous (pluralistic realism)	Psychobiological reality is complex and fuzzy
Explanatory pluralism	A range of explanatory approaches are needed to account for nature	Encompasses different accounts of disorders, e.g. proximal and distal
Erklären and verstehen	In the human sciences, both erklären and verstehen are useful	Psychiatry is concerned with both mechanism and meaning
Epistemic humility	Science entails epistemic virtues such as transparency and fallibility	Psychiatry entails epistemic values such as transparency and fallibility
Practical wisdom	Science is a practical activity that requires practical judgement (phronesis)	Psychiatry is a practical activity that requires practical judgement (medical phronesis)

TABLE 4 Explanations in biology.

	Static focus of study	Dynamic focus of study
Proximal level of explanation	Mechanism (causation) i.e. causal explanation cf. efficient cause	Ontogeny (development) i.e. history of the individual cf. material cause
Distal level of explanation	Adaptive value (function) i.e. evolutionary explanation cf. final cause	Phylogeny (evolution) i.e. history of the species cf. formal cause

biology involves the study of phenomena that range from molecular interactions, through to physiological systems, and on to group behaviour (whether of animals or humans). Indeed, psychobiological phenomena are complex; they encompass a range of structures and processes that operate at different levels and that intersect. Therefore a range of categories and frameworks for describing these phenomena may be useful, and psychiatry should welcome such **conceptual pluralism**.[84]

Second, in thinking about and investigating underlying causes, biology involves the use of a number of different kinds of explanation, and so therefore must psychiatry. The evolutionary theorist Theodosius Dobzhansky noted that nothing in biology makes sense except in the light of evolution: a point that emphasizes the importance of understanding evolutionary mechanisms. These 'distal mechanisms' can be contrasted with the more 'proximal mechanisms' that physicians often think about: doctors think more about how the heart pumps blood than about how the heart evolved. The Nobel Prize winning ethologist Nikolaas Tinbergen listed four kinds of explanations of behaviour (Table 4). These have been compared with the four kinds of explanation outlined by

84. See earlier footnotes on Meyer's use of the term 'psychobiology', on Dewey's use of the term 'psychophysiology', on a 'soft naturalism' that encourages conceptual and explanatory pluralism, and on varieties of 'pluralistic realism'. The term **'biopsychosocial'** was coined by Roy Grinker [324]. George Engel criticized the use of reductionist medical models in psychiatry, but stated that this is a problem in the whole of medicine [325]. Derek Bolton and Grant Gillett provide a useful update of the debate about the term 'biopsychosocial' in their 'The Biopsychosocial Model of Health and Disease' [326]. They note that key criticisms are that the biopsychosocial model lacks conceptual coherence and scientific content. S. Nassir Ghaemi's volume on the biopsychosocial model, for example, argues that the model is overly vague, replacing biomedical reductionism (the so-called medical model) with additive eclecticism (simplistically assuming that 'more is better') [242]. Similarly, David Pilgrim has criticized the model from a critical realist perspective [327]. Bolton and Gillett counter that the value of the model derives from its application to specific health conditions, but that new approaches to thinking about psychosocial causation are needed [326]. We'll return to this point.

Aristotle (also listed in Table 4). Given the complexity of psychobiological phenomena a range of 'difference-makers' need to be considered. Put differently, psychiatric science needs to address a range of different causal mechanisms and to employ a range of different explanatory accounts. In this way it can replace simplistic and dualistic accounts with complex and integrative ones, and foster **explanatory pluralism**.[85]

Third, psychiatry necessarily involves an appreciation **not only of mechanism but also of meaning**, of both erklären and of verstehen. One of the first to emphasize this point was the physician–psychiatrist Karl Jaspers, of 'Axial Age' fame (Table 2).[86] We need to advance both our knowledge and our understanding of people and of the psychiatric disorders from which they suffer. Put differently, we need to understand **both disease mechanisms and illness experience**. This dual focus is consistent with an integrative approach to the cognitive-affective sciences as a whole. We'll go into more detail about several key aspects of an integrative perspective to the brain and mind in the next chapter, as we start to tackle the big questions.[87]

Fourth, in psychiatry, and perhaps also when tackling the big questions and hard problems, even as we make progress in providing more robust explanation and sharper understanding, we need to be open about how much we don't know or appreciate, and be willing to revisit and revise our models. If psychiatry is

---

85. See Hladký and Havlíček [328]. For more on proximal versus distal causes in particular, see E. Mary [329] and D. Haig [330]. For Lakoff and Johnson, the category of causality is even larger, with multiple metaphors available [159]. For more on scientific explanation, see earlier footnote. For more on **explanatory pluralism** or causal pluralism, see earlier footnote on the erklären–verstehen divide, as well as Sandra Mitchell [331], Stephen Kellert, Helen Longino, and C. Kenneth Waters [332], L. De Vreese and colleagues [333], Nikolaj Pedersen and Cory Wright [334], Pierre-Alain Braillard and Christophe Malaterre [335], Chryostomos Mantzavinos [336], J. P. Vandenbroucke and colleagues [337], Stéphanie Ruphy [338], and W. Viet [339]. Some of the first to emphasize explanatory pluralism in medicine were the South African-born couple Mervyn Susser and Zena Stein [340]. The term 'difference-maker' is from David Lewis's work on causality [341], and is drawn on by Kenneth Kendler, a proponent of explanatory pluralism in psychiatry [342]. See also Helen Beebee, Christopher Hitchcock, and Huw Price [343], earlier footnote on explanatory virtues, and later footnotes on the Stanford school and the disunity of science as well as on metaphors of causation.

86. **S. Nassir Ghaemi** argues that Karl Jaspers' largest contribution was not so much the introduction of the erklären–verstehen distinction into psychiatry, but rather his emphasis on a pluralistic scientific model [344]. We later note Kenneth Kendler's idea of 'explanation-aided understanding'—scientific explanations of mental disorders may well help us better understand the experience of suffering from them—a point that is helpful in ensuring that the erklären–verstehen distinction is not itself reified [345].

87. A distinction between **biomedical explanation and illness experience** has been drawn by a number of writers, including Leon Eisenberg [346] and Arthur Kleinman [347]. For a review, see Bjørn Hofmann [348]. S. Nassir Ghaemi suggests that a view that the physician should treat disease in the body, while also attending to the person with the disease, has its roots in Hippocrates, and was modernized by **William Osler** [349]. From a different perspective Michael Schwartz and Osborne Wiggins, in an influential article, also emphasize the importance of including both science and humanism in the medical encounter [350].

to be a science, such **fallibilism** is key. Certainly, in the case of psychiatric disorders it is typically easier to refute overly simplistic or blatantly erroneous theories than to provide fully comprehensive answers. While we have certainly made progress that we should be proud of, the fact that have so much further to go also demands a sense of humility. As we tackle the big questions, we'll again come back to this metaphor of a journey that remains underway, and the issue of whether optimism or pessimism about the trip is the more appropriate attitude.[88]

Fifth, since the time of Aristotle it has been recognized that good medicine involves **practical judgement** or wisdom (i.e. phronesis). Such judgement is necessary not only for optimal diagnosis and assessment, and for explaining and understanding any particular patient, but also for deciding on optimal intervention. As we'll later discuss, a decision to provide no treatment is always a particularly important consideration in medical practice. In addition, in psychiatry the choice between pharmacotherapy and psychotherapy requires a weighing up of a range of considerations, including the nature and context of the disorder, the evidence base related to each modality, and the patient's explanatory models and treatment preferences.[89]

1.3 The big questions and hard problems

What specific big questions about life and hard problems of living may be most usefully informed by ongoing developments in philosophy and psychiatry? This volume is structured around a series of seven large issues, some of which have long been in the domain of philosophy (e.g. the body–mind question), some of which have long been in the domain of psychiatry (e.g. how best to think about pain and suffering), but all of which likely benefit from a dialogue between philosophy and psychiatry. The seven issues are as follows:

1. What is the best way of thinking about the brain–mind?
2. How should we best balance reason and emotion?
3. What is happiness, and how can we bring it about?
4. How should we best think about pain and suffering?
5. How should we conceptualize morality, e.g. good and evil?
6. What is truth, how do we know what is really true?
7. What is the meaning of life, what makes a meaningful life?

I have chosen seven issues that are not only 'big questions' or 'hard problems', but ones where there have also been important advances in cognitive-affective

88. See earlier footnote on fallibilism. For work that documents the inappropriate **hubris of psychiatry**, see, for example, Edward Shorter [351] and Anne Harrington [352].

89. For more on **epistemic virtues in medicine**, see Onora O'Neill [353] and B. Kotzee [354]. For more on medical decision-making as requiring practical judgement, or **medical phronesis**, see my volume on philosophy of psychopharmacology [53]. See also B. Kotzee and colleagues on medical phronesis [355], as well as later footnotes on evidence-based and values-based practice and a flourishing-oriented medicine, and on virtue ethics in psychiatry.

science that may shed some light on the answers. Some chapters fall within the realm of metaphysics or ontology, i.e. what is the nature of reality and of being? Others fall within the realm of epistemology, i.e. what is the truth, and what is the basis of knowledge? And finally, some chapters fall within the realms of ethics, i.e. what is the right way to live? While such a broad array of questions seems wholly overambitious for a single project, in the real world these matters intersect, and the approach here is in line with the current return of some philosophers to the practical and important question of how best to live.

As we discuss these seven large issues we will also cover a range of subsidiary questions, including 'what is science?', 'what is a psychiatric disorder?', and the nature of pleasure. We'll talk about very philosophical issues and very practical issues, including conceptualizations of evil, whether there is progress in the fields of psychiatry and psychotherapy, question about the self and about free will, and the relationship between science and philosophy. So we'll go wide rather than deep, although the approach taken towards this broad range of questions will also be a unified and integrative one, and so will hopefully have some degree of penetration.

In discussing these seven large issues, we will first consider the perspectives of philosophy and of psychiatry, and then note how science is putting forwards more encompassing answers than ever before. One hope is that the exciting disciplines of neurophilosophy and neuroethics, together with developments in the cognitive-affective sciences and in neuropsychiatry, provide resources that may be useful in thinking about new ways to approach age-old questions. Certainly, in the age of neurophilosophy and neuropsychiatry it is difficult not to take a naturalistic view, although we'll also need to consider the limitations of this perspective. We begin with the key question: what is the best way of thinking about the 'brain–mind'?

Chapter 2

Brain–minds: What's the best metaphor?

During high school I studied computer science. This was a newly introduced course with no teacher, I was the only student who took it (after hours), and the reading material focused on ideas and discoveries that today might seem archaic. Nevertheless, I learned the basics of programming, and a whole new world opened up to me. From a practical point of view I wrote programs that did all sorts of interesting things, including playing games like chess (an effort that took many weekends; I worked on a mainframe computer where each line of code had to be punched on a separate card). From a conceptual point of view, I became immersed in the literature on how computers were and were not like minds. At around that time the philosopher Herbert Dreyfus, who I would classify as a critical thinker, wrote that computers would never beat humans at chess. I wasn't so sure.[1]

Broadly, this chapter asks the question, "what is the best way of thinking about the brain–mind"? The term 'brain–mind' is an unusual one, but it may be useful, given how difficult it has been to think clearly about brains and minds, and given the premise of this volume that advances in psychobiology are an important resource for understanding brains and minds. In particular the term 'brain–mind' may help us to avoid positions in philosophy or psychiatry that reduce mind to brain, or that ignore the grounding of the mind in the brain. In day-to-day clinical work, these brainless or mindless positions turn out to be a really bad idea.[2]

We'll begin the chapter by discussing the mind–body problem in philosophy, we'll then go on to discuss different metaphors of the mind that have been formulated in psychology, before proposing an alternative metaphor of the brain–mind that is consistent with evidence from modern psychobiology and from the psychiatric clinic. Over the course of the chapter I'll explain why the

1. **Herbert Dreyfus** argued in a 1965 paper 'AI and alchemy' that a computer would not be able to beat a 10 year old at chess. In 1967, in a match arranged by Seymour Papert, he lost to a computer. Nevertheless, he expanded his paper to book length, and argued that a computer would never beat a world chess champion [356].

2. See earlier footnotes on the use of the terms 'brainless' and 'mindless' by Lipowsky, and the use of the term 'body–mind' by Dewey. Authors who have used the term **'brain–mind' or 'mind–brain'** include Patricia Churchland [235], Carl Carver [357], and Jaak Panksepp [358].

Problems of Living. https://doi.org/10.1016/B978-0-323-90239-7.00005-5

idea that the mind is like a computer has had less and less appeal to me over the years. Certainly, early serial processing programs (such as IBM's Deep Blue, which beat world champion Gary Kasparov) were doing something quite different from human 'chess playing'. Later connectionist approaches (such as Google's AlphaZero, which is even stronger) may provide better models of human cognitive-affective processing, but the hardware–software metaphor of the brain–mind remains misleading in key ways.[3]

2.1 The mind–body problem in philosophy

In thinking about brain–minds, an immediate difficulty is that we have quite different ways of talking about the brain and the mind, despite our growing understanding of how the mind is grounded in the brain. We talk about brains in terms of brain cells or neurons, while we talk about mind using constructs like 'thinking', 'feeling', and 'wanting'. We might say that a particular neuron fires (or more technically, is depolarized) at a particular point in time, and we might say that a particular person feels pain, or perceives the colour green. We say that neurons that 'fire together, wire together', but we say that a particular person is conscious.[4]

The question of how best to conceptualize and to explain the relationship between neurons firing, and mental phenomena, has been called the mind–body problem in philosophy. Put differently the mind–body problem grapples with the question of how the brain, which is material, gives rise to phenomena such as consciousness, which is seemingly immaterial. One response is physicalism, which states that all mental phenomena are brain phenomena. But this reductionistic view faces a number of problems. First, there would seem to be some slippage between brain and mental phenomena: it seems highly unlikely that the perception of the green involves exactly the same neurons firing in all individuals. Second, and this is the hard brain–mind problem, even if we can fully explain a particular brain process, it is not clear that this accounts for phenomenal or subjective experiences or describes what these 'qualia' feel like.[5]

---

3. IBM's **Deep Blue** defeated Gary Kasparov in 1997. Pocket Fritz, running on a cell phone, won a major tournament, receiving a higher chess ranking than Deep Blue in 2009. Google's **AlphaZero** uses neural networks and plays at an extraordinarily high level [359]. While Dreyfus may have been wrong on this specific prediction, it could perhaps be argued that his criticism of symbolic cognition is ultimately supported by the shift to connectionist models.

4. The phrase that neurons that '**fire together, wire together**' is drawn from an article by Siegrid Löwel and Wolf Singer [360], and is often used to summarize Donald Hebb's cell assembly theory [361].

5. There is a vast literature on these issues. See, for example, Tim Crane and Sarah Patterson [362], who note a number of key components of the **mind–body problem**; body and mind seem to interact causally, and the distinctive features of consciousness. For more on mental causation, see D. Robb [362a], James Woodward [363], C.F. Craver and W. Bechtel [364], and Nancey Murphy, George Ellis, and Timothy O'Connor [365]. For a useful discussion of descriptive, explanatory, and functional questions about consciousness, see R. Van Gulick [366]. Importantly, there are a range of

Dualism argues that brain phenomena and mental phenomena are entirely different kinds of things. This position has been called substance dualism or Cartesian dualism, as it rests on the contrast between physical and mental substances. This is not to say that the mental is necessarily nonphysical, rather mental 'stuff' has properties which differ from those of physical 'stuff'. However, such a view again seems unable to provide a rigorous conceptualization of how the brain and mind interact. And certainly they do seem to interact in a causal way: stimulation of neurons may lead to a range of mental phenomena, and conversely a range of mental phenomena appear to lead to changes in neurons.[6]

Many philosophers argue for a position of emergent materialism (Table 5). This stance acknowledges that minds are grounded in brains, agreeing with physicalism that there is no 'ghost in the machine' or 'soul' that accounts for mental phenomena. At the same time this approach acknowledges that brains and minds are different kinds of entities, agreeing with the dualist view that minds have properties that brains do not. This sort of nonreductive naturalist position may not solve the hard problem of explaining how consciousness feels, but it is consistent with the view that consciousness emerges from the brain,

---

'**explanatory gaps**' between our explanations of the physical world and our understanding of consciousness. This phrase was introduced by Joseph Levine [367], who pointed out that the sentence 'Pain is the firing of C fibers', although physiologically correct, does not tell us how pain feels. However, the issue dates back at least as far as Descartes, and similar phrases date back at least to Dewey, who referred in his 'Soul and Body' to an 'an inexplicable mystery, a gulf' [117]. Many have contributed to the literature on the explanatory gaps between our explanations of the physical world and our understanding of consciousness, e.g. Thomas Nagel's paper on 'What Is it Like to Be a Bat?' [368], and David Chalmers's differentiation of the 'hard problem' of consciousness from the 'easy problems' of explaining the mechanisms underlying mental states [369]. For mysterians (a term coined by Owen Flanagan, based on the rock group 'Question Mark and the Mysterians' [370]), this gap will never be closed [371]. For further discussion of **qualia** and related issues, see Nicholas Humphrey [372], Michael Tye [373], and Daniel Dennett's paper on the mystery of David Chalmers [374].

6. For discussion of the history of **dualism**, going back to Plato and Aristotle, and including the work of Karl Popper and John Eccles, see Andrea Lavazza and Howard Robinson [375]. Plato is often viewed as a dualist, but there is ongoing controversy about Aristotle's position. Aristotle seems to emphasize the difficulty of understanding the soul, to take a naturalistic approach, and to hold a position that is neither materialist nor dualist [316, 376]. Although Descartes and substance dualism are synonymous, a more recent literature argues that he has been misread [377]. It's notable that his scientific approach focused on mechanism [273]. Important philosophical criticisms of dualism include those of Spinoza [378], Hume [379], and more recently and influentially, Gilbert Ryle, who argued that Cartesian approaches to the mind make a 'category mistake', because they analyse the relationship between 'mind' and 'body' as if they are terms from the same logical category [380]. Ryle also argued against perceptions of a gap between theory ('knowledge that') and practice ('knowledge how'), an argument redolent of Dewey's reflex arc concept [109], and a forerunner of work on embodied cognition [381]. See also later footnote on the continuity between scientific and everyday knowledge.

TABLE 5 Classical, critical, and integrative positions on the brain–mind.

	Classical position	Critical position	Integrative position
Philosophical model	Mind can be reduced to brain	Mental phenomena require a nonscientific understanding	Mind emerges from brain
Freudian model	Laws of the unconscious explain behaviour	The narrative of the unconscious can be interpreted	The cognitive-affective unconscious needs entirely new metaphors and models
Computational model	Mental processes involve information processing, i.e. symbolic cognition	Mental processes occur in particular contexts, i.e. situated cognition	Wetware is a useful metaphor to explain embodied and embedded cognition
Turing Test	A computer that produces texts like a human can think	The Chinese Room shows that computers have no understanding	Wetware mechanisms lead to meaningful thoughts and feelings

and with knowledge that brain alterations lead to changes in consciousness, and vice versa.[7]

The advent of modern computers inspired a position, known as functionalism, that has some parallels with emergent materialism, insofar as it explains mental states in higher-level terms (mental states have functions or roles) rather than in terms of their particular physical substrates. In this view, any particular mental state can be instantiated using a range of physical substances, ranging from neurons through to silicon. This view suggests that the metaphor that

---

7. The 'ghost in the machine' phrase is from Gilbert Ryle [380]. For a pithy summary of emergent materialism, see Steve Fogel's book title 'The Mind is What Your Brain Does for a Living' [382]. Mental phenomena are, however, not only embodied but they are also embedded; it's not so much minds that think, as people that think—a point we will make in more detail shortly. Dewey's pragmatic realism, Bhaskar's critical realism, and Lakoff and Johnson's embodied realism are all consistent with **emergent materialism**. For additional background, see M. Polanyi [383] and T. O'Connor [384]. For useful collections of work on emergent materialism, see Ansgar Beckermann, Hans Flohr, and Jaegwon Kim [385], Mark Bedau and Paul Humphrey [386], and Antonella Corradini and Timothy O'Connor [387]. For more on emergence in psychology, see R.K. Sawyer [388]. This may also be an appropriate time to mention the work of South African statesman, **Jan Smuts**. Smuts coined the term 'holism', emphasizing how the whole is greater than the sum of its parts [389]. His thinking drew on ideas expressed in **Alfred Whitehead's** 'process philosophy', and in turn influenced a number of important psychiatrists and psychologists, including Adolf Meyer [390]. For more on the relationship between Whitehead and pragmatic realism, see Guy Debrock [391] and Brian Henningh, William Myers, and Joseph John [392].

describes 'computers as thinking' is an accurate one: our minds are computer software running on brain hardware. Modern users of Google Translate are, however, under no impression that when they talk into their cell phones, and their phones translate these words into, say, Chinese, that their phones really speak Chinese (a prediction that the philosopher John Searle made long before the availability of modern phones).[8]

Clearly, we have to pay close attention to the metaphors we use to describe mental phenomena, and the models we use to explain them. The next section will discuss metaphors and models that psychologists have often employed. For now, I wish to return to my suggestion that it may be valuable to refer to the 'brain–mind'. This term may not entirely resolve the body–mind problem. But it helps emphasize the complexity of psychobiological reality; for example, different aspects of the brain–mind may involve different properties—a single neuron firing may have specific effects in a reflex arc, while a broad neural network may ground the more complex phenomenon that is the perception of green. And it helps emphasize the need to discover appropriate causal mechanisms to explain thoughts, feelings, and behaviours; there is no 'ghost in the machine' that can account for these phenomena. The construct of the 'brain–mind' needs, however, to be supplemented by novel metaphors and models that help integrate neuroscience and psychology, and that help conceptualize interactions between different components of the brain–mind; we'll come back to this point when we talk about interactions between reason and emotion in the next chapter.[9]

2.2 The mind–body problem in psychology

Conceptual approaches in psychology have closely reflected the different positions that philosophers have taken on the mind–body problem. Debates on the nature of the soul date back at least to the ancient Greeks. In earlier times scholars were 'natural philosophers', concerned with both conceptual and empirical issues, including those pertaining to the soul. Advances in fields such as physics and chemistry led to the separate emergence of the empirical sciences, but not immediately to a science of the soul. With emerging knowledge of neuroanatomy, in the 17th century René Descartes postulated that the brain and the soul interacted at the pineal gland. While Descartes' contributions to the mathematical sciences remain valid, this sort of idea has not proved productive: instead, focus very gradually shifted to a science of the mind.[10]

---

8. Searle's Chinese Room thought experiment is discussed in more detail later in this chapter.

9. See previous footnote on the use of the term 'brain–mind' by other authors. Along these lines, it's notable that commentators on **Maurice Merleau-Ponty**, who prefigured research on embodied cognition, have emphasized the notion of the 'body-subject'. Although this is not a phrase that he himself used when discussing embodied subjectivities, it also emphasizes the indivisibility of the mental and the physical [393].

10. For more on the **history of the sciences of brain and mind**, see again Carl Zimmer [17] and George Makari [18].

Early psychologists took different approaches, as exemplified in the contrast between Wundt's experimental psychology and James' practical focus. In keeping with positions in philosophy which have emphasized the importance of 'verstehen', one approach within psychology has focused on the level of the mind, attending to central mental phenomena such as human understanding, and key mental constructs such as agency and intentionality (Table 2). Existential philosophers and psychologists, for example, have underscored the lived experience of human subjects, humans' characteristic meaning-making, the agency that individuals have over their thoughts and feelings, and the way in which these are intentionally directed towards the world. From this perspective, we are homo symbolicus; psychology's focus should be on fully understanding human experience and expression, and this requires deep immersion in human narratives and thorough exploration of their meanings.[11]

This perspective has important implications for clinical work. The psychiatrist Victor Frankl was an Auschwitz survivor who placed humans' struggle for meaning at the centre of his system of psychotherapy, called logotherapy. Later developments in psychotherapy, such as existential therapy and narrative therapy, have continued to emphasize the centrality of human experience and expression. More generally, a good deal of current evidence-based psychotherapy also emphasizes the importance of the therapist–client relationship; there is a strong correlation of a good working alliance between therapist and client, with good treatment outcomes. At the same time, a focus on the mental, to the exclusion of the brain, runs the risk of ignoring a whole realm of scientific discoveries about how neuronal activity grounds experience and expression, not only in humans, but across the animal kingdom. Furthermore, there is the key issue of whether some narratives may be better than others; while humans certainly have a penchant for meaning-making, it also turns out that our meaning-making can go horribly awry.[12]

In keeping with a physicalist position in philosophy, an entirely different school of psychology has emphasized the principles of behaviourism. Philosophers working in Vienna at the turn of the 20th century were heavily

---

11. The German philosopher **Ernst Cassirer** who early on argued for the objectivity and validity of both the natural and cultural sciences, noted that "... instead of defining man as animal rationale, we should define him as an animal symbolicum. By so doing we can designate his specific difference, and we can understand the new way open to man – the way to civilization" [394]. See also E.H. Henderson [395] and Christopher Henshilwood and Francesco d'Errico [396].

12. For more on **logotherapy**, see Frankl [397]. For more on later **existential psychotherapy**, see Rollo May [398] and Irving Yalom [399]. For more on **narrative therapy**, see Martin Payne [400] as well as R. Charon and colleagues [401]. Importantly, narrative and meaning has also been emphasized in cognitive and developmental psychology by Jerome Bruner [402] and Jerome Kagan [403]. However, in line with cognitive realism, both Bruner and Kagan see the world as 'given' as well as 'made', with Bruner noting that 'we are natural ontologists but reluctant epistemologists'. For more on how narrative goes awry, see Mary Midgley [404] and Alex Rosenberg [405], as well as later footnotes on cognitive-affective biases and on the neuroscience/evolutionary theory of meaning-making.

influenced by advances in physics, which they saw as a model for both philosophy and all of the different sciences (Table 1). Psychologists such as the Americans Watson and Skinner similarly argued that behaviour needs to be carefully operationalized and observed, in order to uncover the laws that govern the associations between stimuli and responses. Importantly, such work was 'translational': beginning with the Russian Pavlov's work on learning in dogs, it attempted to discover behavioural laws across different species.

The potential scientific rigour and clinical value of behaviourism is exemplified in the work of Wolpe, discussed in the previous chapter. This research became a crucial part of the foundation of behavioural therapy, the first evidence-based psychotherapy. But an immediate problem with evidence-based medicine is the possibility that it doesn't sufficiently address the question of values. The majority of those with, say, phobias want to get rid of them. But for many people with health problems things are more complex: they may need to choose which symptoms are most important to address, and what sorts of treatment are best, taking into account the particular context of their lives. Evidence-based practice therefore needs to be complemented with values-based practice.[13]

Some behavioural approaches have attempted to integrate their concepts with findings in psychobiology and cognitive-affective science. Indeed, there is a risk that a purely behavioural perspective leads to a downplaying of both the brain basis of behaviour, as well as many more complex aspects of human psychology, including human agency and intentionality, which cannot be reduced to observable stimuli and responses. This was part of the great debate between early behaviourists and a second generation of cognitive-behaviourists, a debate in which the cognitive-behaviourists seem to have triumphed. Again, there is a need for metaphors and models that allow the sort of rigorous empirical science pursued by behaviourism, but that also foster integration with a range of findings in cognitive-affective neuroscience.[14]

2.3 Two key mind–body metaphors

While psychology may rely on philosophical concepts, like other branches of science it also builds models of the world and tests them empirically. This is the naturalist approach to psychology: psychology is like any other science insofar

13. There is an argument that **evidence-based medicine** has always referred to the integration of clinical evidence with both clinical expertise and patient values [406]. That said, just as one needs specific tools to gather and assess the relevant evidence, so there may be a need for specific ways of gathering and assessing the relevant values. Bill Fulford, a pioneering psychiatrist-philosopher, has made this argument particularly forcefully in his work on **values-based practice** [407]. Person-centred health care may be considered as integrating evidence-based and values-based practice. Note, however, critique of specific aspects of Fulford's work, and the proposal for a '**flourishing-oriented medicine**' [408]. Along these lines, for a useful discussion of virtue ethics and epistemic virtue in relation to evidence-based practice, see H. Berg [409].

14. For more on the debate between **behaviourism and cognitivism**, see Joseph Wolpe [296] and Albert Ellis [410].

as it uncovers structures, processes, and mechanisms which account for real-world phenomena. Two major models of the mind deserve discussion. The first is the idea that minds can be conceived in terms of energy flow; this is the model of psychoanalysis and of some alternative medicine practitioners, and it retains a great deal of influence. The second was given impetus by the computer age: with exposure to software and hardware, the idea that minds are computer programs came to the fore. As noted earlier this has been a widely influential idea; it is consistent with a functionalist approach to the mind–body problem in philosophy, and is at the heart of much modern cognitive-affective science as well as of much cognitive-behavioural therapy.

Freud, who worked as a neurologist and conducted basic research in neuroscience, is seen by many as one of the first great naturalists of the mind. His contribution was to argue that the mind could be conceptualized in terms of hydraulic forces. Drawing on the physical science of his day, he conceptualized the mind in terms of psychic energy, and the way in which this was suppressed or expressed. While Freud was not the first to write about the importance of the unconscious, his hydraulic models provided a fascinating account of a range of psychological phenomena, including jokes, dreams, and neurotic symptoms.[15]

We earlier raised the important question of whether psychoanalysis is a science. Even on its own terms, however, there have been significant issues with Freud's model. First, while there has been occasional empirical evidence for some psychoanalytic constructs, in general they have not found clear support. There have also been some claims that psychoanalytic therapy is effective, but the evidence base is comparatively small.[16] Second, although Freud himself believed that sciences such as neuroscience would eventually support his views, the integration of psychoanalytic models with neuroscience remains rudimentary. Work in the new field of neuropsychoanalysis has made a concerted effort to move such integration forwards, but its future prospects remain somewhat unclear.[17]

Several other attempts have been made to salvage psychoanalytic theory. One approach has been to argue that psychoanalysis is a science, and to reformulate

15. See earlier footnote, with reference to the work of Henri Ellenberger [246] and Frank Tallis [247].

16. The Freud wars are not over, perhaps, and so not everyone would agree with this assertion about the **evidence base for psychoanalysis**. For a recent review arguing that psychoanalytic constructs and interventions are evidence based, see J. Yakeley [411]. Still, the evidence points to fundamental errors in Freud's thought that are difficult to salvage; see, for example, work on mechanisms underlying incest avoidance [412].

17. **Neuropsychoanalysis** is a field with strong South African roots: Mark Solms, a key advocate, was trained in South Africa and later returned to the country. I would like to be complementary as Mark is a colleague with whom I have worked and published [413]. However, my sense is that the constructs and findings of neuropsychoanalysis have not yet impacted significantly on contemporary neuroscience research or psychotherapy practice. For a more sanguine view, see the writing of Nobelist Eric Kandel [414].

the model using more modern constructs. Some schools of psychoanalysis have, for example, moved away from drive theory towards a focus on interpersonal relationships. Once again, these have had several advantages, providing persuasive accounts of phenomena that emerge during psychotherapy. Attachment theory, which focuses on the psychobiology and impact of early mother–infant relationships, provides a particularly strong interdisciplinary conceptual basis for many components of psychotherapy, but as this work becomes increasingly empirical it becomes more removed from classical psychoanalytic theory.[18]

Freud himself very much thought of psychoanalysis as a science, discovering objective knowledge (erklären) about the mind and its aberrations. Continental philosophers, though, have argued that psychoanalysis is about the understanding and interpretation (verstehen) of narratives; these are hermeneutic methods used in the humanities rather than the sciences (Table 5). Psychoanalytic constructs have certainly been enormously influential in a range of scholarship. However, if a psychoanalytic interpretation is no more scientifically valid than any other kind of interpretation, it runs the risk of becoming just one more narrative. 'Verstehen' or understanding of humans needs to be complemented by 'erklären' or knowledge of their brain–minds: or put differently, psychological science needs to address both meaning and mechanism (Table 2).[19]

Just as 19th-century physics influenced psychoanalytic theory, so the advent of modern computers was key in driving the 'cognitive revolution' in the psychological sciences. Work on computational metaphors of the mind drew on the rigorous empirical efforts of behaviourism, but allowed psychologists to explore a broader range of cognitive phenomena. Early work conceptualized the mind as a serial processor. John von Neumann, Claude Shannon, Alan Turing, and other pioneers in computer science developed new ways of thinking about information, and this provided a solid basis for the metaphor of the mind as an information processor.[20]

The information processor model of the mind was useful in conceptualizing and exploring a range of mental phenomena, including working memory. Nevertheless, the metaphor of mind as information processor is problematic in

---

18. See later footnote on interpersonal schools of psychoanalysis. For more on **attachment theory**, see John Bowlby's 'Attachment and Loss' [415], David Wallin [416], Daniel Spiegel [417], and Jude Cassidy and Phillip Shaver [418].

19. As previously noted, the contrast between 'erklären' and 'verstehen' emerged during 19th century debates between authors in philosophy, sociology, and other fields, about whether the natural sciences and the social sciences require different methods. In the case of psychoanalysis, key authors who have regarded **psychoanalysis as a hermeneutic practice** include Hans-Georg Gadamer, George Klein, Paul Ricoeur, and Roy Schafer.

20. Warren McCulloch and Walter Pitts were perhaps the first to view the mind as a Turing machine [419]. Wilfrid Sellars's account of mental states as 'theoretical entities' is also considered a forerunner of **functionalism** [420]. Hilary Putnam subsequently introduced (and later rejected) a computational model of mind into philosophy, contrasting behaviourism (mental states are behavioural dispositions) and physicalism (mental states are brain states) with functionalism (mental states are characterized by their causal roles) [421]. See also Michael Rescorla [422].

several respects. First, this metaphor ignores the unique substrate of the mind: the brain. Information processing based on computer digits is entirely different in nature from the way in which the brain–mind perceives and acts within the world. Similarly, the functionalist metaphor of the mind as software, running on the hardware of the brain, just isn't the way that the brain–mind works. Second, if early psychoanalytic theory overly emphasizes emotion, the information processing metaphor seems to downplay affect. Third, this metaphor gives short shrift to the context in which information processing is taking place; unlike the relatively constrained algorithms running on computer chips, human cognition and emotion take place within relationships and societies.[21]

Ongoing developments in computational theory and computer chips saw the metaphor of mind as a linear information processor replaced by the metaphor of mind as a parallel processor.[22] Neuronal networks could be built to model such parallel processing in a rigorous way, and ongoing modifications to such neural networks provided more and more robust models. This perspective, which became known as connectionism, had a number of clear advantages for cognitive-affective science; it seemed closer to the way neurons work, and it provided sophisticated models of a broad range of cognitive phenomena, including facial recognition, memory retrieval, and visual processing.

When I was a resident in psychiatry at Columbia University, connectionism was just emerging in the cognitive sciences. David Forrest was one of our lecturers and fond of drawing our attention to the most avant-garde of notions. He was enthusiastic about the idea that neural network theory could be used to model different mental disorders, with 'lesions' to networks resulting in phenomena that paralleled clinical symptoms. Given my interest in computing, I was intrigued, but I suspect most of my fellow residents regarded Dr. Forrest as entirely eccentric. It's remarkable, though, how rapidly and robustly this field has developed in subsequent decades. Neural networks now lies at the heart of many modern artificial intelligence applications, including AlphaZero's chess-playing. And the current Director of the National Institute of Mental Health, Joshua Gordon, has ardently promoted the field of 'computational psychiatry'.[23]

---

21. **Richard Rorty** argues that while the hardware metaphor may apply to brain, the software metaphor is best thought of in terms of culture [74]. Drawing on Ryle and Wittgenstein, his view is that it is a mistake to think of mind in terms of mechanisms.

22. **Connectionism** has a long history in both philosophy and psychology, including contributions from Aristotle, William James, Warren McCulloch and Walter Pitts, Donald Hebb, and many others [423].

23. Lesions are typically used in animal neuroscience research, hence the use of quotation marks here. Influenced in part by David Forrest, when I returned to South Africa, I teamed up with computer scientist Jacques Ludik and edited a volume on '**Neural Networks and Psychopathology**' [424], which collected a range of these sorts of models. It turns out that Josh Gordon was recruited to the National Institutes of Mental Health from Columbia University, and see his more recent volume with A. David Redish, 'Computational Psychiatry' [425]. See earlier footnote on AlphaZero's remarkable achievements in playing chess and other games.

A connectionist perspective is consistent with a general shift in cognitive-affective science from a focus on symbolic cognition to one on situated cognition. Just as interpersonal schools of psychoanalysis argued that there was a need to go beyond models that emphasized psychic drives to models that emphasized interpersonal relationships, so situated cognitivists have been concerned to understand cognition within the context of human bodies in interaction. With this turn to the body, cognition and affect are increasingly seen as embodied (in the brain–mind) and embedded (in interaction with the world). The literature on embodied cognition has become quite fond of 'E's', with contributors adding terms such as enactive, extended, emotional, evolutionary, and exaptative to characterize this model of the brain–mind. Notably, this approach is consistent with some strands in continental philosophy (such as the work of the French philosopher Maurice Merleau-Ponty), and also with cognitive science research which has focused on metaphor as key in cognitive-affective processing (such as the work of George Lakoff and Mark Johnson).[24]

There have also been a number of interesting attempts to integrate psychoanalytic and cognitivist views of the mind. For one thing, cognitive-affective scientists fully accept the notion that much mental processing takes place outside of consciousness, and indeed have developed a large body of empirical work on the 'cognitive unconscious' and the 'emotional unconscious'.[25] Conversely, a number of psychodynamic theorists have emphasized the relevance of cognitive constructs in their work. Mardi Horowitz, a psychoanalyst based in San Francisco, made useful contributions to our understanding of trauma using such models. A number of psychotherapy treatments, such as cognitive-analytic therapy, emotion-focused therapy, and schema therapy, have also brought together

---

24. For early work on the **symbolic vs situated cognition** contrast, see, for example, D.A. Norman [426]. Note that James Gibson's concept of 'affordances' plays a key role in work on situated cognition and has overlaps with the work of Dewey, Merleau-Ponty, and Ruth Millikan [427, 428]. There are also resonances between a focus on situated cognition and what Rom Harré termed the second cognitive revolution [429]. For more on the extended mind in particular, see Andy Clark and David Chalmer [161]. The term '**4E**', referring to embodied, embedded, enactive, and extended cognition, emerged from discussion in 2006 and from a conference of this name organized by Shaun Gallagher in 2007 [430, 431], and can be considered to represent a third cognitive revolution. Mark Johnson usefully put forward three additional Es, namely emotional, evolutionary, and exaptative [432]. See earlier footnotes for references to the corporeal turn and to embodied realism.

25. John Kihlstrom has provided a number of useful reviews of the '**cognitive unconscious**' [433] and the '**emotional unconscious**' [434]. He has summarized four lines of research in this area: the distinction between automatic (unconscious) and controlled (conscious) mental processes, dissociations between implicit and explicit memory, interest in 'subliminal' perception, and research on hypnosis. Dual process theories often distinguish between implicit unconscious processes and explicit conscious processes (see also Nobelist Daniel Kahneman's delineation of a System 1 and System 2 way of thinking in his volume on 'Thinking Fast and Slow' [435]). Notably, William James made an important early contribution to work on the cognitive unconscious, and his notion of a 'fringe of relations' is redolent of schema theory [436]. More recently, Lakoff and Johnson have provided a particularly useful approach to the cognitive-affective unconscious from the perspective of embodied cognition [159].

ideas from both psychoanalytic theory and cognitive-behavioural theory; these may be particularly attractive to clinical trainees and practitioners who wish to integrate different theoretical constructs in psychotherapy.[26]

Schema therapy has become an increasingly sophisticated and evidence-based approach, and it is worth saying something about this in particular. I earlier mentioned watching Jeffrey Young work with patients; I was hugely impressed by his ability to quickly develop a connection, understand personality traits, and push for change, while still maintaining rapport. Jeff's working model is based on the concept of early maladaptive schemas or dysfunctional ways of thinking and feeling that arise in early childhood. For example, early exposure to abuse may lead to a 'mistrust schema', which sees others as untrustworthy, and which hampers the development of loving relationships. The schema construct has a rich basis in cognitive-affective science, and is redolent of Lakoff and Johnson's concept of cognitive-affective maps (they use the term 'image schema'). Jeff's first volume, published in 1989, launched schema therapy. I was inspired by his ideas and signed up for advanced training in schema therapy under his direction in New York. Schema therapy has since spread across the globe and has developed a growing evidence base.[27]

Jeff's work falls squarely in the cognitive-behavioural tradition, with rigorous operationalization of constructs and a collaborative therapeutic alliance. At the same time, schema therapy pays close attention to how ways of thinking-feeling emerge in childhood, it gathers a range of data about thinking-feeling (including from dreams), and it pays close attention to how ways of thinking-feeling impact the therapeutic alliance. Taken together, then, it draws together concepts from attachment theory (which has a strong neuroscientific basis), psychodynamic psychotherapy, and cognitive-behavioural therapy. It also provides useful operationalizations of its constructs (e.g. self-rated measures of early maladaptive schemas), together with an approach that can be manualized and tested. In my mind it exemplifies the best of current work in integrative psychotherapy.[28]

---

26. As discussed later, writers committed to integrating psychoanalytic ideas with more modern ones tend to use the term '**psychodynamic**', indicating a wish to move beyond classical Freudian theory. See, for example, Mardi Horowitz [437]. Witnessing the battle of ideas between Roger MacKinnnon and Jeffrey Young led to me grappling with these ideas; see, for example, my volume on 'Cognitive Science and the Unconscious' [438]. See later footnote for references to the literature on integrative psychotherapy.

27. I have been privileged to work with **Jeffrey Young** on a couple of edited volumes [439, 440].

28. Over the years, there has been a growing literature on **psychotherapy integration**, including work attempting to integrate constructs and methods from different schools of psychotherapy, including psychoanalysis and cognitive-behavioural therapy; see, for example, Paul Wachtel [441] and Hal Arkowitz and Stanley Messer [442]. For a recent review of work in this area, see L.G. Castonguay and colleagues [443]. Such work may have implications for work in global mental health [298]. The construct of 'thinking-feeling' raises the question of the relationship between cognition and emotion, the topic of the next chapter.

2.4 Brain–mind as wetware

Are developments in cognitive-affective and psychiatric science useful in approaching the fundamental and parallel debates about the body–mind problem in philosophy and in psychology? For one thing, it may be helpful to employ a metaphor for understanding brains and minds that goes beyond standard Freudian and computational metaphors. In this section of the chapter, I put forward the suggestion that it would be advantageous to talk about brain–minds in terms of 'wetware'.[29] Wetware is neither hardware (i.e. computer chips) nor software (i.e. computer programs); it involves an entirely different set of processes, grounded in an entirely different 'stuff', that has evolved in the animal kingdom. Like my preference for terms like 'brain–mind' and 'psychobiology', this metaphor cannot entirely resolve perennial contrasts in philosophy and psychology. However, the notion of wetware emphasizes that the brain–mind is an intricate psychobiological entity, and so reminds us to avoid a reductionism and scientism which ignore the complexities of the brain–mind, including its social context.[30]

The **wetware metaphor may be particularly useful in integrating mechanisms and meanings, and in emphasizing how the brain–mind is embodied and embedded** (Table 5). The hardware metaphor of brain is useful in emphasizing how mind is grounded in brain, but the brain works very differently from any machine. The software metaphor is useful in appreciating algorithmic aspects of the mind, but it doesn't sufficiently address how the mind is so closely bound up with meaning and narrative. A wetware metaphor doesn't fall prey to either of these problems, but is it also productive in any way? How does wetware allow for the possibility of thoughts and emotions, which belong to a particular person and are simultaneously about something in the world?

It may be helpful to consider the Turing Test here (Table 5). The metaphor of the mind as a computer led to debate about whether computers actually think. Alan Turing, a seminal figure in computing and cryptography, began his answer by describing an 'imitation game', in which an interrogator is allowed to text two people sitting in different rooms, and then has to decide which is the man, and which is the woman. He then proposed that if a person texting a machine and a person could not distinguish the two, then it was clear that machines can think. This is very much a behaviourist approach, and several counter-arguments have been raised.

---

29. The term '**wetware**' has been used for some decades now, and by 1988 was the title of a novel [444]. The neuroscientist and computational biologist, Dennis Bray, has defined wetware as "the sum of all the information rich molecular processes inside a living cell" [445]. For the purposes of the current volume, the term alludes to the cognitive-affective processing of the psychobiological brain–mind, while simultaneously indicating that this is distinct from the computational processing undertaken by electronic hardware and software.

30. See earlier footnote on reductionism and scientism.

The best known of these is Searle's Chinese Room thought experiment. In the days before Google Translate, he pointed out that computers might someday be able to convert English text into Chinese characters, so well that an observer might think that the computer understood Chinese. He asks us to imagine a room in which English text is put in at one end, where he has thousands of pieces of paper with English text and their Chinese translation, so that he is able to output Chinese text over time. Searle then argues that even if an observer saw him writing fluent Chinese, he would not actually understand or speak Chinese. This thought experiment underscores important differences between digital processing algorithms and real human understanding.[31]

Steve Harnard has in turn proposed a Total Turing Test. In this test, the interrogator can also examine the perceptual abilities of the subject and the subject's ability to manipulate objects. This test, then, is consistent with recent cognitive-affective science which emphasizes that the brain–mind is embodied and embedded. If a machine were sufficiently human-like that it processed visual stimuli and moved around a room in a way that was indistinguishable from a human, then one would be hard-pressed not to describe it as 'thinking'. Still, from a neuroscience point of view, the kind of thinking of this machine would be quite different from human thinking.[32]

Consider, for example, a computer program which provides psychotherapy. Joseph Weizenbaum, working at MIT, developed ELIZA, an early program that simulated conversation by text using a range of algorithms. The most famous set of algorithms was DOCTOR, which used algorithms based on the principles of the humanistic psychologist, Carl Rogers. If a user typed 'I am feeling bad', DOCTOR might respond, 'I hear that you are feeling bad'. Although some dialogue mimicked a real conversation, an interrogator that understood

---

31. For more about the **Chinese room** thought experiment, see John Preston and Mark Bishop [446]. Lakoff and Johnson argue, however, that Searle himself is using a machine metaphor of the mind when making this argument [159]. So while they agree with his point that computers cannot understand Chinese, they see the Chinese Room debate as misguided in a key way. In his critique of functionalism, **Searle** uses the term 'biological naturalism' to emphasize that mental states are emergent and are as real as other biological phenomena [447]. Critics have argued, however, that he is really presenting a 'biological dualism' [448]. Rafael Núñez, who trained with Lakoff and Dreyfus, has suggested instead using the term 'ecological naturalism' [449], which captures Lakoff and Johnson's point that cognition is embodied and embedded.

32. Harnad later produced a hierarchy of **Turing Tests**: T1 (a test that is easier than the Turing test), T2 (the Turing Test), T3 (the Total Turing Test), T4 (everything is indistinguishable, including neural circuitry), T5 (everything down to the last electronic is the same). There is much debate about such modifications. Robert French has argued, for example, that because computers can't easily answer a range of 'subcognitive' questions that tap into our specifically human ways of engaging with the world (e.g. which smells better, freshly baked bread, or newly mowed lawn?), T2 is set too high [450, 451]. A related question concerns the "philosophical zombie": a being which behaves like a human, but which doesn't experience consciousness or qualia. David Chalmers has argued that the possibility of such zombies refutes physicalism [452], but this convinces few emergent materialists [453, 454]. For more on the differences between computational and biological thinking-processing, see R. Brooks and colleagues [455].

the principle behind the algorithms could quickly see that he or she was talking to a computer. Still, Weizenbaum found that some users attributed feelings to the computer.[33] Nowadays, chatbots employ much more sophisticated computations, including those based on neural networks, and provide much more realistic simulations, but it is nevertheless quite clear that they are not thinking-feeling therapists.[34]

Let's consider some human activities that have huge therapeutic value: dancing and singing. As an African I'm aware of the long history of dance and music on the continent, and how it has hugely influenced the world's dance and music. Early Khoi and San paintings depict ritual dances; we now think these were accompanied by trance. Contemporary dance and music may involve a range of other cognitive and affective processes. It's so obvious that computers don't dance, that it barely bears mention. A robot that passed the Total Turing Test would have to dance, which would make it scarily human-like. My strong intuition, however, is that wetware is the only sort of stuff that can produce really complex brain–mind phenomena, from Khoi and San dance to the more standard contemporary experience of emotional joy in listening to a favourite song.[35]

Findings in psychopharmacology, such as the responsiveness of our brain–minds to 'happy pills, smart pills, and pep pills', raise all sort of deep philosophical issues. It's fascinating, for example, that people may have a positive response to chatbots or to inert medications: the placebo response. Classical approaches might emphasis that the placebo response is a conditioned response (i.e. simply talking, whether to a chatbot or a therapist, reliably provides nonspecific relief) or an expectancy response (i.e. we expect medications to work and so they do). A critical approach may emphasize that a placebo response is possible because of the power of healing narratives. An integrative approach may foreground both mechanism and meaning; the placebo response can be characterized as embodied in our particular wetware

33. See **Joseph Weizenbaum** [456]. For more on **Carl Rogers'** approach, see his volume on 'Client-Centred Therapy' [457].

34. **PARRY** was another early computer program; it was developed by psychiatrist Kenneth Colby as a model of paranoid behaviour. PARRY's algorithms were sophisticated, and produced text that really was reminiscent of paranoid patients. ELIZA and PARRY actually 'talked' on a number of occasions. With ongoing advances in artificial intelligence, chatbots can simulate humans to a remarkable degree, and there is ongoing empirical research on their efficacy in the treatment of psychiatric disorders.

35. See the work of the South African archaeologist, David Lewis-Williams, which emphasizes the importance of altered states of consciousness for understanding **rock art**, exploring both neurobiological aspects (e.g. neuropsychology of hallucinations) and cultural aspects (e.g. role of shamanism) [458]. There is ongoing work at the intersection of **music** and embodied cognition, including work on health [459]. While computers may not yet experience joy in dancing, the Turing test for music performance has been passed [460]. Referring to one's **intuitions** is not uncommon in philosophy [461, 462]. While current computers may not experience joy in dance and song, a number of animal species may well [463–465]. See also later footnote on play, dance, and music.

(and underpinned by a range of specific psychobiological mechanisms), and as embedded in our interactions with healers (which are powerfully meaningful for us). We might therefore propose a Placebo Turing Test: a computer may be considered to think and feel if it demonstrates a placebo response. Put differently, a key difference between wetware and hardware–software is that wetware entails the sort of structures, processes, and mechanisms that can produce a placebo response.[36]

Does a view of the brain–mind as wetware help address philosophical questions about the body–mind?

First, this metaphor goes beyond classic reductionistic metaphors of the mind as merely neurons, as well beyond critical views that mind is something that falls outside the sphere of science. More specifically, a view of the brain–mind as wetware is consistent, instead, with a range of 'emergent materialist' positions in philosophy of mind. I may understand a lot about the molecular chemistry of food, but if I want to be a good cook I also need to know a lot about baking. Similarly, neuroscience is crucial to understanding the brain–mind, but other levels of analysis are needed to understand mental phenomena, ranging from colour perception through to thinking and feeling.

The philosophy of colour is informative here. Over the centuries, philosophers have reflected (ha!, couldn't resist!) on the nature of colour.[37] In 1988 Clyde Hardin published a volume 'Color for Philosophers' in which he brought new neuroscientific knowledge to bear on these conceptual issues; this revolutionized the field. Analytic philosophers had, for example, explored the issue of colour incompatibility (there cannot be reddish greens and greenish reds) in terms of the logic of language. Hardin emphasized, instead, that subjective aspects of colour perception were explained by vision wetware (our perception that red and green are maximally dissimilar is due to the way in which the neural pathways underlying colour processing are wired up).[38]

Second, a wetware metaphor may be useful in addressing questions about the relationship between the brain–mind's internal processing and the external

---

36. The point of this kind of Turing Test is not to argue that only machines that have the same electrons as humans can think (see Harnad's level 5 Turing Test), but rather to try to think through the limits of artificial intelligence (AI) and computational models of mind. Key criticisms of AI (and of computational models of mind) include the work of Kurt Gödel, Herbert Dreyfus, and John Searle (see previous footnotes). Phil Hutchinson provides a useful account of the **placebo response** in terms of embodied cognition, and draws a parallel between philosophical approaches to placebo and approaches to emotion [466]. See also my volume on 'Philosophy of Psychopharmacology' [53].

37. Philosophers who have written about **colour** include Descartes, Locke, Mill, Wittgenstein, Wilfred Sellars, Paul Churchland, and Ronald Giere. See again Howard Gardner [2] and also Barry Maund [467].

38. Although Hardin discusses the neurophysiology of colour perception, his work is aligned with theories of colour which emphasize its mental aspects (colour perception is subjective) [468]. This contrasts with theories of colour which emphasize its physical aspects (spectral reflectance is objective) [469].

context. In the case of colour perception, for example, a wetware metaphor helps emphasize that colour perception involves both objective reality and subjective experience. Evan Thompson, one of the co-authors of the pioneering volume 'The Embodied Mind', proposes an ecological perspective, in which "being coloured a particular determinate colour or shade … is equivalent to having a particular spectral reflectance, illuminance, or emittance that looks that colour to a particular perceiver in specific viewing conditions". This relational or enactive position is consistent, then, with a view of reality as highly fuzzy and complex (the word 'dappled' in particular comes to mind here!), and of wetware as embodied and embedded.[39]

The wetware metaphor does not perhaps entirely resolve the hard mind–body question of qualia. Hardin himself wrote a paper called 'Qualia and materialism: Closing the explanatory gap', where he again notes how neuroscientific data explain aspects of the subjective perception of colour. Still, such explanations go only so far. I'm reminded of John Searle's point that water is just H_2O, but in the liquid state it just feels wet—just as colour perception or even consciousness just has certain qualia. While work on the wetware underlying colour perception may well start to address the hard question of how a qualia feels, my view is that only animals with wetware can fully appreciate what it is like to see a particular colour.[40]

A third advantage of focusing on a wetware metaphor is that it moves away from an overly narrow understanding of mental causation. Critics of physicalism worry about how mental causation works: how does higher level consciousness, for example, lead to brain changes? From a wetware perspective, where mind is embodied in brain and embedded in society, this is an issue about how we phrase things: certain kinds of wetware, for example, underlie visual processing, and during visual processing particular visual stimuli may then trigger a range of wetware responses (e.g. seeing a lion leads to a flight-fight response).

39. See Evan Thompson [470]. Similarly, Lakoff and Johnson argue that colour concepts are 'interactional'; they are neither purely objective nor purely subjective, rather they arise from the interactions of bodies, brains, the reflective properties of objects, and electromagnetic radiation [159]. See previous footnote on the varieties of 'pluralistic realism'. For an argument that explicitly emphasizes the concept of **colour pluralism**, see V. Mizrahi [471].

40. For details of this work on 'Qualia and materialism: Closing the explanatory gap', see Hardin [472]. Searle uses this metaphor in different writings [447], arguing that materialism denies the existence of a mind–body problem, while dualism makes the problem insoluble. Note that animals that are wired differently likely perceive colours differently. Note also that the literature on qualia and colour overlap in work on **'Mary's Room'**; this is a thought experiment in which a scientist who lives in a monochromatic world, and who knows everything about the neurophysiology of colour, is exposed for the first time to colour. The question is then asked, does she learn anything new? The thought experiment has earlier roots, but was put in this particular form by Frank Jackson [473]. It is likely that Mary says 'Wow, so that's what red looks like!', consistent with the position of emergent materialism.

Talking about how consciousness of the visual stimuli is a mental cause is a reasonable metaphor for what is happening.[41]

A fourth advantage of the wetware metaphor is that it partly helps address the question of animal minds. Philosophers and scientists have long thought of animals as machine-like, without minds or souls. The discovery that the mind is based in the brain is fairly recent, and the idea that animal brains are sufficiently complex to support minds that can experience a range of pleasures and pains is surprisingly new. Given how much neuroscience supports similarities in brain–minds across the animal kingdom, the philosophical question of whether animals have minds seems increasingly outdated. Certainly brain–minds exist across numerous species, and brain–minds in primates are similar in many ways. This is less and less surprising in terms of ongoing findings in genetics: humans and chimpanzees share around 99% of their DNA.[42]

Does this view help address psychiatric issues? Well, yes, by moving beyond simplistic approaches, this view may be helpful in the clinic in a number of different ways.

First, by emphasizing that whenever we talk about the brain–mind we use metaphors. Thus when a psychodynamic psychiatrist indicates that we are bringing a thought to consciousness, he or she is using a metaphor of the mind as divided into parts, some of which are hidden. Or when a cognitive-behavioural therapist indicates that thoughts underlie emotions, he or she is using a metaphor

---

41. There are several philosophical debates to unpack here. First, there is the question of how to understand **causation**. As noted earlier, Lakoff and Johnson have described multiple metaphors of causation [159]. This sort of view of causation may have been first proposed by Elizabeth Anscombe [474]. In biology and neuroscience, the metaphor of 'mechanism' is often used, a point we will return to later. Second, there is the more specific issue of **mental causation**. In psychiatry, we need ways of describing both brain → mind and mind → brain causality [475]. For more on mental causation, see D. Robb [362a], James Woodward [363], C.F. Craver and W. Bechtel [364], and Nancey Murphy, George Ellis, and Timothy O'Connor [365]. For a view that emphasizes embodied-embedded cognition, see Wim de Muijnck [476]. Third, there is the related issue of the **causal status of reasons**. One camp, exemplified by John Mackie, Roy Bhaskar, and Derek Bolton, sees reasons as causal, and views psychology as one of the sciences (see, e.g. Mackie's volume on 'The Cement of the Universe'; the title is from Hume, and Mackie explores Hume's view of causation [477]). A contrasting camp, exemplified by Ludwig Wittgenstein and the successor to his Chair, Georg von Wright, holds that reasons cannot be causal, and given that brain and mind have entirely different properties, posits that there is an unbreachable gap between explaining brains and understanding minds. (Wittgenstein writes, "the investigation of a reason entails as an essential part one's agreement with it, whereas the investigation of a cause is carried out experimentally", cited by Jacques Bouveresse [249]). See earlier footnote on the Pittsburgh School's distinction between the space of causes and a norm-governed space of reasons (and see Chauncey Maher [187]). For Lakoff and Johnson, seeing reasons as causes is a useful metaphor (based on a view of both causes and reasons as forces) [159]. A partly similar position in the philosophy of folk psychology emphasizes the narrative structure of reasons [478].

42. For more on **animal minds**, see Don Griffin [479], Kristin Andrews and Jacob Beck [480], and Peter Carruthers [481]. My Lasker Award-winning collaborator, Mary-Claire King, and her PhD supervisor Allan Wilson, did pioneering work in the area of chimpanzee genetics [482]. See later footnote for references on animal emotion.

of the mind as a symbol processor, where certain thoughts lead to certain outcomes. There is nothing wrong with such metaphors and models per se. But we need to realize how far from reality they often are. When we tell a patient, for example, that we are prescribing medication in order to address their 'nerves' or to right an 'imbalance', we use metaphors that may be useful, but that can also be characterized as neuromythology.[43]

Second, by providing a sharper way of thinking about a range of brain–mind disorders that are seen in clinical practice. Consider, for example, patients with dissociative disorders: current textbooks distinguish epilepsy as a brain disorder, from psychogenic nonepileptic seizures as a mental disorder. In reality, however, things are much more complex than this. Many patients with epilepsy have psychogenic nonepileptic seizures, as well as a range of other psychiatric symptoms. Brain lesions have been found to underlie both epilepsy and psychogenic nonepileptic seizures, and the mind is involved in the expression and experience of both epilepsy and psychogenic nonepileptic seizures. Both epilepsy and psychogenic nonepileptic seizures are best understood as brain–mind disorders.[44]

Third, a wetware metaphor helps emphasize that we can alter brain–mind disorders by targeting a variety of underlying mechanisms, using a number of different interventions, ranging from psychotherapy to pharmacotherapy. An appreciation of explanatory pluralism may help lay a foundation for thinking through the classification issues that face psychiatry; wetware is involved in a broad range of issues from 'problems in living' through to severe psychiatric disorders, and we must accept that there is rarely a one-to-one relationship between any specific psychiatric symptom and any particular underlying causal mechanism.[45] This perspective may also help encourage new ways of thinking about prevention and intervention; in particular, we need to go beyond simply talking of changing hardware (with medication) and software (with psychotherapy). Interestingly, the term 'biohacking' has recently become popular; this

43. See earlier footnote on neuromythology. See later footnote for a debate about use of the 'chemical imbalance' metaphor and inappropriate medicalization. For more on metaphors in science, see Haack [79].

44. There have been attempts to argue that psychiatric disorders are brain disorders; we'll return to this point when we discuss the future of biological psychiatry. For more on what have been termed 'conversion symptoms' (by psychiatrists) and 'functional neurological symptoms' (by neurologists), see J. Stone and R. Davenport [483]. While there have been attempts to do away with the **organic-functional distinction** in psychiatry, it persists [342, 484, 485].

45. While an approach that emphasizes multiple causes of and multiple interventions for brain–mind disorders makes good sense from a scientific perspective, it has important disadvantages. First, while sensitive and specific **biomarkers** may be found for brain disorders, they are much less likely to be found for brain–mind disorders. Second, while **silver bullets** can be found for some medical conditions (e.g. vaccines for polio, or bed nets for malaria), there are likely fewer silver bullets for brain–mind conditions (making it harder to persuade funders to invest) [486].

word emphasizes how science and technology can perhaps be used in a range of creative ways to improve our own brain–minds.[46]

Fourth, I'd suggest that the wetware metaphor is consistent with a schema model of the brain–mind. Although the schema model is often used at a more mind level rather than at a more brain level, there has been considerable basic and clinical research on schemas and how they are embodied and embedded in the brain–mind. Basic neuroscience research, for example, has mapped out how the brain grounds body schemas, with different brain regions mapping different parts of the body. Clinical work on schema therapy is also consistent with a wetware metaphor; for example, early maladaptive schemas likely entail specific psychobiological changes as well as particular 'modes' of thinking-feeling about the world. Put differently, work on schemas is consistent with an emphasis on both mechanism and meaning. We'll expand on this idea in the next chapter.[47]

2.5 Conclusion

We've seen that philosophy and psychology are characterized by a number of long-standing and overlapping debates about the nature of the mind. Philosophers have argued about physicalist and dualist answers to the body–mind problem, and psychologists have developed different schools including behaviourism and existentialism. This chapter has suggested that the **metaphor of the brain–mind as wetware is consistent with current cognitive-affective and psychiatric science, and provides an integrative approach that goes beyond a number of earlier failed metaphors and models.**

More specifically, if the body–mind is wetware, and if it is embodied and embedded, then we can study it scientifically (drawing on the best empirical traditions of positivism and behaviourism, and avoiding skepticism) in all its complexity (drawing on the best hermeneutic traditions in philosophy and psychology, but avoiding scientism). Work on schema models of the brain–mind, for example, is consistent with the wetware metaphor, and exemplifies research that is both scientific and complex. Insofar as the wetware metaphor draws on

46. For more on the application of ideas regarding **embodied and embedded cognition to psychiatric disorders**, see T. Fuchs and J.E. Schlimme [487], F. Röhricht and colleagues [488], S. Gallagher and S. Varga [489], L.J. Kirmayer and A. Gómez-Carrillo [490], S. de Haan [491], and D.J. Stein [492].

47. Head and Holmes first described the body schema on the basis of a study of cerebral lesions [493]. A neurobiological line of research has become increasingly sophisticated [494]. The **construct of schemas** has, however, been used in range of ways, including by Kant, Frederic Bartlett, and Jean Piaget (see earlier footnote on Kant's schemata). It has been asserted that the schema construct is not consistent with Aristotle's ideas, but Aristotle's theory of mind is open to different interpretations [316, 495]. For a discussion of schemas in relation to embodied cognition, see Shaun Gallagher [496] and Raymond Gibbs [497]. For reviews of some more recent uses of the term schema, see D.J. Stein [498] and K.L. Plant and N.A. Stanton [499].

current science, this metaphor may facilitate a more fine-grained approach to reality than earlier metaphors, including a better response to the dictum of 'Know thyself', as instructed by thinkers from Socrates to Freud.

Of course, we could be better off not knowing ourselves. This is a view that has also been put forward by a number of philosophers and psychologists, including some of those doing research in the area of self-deception (who show how prevalent and important self-deception is in *Homo sapiens*), as well as by those who argue that the meaning of life does not rely entirely on self-insight. We will come back to this debate later on. For now, given that wetware has some very unusual and intriguing features, we go on in the next chapters to consider some of these fascinating characteristics in more detail. We began with a particularly important fact about our brain–minds; they not only think, but they feel.

Chapter 3

Reason and passion

In the digital age, when we think about brain–minds, we typically think about 'cognitive stuff' such as information processing. In my work as a psychiatrist, though, I and my patients are often focused on quite different issues, such as those involving stress and trauma, where our responses include anxiety and sadness. So this work immediately involves 'emotional stuff'. One of the most interesting features of wetware is that it produces not only thoughts but also emotions. Classical metaphors of the mind as a serial computer don't seem up to the task of being able to fully explain this: will a wetware metaphor perhaps be more successful?

Philosophers and psychologists from Plato onwards have contrasted reason and passion.[1] This contrast involves conceptualizing the mind in terms of its different, sometimes conflicting components (using the metaphor of mind as comprised of different parts). Cognitive-behavioural psychotherapy, the most evidence-based form of psychotherapy, relies on a model of thoughts leading to emotions. Additionally, throughout Western thinking, one of the ways in which we think through conflicts between reason and emotion is by using the metaphor of emotion as a force. Psychoanalytic theory has continued to rely on this sort of metaphor.

In this chapter we'll begin by considering some philosophical approaches to the question of how reason and passion intersect, before going on to consider some psychological approaches to the same question. We'll then draw on the resources of psychobiology, to argue that like the brain and mind, cognition and emotion are deeply linked. With these resources in hand, we'll conclude that the wetware metaphor in general, and a schema model of cognitive-affective processing in particular, are particularly useful for thinking through how the brain–mind integrates cognition and emotion, and that this view has useful clinical applications.

1. In the 'Republic', for example, Plato described a tripartite model of soul, comprising reason, spirit (including emotions), and appetite (or desire) [376]. Parallels between **Plato and Freud** have been drawn [500, 501]. However, such parallels may simply reflect the existence of long-standing metaphors of the mind.

Problems of Living. https://doi.org/10.1016/B978-0-323-90239-7.00006-7

3.1 Philosophy of reason and passion

Western philosophy has often valued reason over passion. The Stoics, for example, held that emotions often entail false judgements. In another common view, human reason is seen as distinctive: humans are the only animals able to tame their animal urges. This position is also taken in a good deal of theological thinking, where the conflict between reason and passion is also one between good and evil. René Descartes' view is that the essence of human nature is reason ("I think therefore I am"); in his perspective, reason is disembodied and spirit-like, in contrast to emotion, which can be understood in bodily and mechanical terms.

A long-standing tradition in philosophy of emotion has emphasized that emotions are primarily subjective feelings, with a modern version of this tradition arguing that emotions are feelings caused by perceptions of changes in the body. This latter view was first put forward by James, who focused on perceptions of changes in the autonomic and motor functions, and who held that "our feeling of [bodily] changes as they occur IS the emotion". There are also views in philosophy of emotion that see emotions as well-differentiated essential natural kinds, perhaps entailing particular beliefs and desires, and corresponding to specific physiological signatures.[2]

At the same time, there has been an argument that emotions tell us something important about the world, and our role in it. Within the realm of theology there is a view that our connection with a higher being goes far beyond the purely rational. Hume argued that reason alone cannot move people to action; rather "Reason is, and ought only to be the slave of the passions and can never pretend to any other office than to serve and obey them". In France, Jean-Jacques Rousseau's romantic view emphasized the importance of compassion as providing a foundation for a moral society, and Jean-Paul Sartre's existential position emphasized emotion as a key form of consciousness and understanding. In contrast to Descartes, the pioneering affective neuroscientist Jaak Panksepp argues, "I feel therefore I am".[3]

Furthermore, a long-standing tradition in philosophy of emotion has emphasized that emotions are evaluative, and that they have a key role in guiding

---

2. The full citation is "Our natural way of thinking about emotions is that the mental perception of some fact excites the mental affection called the emotion, and that this latter state of mind gives rise to the bodily expression. My thesis on the contrary is that the bodily changes follow directly the PERCEPTION of the exciting fact, and that our feeling of the same changes as they occur IS the emotion" [502]. For more on James and **neo-Jamesian theories of emotion** such as that of Antonio Damasio, see Edmund Rolls [503] and A. Damasio and G.B. Carvalho [504].

3. Note that while Descartes foregrounds reason, his volume on 'The Passions of the Soul' also details how emotions have important functions [505]. For Hume's sentence, see his 'Treatise of Human Nature' [506]. For more on Rousseau and emotion, see Martha Nussbaum [507]. For more on Sartre and emotion, see his 'Sketch for a Theory of Emotions' [508]; for more on the **passions of the existentialists**, see Sarah Bakewell [509]. For Panksepp's slogan, see his 'The Archeology of the Mind' [358].

our responses. In this view emotions have intentionality, involve cognitive evaluations, and can be more or less appropriate. Some in the evaluative tradition, including Robert Solomon and Martha Nussbaum, identify emotions with judgements, while others see emotions as also including evaluative perceptions. Paralleling these views, affective scientists such as Magda Arnold and Richard Lazarus have developed the notion of appraisal, which leads to attraction or aversion. There are also views in philosophy of emotion which emphasize that emotions overlap with one and other, and are socially constructed.[4]

In some ways these contrasting views of emotion are reminiscent of the classical versus critical divide as a whole. A classic position places significant emphasis on reason; reason can control emotions, which are essential natural kinds with specific features. The critical position says that this approach has things backward; we should highlight and appreciate the importance of emotions, which cannot simply be specified in formal terms, and which are instead intentional and constructed. The classical position often implicitly employs the notion of a disembodied and independent reasoner, while the critical position often emphasizes how emotion necessarily involves lived experience and is embedded within particular social contexts.[5]

As elsewhere in this volume, I am drawn to an integrative position that brings together the best of both the critical and the classical approaches, recognizing the importance of both reason and passion in good decision-making and other brain–mind cognitive-affective processes. Dating back to Aristotle, and arguably including Spinoza and Hume, there is a tradition acknowledging that reason and passion are deeply interconnected, and that both are key for our flourishing. A complementary tradition, dating back at least to John Dewey, has emphasized that emotions emerge from an interaction of the organism with the world. Put differently, emotions are not merely cognitive calculations nor simply bodily feelings; they are embodied in the brain–mind and they

4. Andrea Scarantino and Ronald de Sousa outline three broad traditions of **theories of emotion**: the feeling tradition, the evaluative tradition, and the motivational tradition [510]. Nevertheless, there is contestation in the literature about where any particular thinker or group of thinkers fit, and they note that with the feeling tradition now including work on evaluative feeling (e.g. Peter Goldie), and the evaluative tradition now including work on evaluative perceptions (e.g. Jesse Prinz), many current theories are hybrids. Thus Prinz's Hume-inspired and neo-Jamesian theory of evaluative perceptions sees emotions as 'embodied appraisals' analogous to perceptions, responsive to external stimuli, and reliant on a specific somatosensory system [511]. Martha Nussbaum argues that the Stoics view all emotions as false judgements [507], while Margaret Graver argues that the Stoics held that emotions are affective responses can be right or wrong [512]. The debate is complexified by differences between different Stoics [99]. Hume is often viewed as falling within the feeling tradition, but he also emphasizes the cognitive antecedents of emotion [513]. Jerry Fodor has gone so far as to claim that Hume's 'Treatise of Human Nature' [506] is the foundational text in cognitive science [514], and see Phillip Reed and Rico Vitz's volume on Hume's moral philosophy and contemporary psychology [515].

5. For Wittgenstein, for example, emotions are not universal forms, but rather are inextricably embedded in 'the stream of life' [39].

are embedded in particular contexts, where they play an integral role in a broad range of intraindividual and interpersonal processes.[6]

Aristotle had two additional key insights. First, there is his view on emotional development; he argued that practical judgement is obtained during a process of teaching and training, directed by a mentor for whom the student has affection, in which appropriate habits of deliberating, feeling, and acting are learned and practiced over time.[7] Lifelong learning and practice leads to practical wisdom, characterized by the appropriate engagement of universal principles with particular situations, so allowing the 'priority of the particular'.[8]

Second, there is Aristotle's view of the expression of emotions, as aiming at 'what is intermediate', the so-called golden mean. Thus Aristotle emphasized the importance of expressing the right degree of emotion at the right time for the right purpose, and in the right way. Determining the right degree of emotion at the right time for the right purpose in the right way is no easy task; a good

---

6. In his 'Nicomachean Ethics', Aristotle suggests that the **relationship between reason and emotion** is not one of polarity but 'like the convex and concave sides of a curve' (cited by Edith Hall [516]). He also suggested that emotions involve a belief together with feelings of pleasure or pain, the topics of our next two chapters (cited by Martha Nussbaum [98]). In the previous chapter we didn't discuss Spinoza's theory of the body–mind, but his work criticizes both physicalism and dualism. Taken together, his position on reason and emotion is an integrative one which anticipates key ideas in embodied cognitive-affective approaches; see H. Ravven [517], Michael LeBuffe [378], and Amy Schmitter [505]. For both Aristotle and Hume, virtue is the state in which reason and passion speak with the same voice [518]. Dewey has been identified with the motivational theory of emotion; his view is consistent with both an evolutionary perspective on emotions [92], and with later work on embodied cognition and emotion [519]. Often overlooked is Hume's strong influence on Darwin, leading to important continuities in their work [520].

7. There is a broad range of work on **habits in philosophy**. Xabier Barandiaran and Ezequiel Di Paulo provide an extensive mapping, and suggest that there are two broad approaches [521]. The first begins with Aristotle, includes thinkers such as Spinoza, Ravaisson, James, Dewey, Piaget, and Merleau-Ponty, and conceptualizes habits as self-organizing structures or schemas. The second begins with Locke and Hume, includes thinkers such as Mill, and conceptualizes habits as automatisms or stimulus–response associations. Related work focuses on habitus or social practices, and includes thinkers such as Bourdieu [522]. Felix Ravaisson's work draws on Aristotle, and has been particularly influential [523]. For more on the philosophy of habit, see Tom Sparrow and Adam Hutchinson [524] and Claire Carlisle [525].

8. Aristotle's **moral particularism** emphasizes that the application of general principles without appreciation of specific contexts may be misleading, see J. McDowell [526], Martha Nussbaum [527], and Edith Hall [516]. The phrase 'priority of the particular' is from Nussbaum, who cites Aristotle as saying "Practical wisdom is not concerned with universals only; it must also recognize particulars, for it is practical, and practice concerns particulars" (in his 'Nichomachean Ethics'). A focus on education of children, and on the use of narrative to convey particulars, resonates with Hebrew texts of the Axial Age [528], as well as with contemporary cognitive-affective science [529]. For more on the development of the virtues, including their empirical grounding, see Nancy Snow [530], Michael Slote [531], Julia Annas, Darcia Narvaez, and Nancy Snow [532], and Matt Stichter [533].

deal depends on the particular details of the situation, and we'll consider some exemplars as we go on. Aristotle's view entails taking responsibility for our emotions.[9]

There are a number of remarkable parallels between Aristotelian thinking and Eastern teachings. Ancient Chinese writings, such as those of Confucius and Mencius, for example, emphasize how the heart–mind (or xin) brings together reason and emotion, the value of training the heart–mind to respond with propriety, and the importance of living with balance. Habits and rituals are key for the cultivation of the heart–mind, and 'flexible judgment' is needed to allow the heart–mind to respond most appropriately and sensitively to any particular situation.[10]

Lakoff and Johnson have pointed out that throughout Western thought, a key metaphor sees reason as a force that pushes against passion. While this metaphor is useful for describing some aspects of our cognitive-affective processing, more nuanced models, which help conceptualize the way in which the embodiment of reason and emotion is closely interdigitated, may be helpful. We'll speak more about this in the next sections, where we expand on the notion of schemas and emotion. In the interim, however, it may be useful to consider the relevance of paying mindful attention to our thoughts and emotions: listening to and exploring our bodies/feelings/thoughts, and then attempting to articulate these in a fine-grained and nonjudgemental way.[11]

The idea of mindful attention to our thoughts and emotions is consistent with an integrative position which holds, on the one hand, that both thoughts and emotions can provide important insights, but which at the same time emphasizes the need for careful consideration before any is wholeheartedly adopted. This sort of mindful attention has a long tradition in both Western and Eastern philosophy, and differs from concepts of mindfulness that focus on detachment

9. Practical judgement and the **golden mean** are key in virtue ethics, which we discuss later [534]. The phrase 'what is intermediate' is from Aristotle's 'Nicomachean Ethics', and has analogues in other ideas of the Axial Age, such as the Confucian 'doctrine of the mean'. For more on how Aristotle's ideas about the golden mean draw on medicine in particular, see Tom Angier [241]. For speculations on the history of the golden mean, and its sociopolitical relevance for Aristotle, see J.D. Pappas [535]. For more on Aristotle and emotion in general, see A.W. Price [536], C.D.C. Reeve [537], and Richard Kraut [538]. The idea of emotion as not simply a passive response, but rather an active choice, has been expanded by Robert Solomon [539].

10. For more on **Eastern teachings**, see Kwong-Loi Shun and David Wong [540] (with a concluding discussion by Alasdair MacIntrye), Stephen Angle and Michael Slote [541], Michael Puett and Christine Gross-Loh [542] (which contrasts the moral particularism of **Mencius** with the utilitarianism of Mozi in Eastern philosophy), Douglas Robinson [543], and Michael Slote [544].

11. For more on **embodied and enactivist theories of emotion** that are consistent with Lakoff and Johnson's view, see Andrea Scarantino and Ronald de Sousa [510]. See also P.M. Niedenthal's review [545] and Phil Hutchinson's volume, which rejects neo-Jamesian approaches (for reducing emotions to biology) and cognitivist approaches (for seeing emotions as propositions), and instead proposes a view of emotions as an embodied response to the meaningful world [546].

or compassion.[12] Spinoza, for example, advises us that "An emotion which is a passion, ceases to be a passion, as soon as we form a clear and distinct idea of it". Put differently, attention to our thoughts and emotions may be one useful starting point for us, as we attempt to train our cognitive-affective responses and express them in a balanced way.[13]

3.2 Psychiatry of reason and passion

Debates in philosophy on the relative importance of reason and emotion have played out again in psychology. Freud's views of reason and passion, like so much of his thinking, have been tremendously influential in psychiatry and clinical psychology, as well as in many other fields. In one key version of his model, Freud distinguished the id, the ego, and the superego. The id comprises a kind of seething cauldron of unconscious drives, particularly sexual and aggressive drives, that forms the basis of our all our motivations. The ego and the superego, on the other hand, regulate these drives, either suppressing them, or transforming them into acceptable behaviours.

Freud's writing is ingenious. In his volume on jokes, for example, he supplies a range of fascinating examples, and concludes that jokes are either hostile (expressing aggression) or obscene (allowing exposure). Laughter, an indication of pleasure, takes place when psychic energy used to repress libido is freely discharged. As noted earlier, Freud's explanation relies on a hydraulic model of mental forces in conflict; this model provides an account not only of jokes, but also of dreams, slips of the tongue, and psychiatric symptoms. The model draws in turn on an age-old metaphor of the brain–mind as made up of different compartments, with different forces ensuring expression or repression of thoughts and feelings.[14]

Perhaps it's this consistency with an age-old metaphor that explains how the idea that we should allow emotions to emerge has found its way into so much psychological thinking. Those who emphasize that emotions are feelings may view the expression of emotion as ultimately allowing better control, while those who emphasize that emotions are evaluative may view the expression of emotion as allowing important insights. Many forms of psychotherapy, influenced by psychoanalysis, encourage clients to allow feelings to emerge, with the idea that 'working through' such feelings will in turn lead to therapeutic changes.[15]

---

12. For more on the distinction between **mindful attention and mindfulness**, see again Michael Puett and Christine Gross-Loh [542]. Both constructs may also differ from 'being present', another state that has long been advised by both Western and Eastern philosophy. Seneca, for example, writes "Lay hold of today's task, and you will not need to depend so much upon tomorrow's. While we are postponing, life speeds by" [547]. We'll later discuss limitations of work on mindfulness in more depth.

13. This citation is from Spinoza's 'Ethics' [548]. Note that Spinoza differentiates affects (emphasizing their positive aspects) from passions (which may be confused).

14. For useful insights into **Freud's own sense of humour**, see Elliot Oring [549]. We will discuss humour in more detail in the next chapter, on pleasure.

15. **Ernest Gellner** points out that where Stoicism demands acceptance of external reality by modifying inner reality, psychoanalysis advises acceptance of inner reality [550].

In cognitive-behavioural therapy this approach is tweaked to be more consistent with the cognitive revolution: the fundamental idea is that thoughts lead to emotions, and that in order to change our emotions we need to alter our cognitions. When closely examined, our thoughts are not always rational. So we need to listen not just to our emotions, but also to the thoughts that underlie them. Given the focus of the Stoics on close examination of the rationality of our emotions, it is not surprising that Albert Ellis, a key figure in cognitive-behavioural therapy, described himself as espousing Stoicism.[16]

One of the most important components of my residency training in psychiatry at Columbia University was 6 months spent on the inpatient personality disorder unit. Each resident had only a few patients, and we were supervised intensively by psychoanalysts. We were to taught to listen closely both to patients' emotions and to the emotions they aroused in us. A patient that made us angry, for example, might well be angry herself, and 'projecting' this emotion on to us. We also learned to look out for key emotion-related themes, such as abandonment, and how they influenced what was said in sessions. I have found these ideas invaluable during my subsequent work as a psychiatrist.

How does this square with my earlier sceptical points about psychoanalysis? Importantly, the model used on the unit was not that of classical Freudian psychoanalysis, but rather was based on subsequent writings that are focused instead on interpersonal theories. The term 'psychodynamic' has been used to refer to these more modern psychotherapy approaches that are committed to integrating better validated constructs and to using evidence-based interventions. In many ways, Jeffrey Young's schema therapy can be considered 'psychodynamic' insofar as it relates particular ways of processing thoughts and emotions (schemas) to early childhood, and insofar as the therapist–client relationship is seen as a key tool for assessing the patient and for bringing about therapeutic change.[17]

How evidence based is the idea that expression of emotion is healthy? One immediate concern is the way in which emotions can easily escalate, spiralling out of control. Aristotle's cautionary note about the need to express the right amount of emotion at the right time is consistent with this key point. It's also a point that's been made by those focused on the psychobiology of emotion; in his seminal contribution on the "Expression of the Emotions in Man and Animals", Darwin noted that "The free expression by outward signs of an emotion intensifies it. On the other hand, the repression, as far as this is possible, of all outward signs

---

16. Aristotle provides a number of **philosophical views consistent with cognitive-behavioural therapy** [551, 516]. Parallels between Stoic philosophy and cognitive-behavioural therapy have increasingly been pointed out [552]. Ronald Pies has noted the overlap between cognitive-behavioural therapy and the writings of Maimonides and others in the rabbinical tradition, who have argued how important it is to use reason to understand and to modulate passion [553].

17. For more on **interpersonal schools of psychoanalysis**, see Stephen Mitchell [554]. For more on psychodynamic psychotherapy, see R.C. Friedman and colleagues [555]. For more on the origins of the term 'dynamic' in psychiatry, see Henri Ellenberger [246].

softens our emotions. He who gives way to violent gestures will increase his rage; he who does not control the signs of fear will experience fear in greater degree".[18]

Another concern that springs to mind is work on trauma debriefing. In the immediate aftermath of a trauma, there is frequently a typical stress response with intrusive thoughts, altered emotion, and hyperarousal. The idea that expression of emotion is healthy suggests that we should talk about the trauma, and indeed this is what debriefing therapy encourages. The only snag is that empirical investigation has shown that people who undergo debriefing therapy after trauma exposure are more likely to develop posttraumatic stress disorder! It turns out that sometimes it's important to bottle up one's thoughts and feelings.[19]

Consider also examples from work on anger and forgiveness. When I returned to South Africa from the United States in 1994 at the dawn of democracy there was a great deal of discussion about how best to address past injustices. Archbishop Desmond Tutu, who led the country's Truth and Reconciliation Commission, used a psychoanalytic metaphor of catharsis: one needed to lance the wound, in order for healing to take place. This may well be true in some cases, but there were also notable objections to the Commission. For one thing, testifying in front of such a Commission may not provide a great deal of help to people with posttraumatic stress disorder. On the other hand, there is an argument that the Commission was an important acknowledgement of suffering, and provided a useful societal approach to addressing distress.[20]

While there may be disagreement about how valuable the expression of emotion is, there is a good deal of agreement amongst psychologists on the notion that awareness of one's thoughts and emotions can be useful. Freud himself argued that free association was a helpful technique, and that the goal of therapy was to replace id with ego. Today's mindfulness practitioners tend to fall into a different tradition: one that combines cognitive-behavioural therapy and ideas from Buddhism. Here the approach is not necessarily to obtain insight into the unconscious, but rather to get in touch with one's mind–body and to cultivate an accepting and compassionate approach.[21]

From a wetware perspective, it seems clear that our sort of wetware produces a range of thoughts and emotions that come and go, in a way that we very likely don't fully understand (after all, the cognitive-affective unconscious is likely playing a large role). There is an argument that we don't really know what we think and feel until we actually pay attention to this coming and going, and then articulate it. For me this is one of the most fascinating aspects of writing: putting text to screen lets me know more about what I think and feel. At the same time

---

18. See Darwin's 'The Expression of the Emotions in Man and Animals' [556].
19. See S. Rose and colleagues [557].
20. For more on the **Truth and Reconciliation Commission** and psychiatry, see D.J. Stein and colleagues [558]. See also Martha Nussbaum [559], and note her view that "if anything it is the therapeutic insistence on accessing buried anger that keeps it fixed and immovable, like a stone". See also L. Allais [560], A.A. van Niekerk [561], and R. Oelofse [562].
21. See, for example, Jon Kabat-Zinn and Richard Davidson's work on **meditation** [563]. There's also a growing field of 'compassion science' [564].

this point unfortunately emphasizes the possibility that my understanding of my own cognitive-affective processes is rather more limited than I might imagine.[22]

3.3 Neurophilosophy and neuropsychiatry

Cognitive-affective science provides a number of important insights into these debates.

First, there is the question of how different cognitive and affective neurocircuitry is? It is thought that subcortical structures play a key role in emotional responses: the amygdala, for example, is immediately activated by threatening stimuli, and rapidly contributes to coordinating a flight or fight response. In contrast, cortical structures play an important role in more cognitive processes: the prefrontal cortex is involved in articulating the thoughts and emotions that occur in the aftermath of exposure to stressors. Darwin's work on the evolution of emotions showed that emotional responses to threat have important parallels across the animal kingdom. Our subjective experience of emotions such as fear, and of controlling our responses to threat, gives rise to the metaphors of emotion as forces, and of mind as made up of compartments.[23]

At the same time, neural circuitry is widely distributed, with continuous relays between cortical and subcortical structures, and with cortical circuitry particularly complex in primates. Joseph LeDoux, a neurobiologist who has made pioneering contributions to understanding the neural circuitry underlying fear, argues that inputs from subcortical networks to cortical networks lead to states of consciousness characterized by emotional feelings, and that there is also top down cognitive and emotional modulation of subcortical circuitry. Thus, **while different brain–mind structures underlie different cognitive and affective processes, they also facilitate integrated cognitive-affective functioning**.[24]

---

22. Several writers have made this point about **writing**. See later footnote on Michel de Montaigne's early thoughts about this issue. More recently Flannery O'Connor stated "I write because I don't know what I think until I read what I say" while Susan Sontag stated "I write to define myself – an act of self-creation – part of my process of becoming".

23. See again Darwin's 'The Expression of the Emotions in Man and Animals' [556]. For more recent work on **animal emotions**, see again Marc Bekoff [464], Carl Safina [565], and Frans de Waal [566].

24. **LeDoux** and Brown differentiate their cortical–subcortical view of consciousness of feelings from those of Jaak Panksepp and Antonio Damasio, other key authors on the neurobiology of emotion [567]. Panksepp regards feelings as "implicit procedural (perhaps truly unconscious), sensory-perceptual and affective states" that arise in mammals from evolutionarily conserved subcortical circuits. Damasio initially emphasized the importance of body sensing areas of the cortex in giving rise to feelings, but subsequently proposed that feelings are products of subcortical circuits that receive primary sensory signals from the body. Note that contemporary neuroscience has found that in some subcortical regions, such as the nucleus accumbens, there is a complex 'keyboard pattern' of valence organization, that underpins a broad range of affects, and that can be 'retuned' by the environment [568]. Further, contemporary neuroimaging data indicate that cognitive-emotional behaviours are based in brain areas with a high degree of connectivity, and involve dynamic networks, none of which should be conceptualized as specifically affective or cognitive [569, 570]. Terms such as 'affective cognition' or 'hot cognition' may be used to differentiate cognitive-affective processing that is accompanied by arousal of feelings [571–573].

Second, there is the question of what light neuroscience sheds on the early development of particular cognitive-affective styles and responses. Both animal and human studies demonstrate, for example, that early adversity is associated with specific changes in the neurocircuitry that underpins cognitive-affective processing, and that this in turn is associated with particular kinds of cognitive-affective approaches later in life. Humans with a history of early childhood abuse, for example, may have schemas based on 'mistrust', which view the outside world as threatening, and which predispose them to responding to external stimuli with higher levels of fear and anxiety. While stressful environments may lead to fear and anxiety in any individual, this response is more likely in those with prior exposure to traumatic events.[25]

Such findings make good sense from an evolutionary perspective; particular cognitive-affective styles and responses may be suited to particular environments. Table 4 emphasized the difference between explanations that focus on proximal mechanisms (e.g. amygdala activation in response to threat) versus those that focus on distal mechanisms (e.g. the adaptive value of threat responses in dangerous environments). Randolph Nesse, a father of both evolutionary medicine and evolutionary psychiatry, views emotions as coordinated states which represent adaptive responses that help individuals maximize opportunities and minimize threats. Although some regard emotions as evolved 'brain modules', Nesse sees emotions as prototypes without sharp boundaries: any emotion has many functions, and any function may involve many emotions. His most recent volume is titled 'Good Reasons for Bad Feelings', and emphasizes that emotions provide important information about the world.[26]

Third, there is the question of what neuroscience might have to say about Aristotle's recommendations regarding the importance of cognitive-affective training, and balanced expression of emotion. It is notable that over the course of development, there are changes in our cognitive-affective processing. In the 'marshmallow test', a toddler is presented with one piece of candy, and told that if they don't eat it, then after some time they will get two pieces of candy. Many toddlers cannot wait, but others have greater self-control, a predictor of success in later life. Similarly, there is good evidence that during adolescence, subcortically driven motivations are more intense, while cortical control mechanisms are weaker, consistent with the impulsivity characteristic of this life stage.

25. The term 'cognitive style' was introduced by George Klein [574]. The notion of **cognitive-affective style** raises that of personality; we discuss antisocial personality disorder and some other aspects of personality later.

26. See earlier footnote on Randy's mentorship of me. For a synthesis of his work in **evolutionary psychiatry**, see his volume on 'Good Reasons for Bad Feelings' [575]. For criticism of the modular perspective within evolutionary psychiatry, see Jaak Panksepp and Jules Panksepp [576]. Relevant also is S.J. Gould and R. Lewontin [577], and their critique of 'just so' stories in evolutionary theory. For more on related **debates about sociobiology**, see Marshall Sahlins [578], Ullica Segerstråle [579], and Mary Midgley [580]. Spinoza anticipated Nesse, arguing that while there may be some primary affects, "the various affects can be compounded with one another in so many ways, and that so many variations can arise from this composition that they cannot be defined by any number" [505].

Under particular circumstances (e.g. extreme hunger, fatigue) one can imagine that almost no one can pass this kind of test.[27]

Different psychological traits and mental disorders may be characterized by different kinds of connectivity between brain structures. It turns out that different individuals have different degrees of prefrontal-amygdala coupling. Social anxiety disorder may be characterized by overactivation of the amygdala, while antisocial personality disorder may be characterized by underactivation of the amygdala. The case of Phineas Gage is a fascinating illustration; as the result of an unfortunate accident, Gage suffered a lesion to his prefrontal cortex: this led to a dramatic personality change, with Gage becoming extremely impulsive. In most cases, however, connectivity reflects more nuanced interactions of nature and nurture. There is a growing body of work on the notion of 'emotional dysregulation', how it is characterized by specific alterations in neuronal circuitry, the causal contributions of genes and environments, and its response to treatment. Such intervention may well parallel an Aristotelian focus on the importance of practice, and on expressing emotion with the golden mean in mind.[28]

Fourth, is there any neuroscientific evidence that paying close attention to thoughts and feelings, and articulating and appraising these, is helpful? Cognitive-affective neuroscience has usefully differentiated our explicit knowledge of the world from our implicit knowledge. Perhaps monitoring the products of our cognitive-affective processing provides a way to better understand our own implicit processes. There is also evidence that with recall and verbalization of past memories, new sets of thoughts and feelings can be put in place, and new ways of viewing and understanding the world can emerge. Reappraisal and regulation of our thoughts and feelings is possible, and neuroimaging studies have investigated the particular circuitry activated by these strategies. Further, there is growing research on the neuroscience of psychotherapy, some of it specifically focused on the role of cognitive-affective learning, and enhanced cognitive-affective appraisal and regulation, during such treatment.[29]

---

27. See Walter Mischel's work on the **Marshmallow Test** [581]. This finding has received some criticism; for example, it has been suggested in low-resource contexts, it makes sense to opt for smaller but sooner rewards [582]. We return to the issue of willpower later on.

28. For more on Phineas Gage, see D.J. Stein and colleagues [583]. For more on emotional dysregulation, see E. Sloan and colleagues [584]. For more on the **nature vs nurture** debate, see Patrick Bateson and Paul Martin [585] and Matt Ridley [586]. From a neuroscience perspective, LeDoux emphasizes that nature and nurture both act to change the wiring of synapses [587]. Although the science of genetics is new, discussion of how we are moulded by society is not, and speculatively a proto-concept of nature–nurture intersection has a long pedigree, including the work of Aristotle [300], Spinoza [517], Hume [588], and Dewey [589]. In his 'Nicomachean Ethics', for example, Aristotle says of the moral virtues that they "are engendered in us neither by nature nor yet in violation of nature; nature gives us the capacity to receive them, and this capacity is brought to maturity by habit". We'll return to the issue of human nature later.

29. As noted earlier, John Kihlstrom has provided a number of useful reviews of the 'cognitive unconscious' and the 'emotional unconscious', including implicit versus explicit processing, and see next footnote for further literature. For more on the **neuroscience of cognitive-affective appraisal**, see J.T. Buhle and colleagues [590]. For more on how psychotherapy alters the brain, see Louis Cozolino [591].

At the same time, we need to be cautious both about our ability to have insight into ourselves, and about whether insight is really what brings about therapeutic change. The evidence for a Freudian unconscious as a cauldron of drives exploding against the barrier of the ego is thin, but evidence that most of our cognitive-affective processing is nonconscious continues to grow. Given the importance of emotion, we should extend the phrase 'cognitive unconscious' to 'cognitive-affective unconscious'.[30] While it may be true that with dedicated Aristotelian practice, begun from an early age, we can mould our cognitive-affective habits, and take responsibility for our thoughts and feelings, there are likely also limits to what we can change. We can be proud of the range of current evidence-based psychotherapies, but their goals are often relatively circumscribed. And sometimes robust action, rather than contemplative mindfulness or even psychological change, is called for.[31]

Taken together, then, the **data from science are consistent with the idea that both reason and emotion are needed for optimal decision-making, that cognitive-affective processes may be impacted by early life experiences, and that emotional expression should be balanced** (Table 6). Put differently, neuroscience can be used to support an integrative position in both philosophy and psychology according to which cognitive-affective processing provides us with useful information about the world, but that cognitive-affective processing also reflects past experiences and may go awry in important ways. There is some science to support the idea that mindful attention to our thoughts and emotions may be useful, although given how important the cognitive-affective unconscious is there are likely key limits to self-understanding, and other ways of effecting change in the world may also be helpful.[32]

3.4 Schemas and cognitive-affective processing

Lakoff and Johnson have listed a range of metaphors that we use to talk about emotions, and about the interaction of emotion and reason. These metaphors

---

30. A large literature has developed on the **cognitive-affective unconscious**; see Matthew Erdelyi [592], Matthew, my volume on 'Cognitive Science and the Unconscious' [439], Lakoff and Johnson [159], Timothy Wilson [593], Shankar Vedantam [594], John Bargh [595], and Joel Weinberger and Valentina Stoycheva [436]. A related literatures focuses on **cognitive-affective biases**; see Leon Festinger and colleagues [596], Thomas Gilovich [597], Massimo Piatelli-Palmarini [598], Robert Sternberg [599], Cordelia Fine [600], Carol Travis and Elliot Aaronson [601], Dan Ariely [602], Ori and Rom Brafman [603], Christopher Chabris and Daniel Simon [604], Joseph Halinan [605], Daniel Kahneman [435], David Dunning [606], Mahzarin Banaji and Anthony Greenwald [607], Nicholas Epley [608], Richard Nisbett [609], and Hugo Mercier and Dan Sperber [610].

31. The point that insight is important but perhaps not as important as action is summarized pithily in **Karl Marx**'s famous epitaph, "The philosophers have only interpreted the world, in various ways. The point, however, is to change it" [611]. From a psychological point of view, we will later refer to Paul Dolan's work on "Change What You Do, Not How You Think" [612]. We will also return to the issues discussed here in the sections on self-deception and self-insight.

32. The term 'awry' suggests psychopathology. Certainly there has been debate in the philosophy of emotion on the **rationality of emotions** [510]. We will discuss mistakes in brain–mind systems, and the nature of mental disorder more fully in subsequent chapters.

TABLE 6 Thoughts and emotions.

	Classical position	Critical position	Integrative position
Philosophical roots	Plato, Descartes, logical positivism	Rousseau, Sartre, romanticism, existentialism	Aristotle, Spinoza, pragmatism[a]
Cognition	Cognition is rational	Cognition can be flawed and misleading	Cognition and emotion are intertwined
Emotion	Emotion is irrational	Emotion can provide important understanding	Cognition and emotion together provide insights
Categories of emotion	Emotions are essential natural kinds, with a specific physiology	Emotions are kinds of consciousness, they are socially constructed	Emotions are embodied and embedded prototypes
Cognitive-affective development	Cognition learns to appropriately control emotion during childhood	Emotional responses of childhood should be cherished	Wise cognitive and affective responses are learned over time, via practice
Psychotherapy	Psychotherapy leads to cognition being able to override emotion	Emotion provides useful insights during psychotherapy	Psychotherapy employs and changes cognitive-affective processes

[a]*For a discussion of how Freud straddles the* ***Enlightenment*** *(focusing on rational methods, as did Hume and his heirs) and* ***romanticism*** *(focusing on the irrational, as did Rousseau and his heirs), see Jerome Bruner [613] and Martin Bergmann [614]. As noted in an earlier footnote, Hume is often viewed as foregrounding emotion, but he also emphasizes the cognitive antecedents of emotion and the importance of sympathy [513]. While Spinoza may fall into the group of philosophers for whom emotion is something to be controlled, as noted in a previous footnote, there are other aspects of his work that push him closer to an integrative and embodied position [517]. As again noted in a previous footnote, for a view on how Dewey's pragmatism also supports positions in embodied cognition and emotion, see Mark Johnson [118].*

include 'Emotional Reaction is Feeling' and 'Emotional Experiences are Physical Forces'. They emphasize that in folk or common-sense psychology, our metaphors suggest that the mind has different faculties, and that emotion can disrupt reason. They point out, however, that in fact human thinking is embodied and emotionally engaged, it's often imaginative and metaphorical, and it is mostly unconscious. Furthermore, human thinking occurs within particular contexts; it's embedded within particular interpersonal relationships.

Are there sharper metaphors and models for thinking about the relationship between cognition and emotion? Earlier we suggested that the metaphor of wetware helps us to address perennial parallel debates in philosophy and psychology. The kinds of work in cognitive-affective and clinical science that have been summarized here suggest a model of wetware-based cognitive-affective schemas, which govern a range of often nonconscious cognitive-affective processes. Indeed, a focus on **wetware-based cognitive-affective schemas may be helpful in integrating different approaches to cognition and emotion, and in emphasizing that cognitive-affective processing is embodied and embedded** (Table 6).

Consider, for example, the schema that Jeffrey Young terms 'unrelenting standards'. He posits that this schema is more likely to develop in families characterized by high levels of parental criticism. In adult life, the person with an unrelenting standards schema may respond to criticism with feelings of guilt and thoughts that he or she is a failure. The schema is instantiated in the person's wetware, where it is responsible for specific cognitive-affective processes and products. The schema is not always active, but it may be triggered by particular kinds of social interaction. In some cases, the schema may be useful, ensuring high performance. In other cases, thoughts and emotions focused on the theme of "I am not good enough" may lead to depression.

The concept of wetware-based cognitive-affective schemas may be useful in integrating different positions in philosophy of emotion.

First, a focus on wetware-based cognitive-affective schemas may be helpful in moving away from a model of the mind as comprised of different compartments and of emotion as a force that disrupts cognition, and towards an integrated model of cognitive-affective structures, processes, and mechanisms that are embodied in the brain–mind and embedded within particular contexts. Consider Aristotle's advice about habit training and emotional balance. As a child is brought up, so his or her wetware instantiates cognitive-affective ways of understanding the self, others, and the world: cognitive-affective schemas. Where the child has an affectionate bond with adults, and where the environment is safe and supportive, such schemas will later likely facilitate efficient and appropriate responses. With practice our wetware develops in such a way that we are able to achieve greater wisdom, greater proficiency in our decision-making and practices, and greater balance in our emotional expression over time.[33]

Second, a model of wetware-based cognitive-affective schemas moves us away from a focus on emotions as essential categories, and towards an emphasis on emotions as graded entities (with more central and more peripheral exemplars), that facilitate adaptive responses to the environment. A comprehensive

33. John Dewey, as noted earlier, emphasized that knowledge was gained through interaction of an organism with its environment. Similarly, as again noted earlier, **Jean Piaget** carefully outlined how during cognitive development there is schema assimilation and accommodation; assimilation uses schemas to interpret new information, while accommodation alters schemas to incorporate new information. Notably, Piaget, who was interested in but critical of psychoanalysis, also attempted to integrate affectivity into his constructs [615]. See also Hans Furth [616], and earlier footnote on emotional embodiment.

explanation of cognitive-affective schemas requires an account both of the specific genetic and environmental mechanisms that shape them, as well as of how they are adaptive from an evolutionary perspective (Table 4). It increasingly seems **that psychobiological mechanisms typically result in dimensional traits, and that evolution does not ordinarily produce essential natural kinds but rather leads to complex and fuzzy phenotypes.**[34]

Third, philosophy of emotion has increasingly intersected with affective science, and we have noted that schema theory is consistent with a range of basic and clinical scientific findings on cognitions and emotions. Basic neuroscience, for example, has demonstrated the neural basis of body schemas, and has investigated the neurocircuitry of particular cognitive-affective styles. There is also more clinically oriented work, such as that on how early traumatic experiences lead to changes in cognitive-affective schemas and in cognitive-affective processing. In those with histories of early adversity, for example, we can start to link changes in neuronal circuitry with alterations in both cognitions and emotions. Importantly, this sort of work can also be undertaken across species, consistent with the current call for better 'translational neuroscience', that moves from the bench to the bedside.

Is a model of wetware-based cognitive-affective schemas also helpful in the clinic? There are a number of reasons to answer 'yes'.

First, we noted earlier that schema models help integrate psychodynamic and cognitive-behavioural constructs. But is such an integration useful in day-to-day work with patients and their emotions? Leslie Greenberg is a South African-born psychologist who is a pioneer of emotion-focused therapy, an approach that draws on psychodynamic and cognitive-behavioural ideas, and that emphasizes emotional schemas. This integration leads, for example, to helpful practical clinical guidelines. Thus Greenberg notes that a first crucial question for the psychotherapist is whether to help clients feel their feelings, whether to distract them from their feelings, or whether to help them regulate their feelings. Put in terms of the standard metaphor of emotion (which it is difficult to escape), this point usefully reminds therapists that it is sometimes important to express emotion, but at other times it is important to contain it.[35]

---

34. See earlier footnotes on natural kinds, on essentialism, and on graded categories. For more on **emotions as natural kinds**, see L.F. Barratt who argues that emotions are not natural kinds [617]) and C.E. Izard (who helpfully suggests that basic emotions may represent natural kinds, but that emotional schemas do not [618], consistent with a view of emotions as fuzzy natural kinds). With regard to the issue of **biological categories as natural kinds**, it is notable that Darwin himself did not see species as essential natural kinds, but rather as fuzzier constructs [619]. For more on the view that evolutionary theory undermines the idea of essential natural kinds, see Steven Gould [620], E. Sober [621], John Dupré [622], and Daniel Dennett [623]. Richard Boyd has argued that fuzzy natural kinds may be characterized by networks of causal mechanisms [624], and see previous footnote on the philosophy of species.

35. See, for example, **Leslie Greenberg** [625]. The importance of weighing up thoughts and reasons, in a deliberative rather than impulsive way, was made by Aristotle [516], and along these lines we will later emphasize the close links between psychotherapy and morality. The psychodynamic psychiatrist **Mardi Horowitz** has also attempted to develop an integrative approach to schemas [626].

Second, a model of cognitive-affective schemas that emphasizes how they may be influenced by both nature and nurture is clinically useful. Jeffrey Young posits that the 'unrelenting standards' schema is more likely to develop in families characterized by high levels of parental criticism. At the same time, neurogenetic findings suggest that cognitive-affective schemas likely have a strong hereditary component. While the psychoanalytic literature has long described perfectionistic traits (hypothesized to be due to strong 'anal drives'!), schema therapy provides rigorous operational definitions of early maladaptive schemas, and empirical investigation of the underlying causal mechanisms that shape their development.

Third, schema theory provides a way of thinking about a comprehensive treatment plan that addresses both mechanism and meaning (Table 5). The idea that schemas are instantiated in the wetware of the brain–mind is consistent with data that the development and activation of schemas are associated with brain–mind changes, and that medications that act on the brain–mind can be useful for changing cognition and emotion. On the other hand, the idea that schemas play out in particular social contexts is consistent with data that schemas are triggered by specific situations, and that psychotherapies that explore our thoughts and emotions and reframe our interpretations of the world can be helpful. Therapeutic change involves laying down new wetware-based cognitive-affective schemas.[36]

3.5 Conclusion

Cognitive-affective science tells us that the brain–mind is configured to use cognitions and emotions in an integrated way, in order to respond adaptively to the world. Furthermore, cognitive-affective science emphasizes that cognitive-affective processing styles develop in early life, and that they are impacted by early life experiences. Put differently, modern cognitive-affective science is consistent with a position which emphasizes that cognitive-affective processing provides us with useful information about the world, but which also accepts that cognitive-affective processing reflects past experiences and may go awry in important ways.

Psychiatry and psychology give us key examples of pathological alterations in cognitive-affective processing. There are, for example, psychiatric disorders

---

36. There are notable similarities between Aristotle's advice to adhere to the **golden mean, and schema therapy's** view of optimal cognitive-affective processing. Edith Hall cites Aristotle's second book of his 'Eudemian Ethics', where he tabulates qualities of character, indicating the virtue of having this quality in an appropriate amount, as well the vices of having the quality in excess or deficient. Thus, for example, self-control is a virtue in comparison to self-indulgence (excess) and self-denial (deficiency), and respectfulness is a virtue in comparison with insolence (excess) and shyness (deficiency). He further notes that "it is clear that all the moral virtues and vices have to do with excesses and defects of pleasures and pains" [516]; pleasure and pain will be the topic of the next two chapters.

characterized by excessive activation of passion, and psychiatric disorders characterized by insufficient cognitive control. The term 'balance' to describe optimal cognitive-affective functioning draws on the age-old and somewhat dated metaphor of cognition and emotion as opposing forces, but seems useful. Clinicians have at times been overly optimistic about the therapeutic value of expressing emotions; more contemporary views emphasize, for example, that sometimes it is useful to express emotions, while other times it is useful to contain them.

This position in psychiatry and psychology is consistent with philosophical approaches that have argued for an integrated approach to emotion, which emphasizes that any particular emotion may provide useful information about the world, but also that emotions may entail false judgements. Aristotle wisely recommended rigorous training of our cognitive-affective responses, so that we can live lives of practical wisdom, expressing the right amount of emotion in the right way at the right time. It seems that we are wired in such a way that we are thinking-feeling beings; this is the one of the key aspects of the human condition, and we would do well to accept and embrace this.

Although we might recognize the potential value of our thoughts and feelings for optimal decision-making, at the same time we don't always fully appreciate or understand our own thoughts and feelings. After all, the cognitive-affective unconscious is always busy and below the surface. We have focused on the potential value of the model of schemas, and the method of mindful attention, to obtain information about and insight into cognitions and feelings. In the next chapter we will build on these themes to address a particularly fast-advancing set of philosophical and psychological work on the brain–mind; that addressing pleasure and reward.

Chapter 4

The pleasures of life

Now that we have a sketch of how to conceptualize the brain–mind, as well as thoughts and emotions, we can move to the question of what brain–minds mostly think-feel about. Aristotle and evolutionary theorists have pointed out that emotions are related to either pleasure or pain; reward circuitry facilitates an approach towards opportunities, while fear circuitry facilitates an escape from threat. The next two chapters will cover pleasure and pain. We'll also cover some issues at the intersection of pleasure and pain, such as the sweetness of revenge. In this chapter, on pleasure, we'll also consider a number of related questions, such as those pertaining to food and drink, to beauty and nature, to play and music, to exercise and running, and to attachment and love.[1]

From adolescence onwards, I've been an absolute sucker for self-help books on happiness. These tend to fall into at least two categories. On the one hand, there are those that promise that if we just stick to a few key rules, we can achieve happiness. Norman Peale's 'The Power of Positive Thinking', which sold 2.5 million copies in the 1950s alone, is a good example of this category. A more recent best-seller is 'The Secret', which also emphasizes the power of positive thinking: if we just focus optimistically on what we want, our desires will materialize.[2]

On the other hand, there are those self-help books which emphasize that life is pretty complex, that no rules are infallible, and that the path to happiness requires embracing some degree of anguish. Greek tragedy emphasizes that life

---

1. For Aristotle, "Emotions are the things on account of which the ones altered differ with respect to their judgments, and are accompanied by **pleasure and pain**: such are anger, pity, fear, and all similar emotions and their contraries" (in his 'Rhetoric'). Jeremy Bentham begins his volume, 'An Introduction to the Principles of Morals and Legislation', saying "Nature has placed mankind under the governance of two sovereign masters, pain, and pleasure. It is for them alone to point out what we ought to do, as well as to determine what we shall do" [627]. We will get to the issue of morality, immediately after the chapters on pleasure and pain.
2. See **Norman Peale** [628] and **Rhonda Byrne** [629]. The phrase 'negative pathway' to happiness is from Oliver Burkeman [630], although see later footnote on others who have advocated for the value of negativity, sometimes using similar phrases.

Problems of Living. https://doi.org/10.1016/B978-0-323-90239-7.00007-9

is precarious and fragile; a 'negative pathway' to happiness accepts this point, emphasizing also the beauty and grandeur of life.[3] As Bertrand Russell, a winner of the Nobel Prize in literature, poignantly puts it, "only on the firm foundation of unyielding despair, can the soul's habitation henceforth be safely built". This sort of near paradoxical approach is captured in phrases such as the 'strength of pessimism' (from Nietzsche) and 'tragic optimism' (from Victor Frankl, referring to finding meaning in even the worst situations).[4]

How can we move forward in this debate, distilling the best of the self-help literature? We'll take the same approach as elsewhere in the volume, beginning by discussing work in philosophy, before going on to discuss ongoing psychological debates. Then we'll use findings from cognitive-affective science to help formulate an integrative position on the question of happiness. We'll see that positive psychology has itself become increasingly interested in the science of happiness, and so draws on a neuroscience evidence base. At the same time, the devil is often in the detail, and we'll need to examine closely the extent to which a positive outlook really is paramount.

---

3. Nussbaum reads Plato as opposing the idea of **the tragic** (arguing that we can overcome the limitations of the body and its passions), and Aristotle as affirming it (arguing that its truths have important value) [527]. She writes, "There is a beauty in the willingness to love someone in the face of love's instability and worldliness that is absent from a completely trustworthy love. There is a certain valuable quality in social virtue that is lost when social virtue is removed from the domain of uncontrolled happenings. And in general each salient Aristotelian virtue seems inseparable from a risk of harm"; we will discuss the virtues later. Harold Bloom's notion of the 'higher unpleasure' that more difficult texts give us is relevant here [631], as is Simon Critchley's argument that looking at tragedy allows us to look further and more clearly [632]. Spinoza, Hume, and Dewey have each been interested in or linked to the tragic, see respectively R. Leo [633], E. Galgut [634], and R.A. Jacques [635].

4. The term '**grandeur**' here is taken from Darwin, who closed his 'Origin of Species', with its rather disconcerting view of how natural selection had led to humankind, by reassuring readers that "from the war of nature, from famine and death, the most exalted object which we are capable of conceiving, namely, the production of the higher animals directly follows. There is grandeur in this view of life, with its several powers, having been originally breathed by the Creator into a few forms or into one; and that, whilst this planet has gone cycling on according to the fixed law of gravity, from so simple a beginning endless forms most beautiful and most wonderful have been and are being evolved" [619]. Russel's sentence is from "A Free Man's Worship" [636]. It's redolent of Blaise Pascal's earlier "Man's greatness comes from knowing he is wretched" [637]. For more on Nietzsche's views (and those of Bernard Williams), see M.P. Jenkins [638]. For Frankl's phrase, see "Man's Search for Meaning" [397]. The sentiments here are also redolent of the work of Stellenbosch University philosopher Anton van Niekerk, who argues with Nietzsche, and against Schopenhauer, that the recognition of life as tragic has important positive consequences [639], and of Rhodes University philosopher Pedro Tabensky who holds that ethical living is necessarily born of struggle [640]. See also Michael Brady on suffering and virtue [533], and later discussion and footnote on optimism and pessimism.

4.1 Philosophy and happiness

Happiness has long been a topic of consideration in philosophy, and it seems to be receiving even more attention at the moment, with courses called 'Happiness 101', and with dozens of philosophically minded volumes on the topic coming out each year. We'll begin by considering some of the views of key ancient philosophers, the Epicureans, the Stoics, and Aristotle. Each of these can be taken to represent key positions in the philosophy of happiness, ranging from those who emphasize that happiness is a subjective mental state, to those who argue that happiness is a form of activity which can be objectively defined.[5]

Epicureans argued that what is of primary importance is the pleasure that one obtains in life; they advocated that moderate pleasures from knowledge, friendship, and a temperate life would lead to a life of tranquillity (ataraxia). In this view, happiness is dependent on subjective experiences. This idea continued in the work of John Stuart Mill who, in addition to helping set off the erklären versus verstehen debate in the 19th century, argued that ethical decision-making is based on the sum of happiness. Jeremy Bentham, who had earlier taken this utilitarian position, primarily conceptualized pleasure in terms of its duration and intensity. However, Mill was also concerned with the issue of quality: he argued that some pleasures are more valuable than others.[6]

Stoics argued that happiness depends on how one thinks about things; we may not have control over the vicissitudes of life, but we can determine our cognitive and emotional response to these, and so achieve ataraxia. Epictetus was an early exponent; he had been brought up as a slave, and he figured that the way to cope with this situation was to change his cognitive-affective responses. As mentioned earlier, this position resonates a good deal with the approach of cognitive-behavioural psychotherapy: the therapist asks the client to explore their thoughts and feelings, and then to consider whether his or her thoughts are really true, or whether they are overly negative and pessimistic.[7]

---

5. For reviews of some of the literature on **philosophy of happiness and well-being**, see Darrin McMahon [641], Lisa Bortolotti [642], Steven Cahn and Susan Vitrano [643], Susan David, Ilona Boniwell, and Amanda Ayers [644], Guy Fletcher [645], Øyvind Rabbås, Eyjólfur K. Emilsson, Hallvard Fossheim, and Miira Tuominen [646], Philip Bosman [647], and Alan Goldman [648].

6. While **Epicureanism** can potentially be considered a form of hedonism, it is also focused on the virtues, and the 'simple pleasures' that it advocates verge on asceticism. The neologism 'epicure' to denote someone devoted to sensual pleasure is based on a misunderstanding of its tenets. **Jeremy Bentham** was one of the first to tackle the questions of how to measure and promote 'well-being'. For Bentham, if they produced the same degree of pleasure, then 'pushpin is equal to poetry' (pushpin was a pub game played in Victorian times). For critics such as Thomas Carlyle, such an approach to utilitarianism was the 'philosophy of swine'. Daniel Kahneman has added that hedonia can be reported in real time or retrospectively, and that there are systematic errors in retrospective evaluations [649].

7. There is a subgenre of self-help volumes that draws on the **work of Stoics**, such as Epictetus, Seneca (earlier than Epictetus, and Nero's tutor), and Marcus Aurelius (later than Epictetus, and emperor), to address the big questions and hard problems: see Anthony Long [650], William Irvine [651], Donald Robertson [652], Massimo Pigliucci [653], and Antonio Marcano [654]. Nancy Sherman's work is useful in pointing out the strengths but also the limitations of Stoic philosophy [655].

An emphasis on happiness as a hedonic state seems, however, to miss something important. I vividly remember the first time one of my patients flipped from depression to mania: an occasional adverse effect of antidepressants, and a reminder of the tremendous impact that these agents can have. Graham was an elderly gentlemen who had been admitted because of severe depression: he had insufficient energy to leave his bed, he was eating too little to maintain his weight and health, and he saw himself and his future in bleak terms. After 2 weeks of antidepressant treatment, he was an entirely different person, getting out of bed, eating and socializing, with hope about the future. But then things went too far; he no longer felt the need to sleep, his energy was entirely focused on flirting with female nursing staff, and he spoke grandiosely about how he would soon fix the world's problems. Notably, Graham had little insight into his flirtatiousness and grandiosity, and so was very loath to take a mood-stabilizing medication that would diminish the pleasant buzz of his mania.

Rather earlier than my experience with Graham, the Roman poet Horace described the merchant Lycas, who imagined himself a spectator in a wonderful amphitheatre—a state of illusory bliss apparently akin to a manic episode. Philosophers have since disagreed about whether or not it would be appropriate to intervene to attempt to remove this state. Clinicians, on the other hand, rarely disagree on whether patients diagnosed with mania deserve treatment, no doubt because in the clinic a manic episode is not merely a state of bliss—it's associated with major disruption to a person's life, with loss of work, relationships, and more. A thought experiment put forward by Robert Nozick may be more persuasive. He asks us to consider a brain that is hooked up to an experience or pleasure machine that ensures a state of bliss. Most of us are loathe to describe such a brain as truly happy, and indeed his work has been taken to provide an important critique of approaches to happiness that rely entirely on subjective mental states.[8]

Wittgenstein, too, was not a fan of hedonism, apparently noting, "I don't know why we are here, but I'm pretty sure that it is not in order to enjoy ourselves". At the same time, formulating an objective account of happiness was challenging. Wittgenstein wrote, "I keep on coming back to this! Simply the happy life is good, the unhappy bad. And if I *now* ask myself: But why should I live *happily*, then this of itself seems to me to be a tautological question; the happy life seems to be justified, of itself, it seems that it is the only right life.

8. Some (e.g. Michel de Montaigne) have argued that it was appropriate to intervene for **Lycas**, others (e.g. Desiderius Erasmus [656] and Madame du Châtelet [657]) have opposed this view [658]. For a consideration of more subtle cases, see my 'Philosophy of Psychopharmacology' [53]. For more on Nozick's thought experiment, see his 'Anarchy, State, and Utopia' [659]. For a defence of hedonism, see Fred Feldman [660]; for counter-arguments to Nozick, see B. Bramble [661]; and for an approach to the experience machine from experimental philosophy, see F. Hindriks and I. Douven [662].

But this is really in some sense deeply mysterious! *It is clear* that ethics *cannot* be expressed! [Cf. 6.421.] But we could say: The happy life seems to be in some sense *more harmonious* than the unhappy. But in what sense? What is the objective mark of the happy, harmonious life? Here it is again clear that there cannot be any such mark, that can be *described*".[9]

An important response to this challenge is provided by Aristotle, who saw human flourishing (eudaimonia) as a form of activity (energiea), as a mode of being that includes physiological and social aspects, and as emerging over the course of a life.[10] Aristotle was perhaps the first to take a naturalistic approach to happiness, rigorously examining the conditions for human flourishing. While a plant might need water and sunlight in order to grow well, a human might need food and shelter, intellectual stimulation, the companionship of family and friends, and so on, in order to flourish. At the same time, Aristotle's approach is far from reductionistic or scientistic; he views flourishing as a complex and fuzzy construct, and he emphasizes that flourishing does not simply emerge from the application of a set of techniques. Rather, his view is that as our knowledge and experience grow, so we continue to sharpen our concepts of flourishing and to improve relevant practices.[11]

An integrative approach to happiness may emphasize aspects of both subjective pleasure as well as of objective activity. Aristotle himself was intently interested in pleasure and what it says about human nature, and held that a flourishing life includes pleasure. At the same time, in line with his emphasis on teleological explanation, he emphasized the importance of individuals having purposes and reaching particular goals. Indeed, while Aristotle held that "knowing yourself is the beginning of all wisdom", his focus was less on self-examination and more on purposeful activity (praxis); virtuous practices lead to flourishing. A number of subsequent authors have suggested that **happiness entails both pleasure and purpose**. Hedonism by itself runs the risk of meaninglessness by prioritizing pleasure for any reason. A focus on purposeful activity alone, on the other hand, runs the risk of a life that is entirely ascetic. At the

9. This citation is from Wittgenstein's 'Notebooks, 1914–1916' [663]. For more on **Wittgenstein's views of happiness**, see M. Balaska [664].

10. The University of Cape Town philosopher Tom Angier draws a helpful contrast between philosophers who view **happiness as a mental state brought on by various techniques**, and **Aristotle's concept of flourishing** [665]. Philosophers who focus on happiness as a mental state include Daniel Haybron [666] and Fred Feldman [667]. Aristotle sees eudaimonia as achieved "in a complete life. For one swallow does not make a summer, nor does one day; and so too one day, or a short time, does not make a man blessed and happy" (in his 'Nicomachean Ethics', cited by Angier in this paper). See also a later footnote on this point.

11. For a reading of Aristotle which emphasizes that **flourishing is an indeterminate or fuzzy construct**, and that **practical wisdom involves more than just a set of skills**, see Martha Nussbaum [527]. See also later discussion of Aristotle's 'sketch for the good', as well as later discussions of moral naturalism and human nature, where it is suggested that for Aristotle human function involves a life of practical judgement [173].

same time, Aristotle's notion of 'flourishing' is purposefully a general one, and further discussion of relevant specifics is needed. We'll get into more detail as the chapter progresses.[12]

4.2 Psychiatry and happiness

Psychiatry and psychology have not always centralized the issue of happiness. A tongue-in-cheek paper in the Journal of Medical Ethics went so far as to argue that happiness is an illness. That said, in the past few years, the field of 'positive psychology' has emerged, with 'positive psychiatry' following in its footsteps. Psychological and psychiatric views of happiness reflect various theoretical positions in the field, so we'll begin with a brief review of this work. We'll then go on to discuss addictions—psychiatric conditions in which there seem to be disturbances in reward systems. We'll close with a brief word of caution about positive psychology.[13]

Theoretical positions in psychiatry on happiness

As we've noted, Freud was very much influenced by the science of his day, and he proposed that unconscious drives including sexual and aggressive ones were always pushing for expression. These drives had to be constrained by reality, and so often expressed themselves in dreams, jokes, or neuroses. This theoretical

---

12. For more on Aristotle's views on pleasure, see D. Bostock [668] and Richard Kraut [538]. For more on Aristotle's views on purpose, see Edith Hall [516]; she cites his words that, "Everybody who is able to live according to their own purposive choice should set before themselves some goal [skopos] to aim at through living in a good way – the goal could be achieving recognition, or distinction, or wealth, or culture – on which they will keep their eyes fixed in everything they do. It is clearly a sign of foolishness not to create order in your life in terms of having an end [telos]" [669]. More recent authors who have discussed notions of '**pleasure plus purpose**' include Derek Parfit [670], Martin Seligman [671], John Kekes [672], Mike Martin [673], and Paul Dolan [612]. The subtitle of Dolan's volume is consistent with Aristotle's focus on activity, and also brings to mind Bertrand Russell's earlier 'The Conquest of Happiness' which was similarly focused on activity rather than introspection, and which spoke elegantly to a range of pleasures and purposes [674]. A number of our favourite philosophers have followed Aristotle in taking an integrative position. Spinoza's 'free man', for example, is one who experiences joy, but who is focused on a virtuous life [675]. Hume wrote several essays depicting different ideas of happiness (including 'The Epicurean' and 'The Stoic'), and advocated for philosophy as a 'medicine of the mind', that promotes a balanced view [676, 677]. Although Hume has been characterized as a moral emotivist (see later), his views of the virtues have much in common with Aristotle [221, 518]. Aristotle focused on flourishing, while in a more American way pragmatists such as Dewey addressed the pursuit of happiness; both emphasized the importance of purposeful activity or virtuous practices [678]. More recently, Pierre Hadot noted that Stoicism and Epicureanism seem to correspond to "two opposite but inseparable poles of our inner life: tension and relaxation, duty and serenity, moral consciousness and the joy of existing" and suggested that the complexity of life requires using simultaneously or successively a range of different approaches [679]. Arguably in line with Aristotle, for Spinoza, Hume, Dewey, and Hadot, philosophy has a therapeutic aim, pointing towards a way of living (see earlier footnote on philosophy as therapy for references).

13. See R.P. Bentall [680] and J. Harris and colleagues [681].

perspective seems somewhat pessimistic, but the use of mature defences, like love and work, could potentially minimize neurosis. Freud himself felt that psychoanalysis could transform 'neurotic misery' into 'ordinary unhappiness', but the extent to which psychoanalysis is in fact an efficacious treatment for mental disorders remains an open question.[14]

Skinner, as a key proponent of behaviourism, put reward and punishment at the centre of his theories about the brain–mind and psychopathology. In line with the maxim "give me the child … and I will give you the man", he saw the child as a 'blank slate', and suggested that rewards were key for ensuring appropriate conditioning during development. At some level, this too seems a rather pessimistic perspective on the world, although in his volume 'Walden Two', Skinner presents an optimistic blueprint for the future. Behaviourism's perspective has been criticized from several angles, but a particularly clear deficit in the age of neuroscience is that the blank slate argument is quite incorrect, insofar as it ignores the inbuilt nature of the reward and threat systems.[15]

We've noted that cognitive-behavioural therapy has much in common with the approach of Stoicism. Cognitive-behaviour therapy holds that emotions are driven by thoughts: we can therefore improve our well-being by rethinking how we approach the world. There are, however, several potential criticisms of this view. First, it's unlikely that cognitive-affective systems work in such a simple linear way. Schema therapy may do a better job of explaining the origins of cognitive-affective attitudes and styles. Second, it's possible that people who are happier are more out of touch with reality, demonstrating positive bias; this phenomenon is known as **depressive realism**, and although the evidence is not entirely consistent, there are some data that support it.[16]

With the advent of modern psychopharmacology, a range of drugs that improve mood became available. There has been some contention about how

---

14. There is debate about this point. In an earlier footnote I referenced a review arguing that psychoanalytic constructs and interventions are evidence based [411]. For a rather more sceptical perspective, see the work of **Stuart Schneiderman**, an analysand of the famous French psychoanalyst Jacques Lacan, and critic of much contemporary psychotherapy [682]. Stuart has been another influential figure in my life, from the time of my psychiatry residency, in one way or another.

15. "Give me the child for the first seven years and I will give you the man" is considered to be a Jesuit maxim. See Skinner's 'Walden Two' [683], and Steven Pinker's 'The Blank Slate' [684]. The '**blank slate**' argument is most often associated with John Locke [685]. See earlier footnote on the nature–nurture dichotomy.

16. Freud noted that the depressive "has a keener eye for the truth" [686]. For more on **depressive realism**, see L.B. Alloy and colleagues [687], D.A. Haaga and colleagues [688], and M.T. Moore and D.M. Fresco [689]. For more on **positive illusions**, optimism bias, negativity effects, and related constructs, see S.E. Taylor and J.D. Brown [690], Julie Norem [691], Suzanne Segerstrom [692], Tali Sharot [693], and John Tierney and Roy Baumeister [694], which includes a discussion of the Pollyanna principle. There is some evidence that moderate optimists may be better decision-makers (see Sharot's volume), but there is also evidence that for some individuals defensive pessimism works best (see Norem's volume). Indeed, depressive realism is relevant to debates about optimistic vs pessimistic approaches [695], which we will discuss in more detail later. In philosophy, see Lisa Bortolotti [696].

effective these drugs are. The technical term is 'effect size': drugs that are much better than placebo have a high effect size, while those that are only a little better than placebo have a low effect size. It's remarkable that placebo is so effective in depression, a fact that underlies how complex the brain–mind is. It is noteworthy, however, that in people without depression, it's much harder to discern any impact of antidepressant use. These drugs may act in part by changing our setpoint for the triggering of sadness, rather than by increasing happiness in everyone.[17]

Addiction and related disorders

Investigation of reward systems in psychiatric disorders is another place to look for psychological contributions to the understanding of happiness. Reward systems may be involved in a range of different psychiatric conditions, including mania. Elevated mood is characteristic of mania and even pathognomonic: mania is present if, and only if, this sort of mood disruption is present. Mania is clearly pathological in that people with mania have multiple dysfunctional symptoms, and there is broad agreement that manic episodes should be treated. People with hypomania, a less severe form of mania, may not be as dysfunctional and may decline treatment, which again raises the threshold issue (what is the best cut-off for defining illness) as well as the ethical issue of when to treat impaired but nonconsenting patients.[18]

Depression is much more common than mania. On the one hand, there is good evidence of abnormalities in the reward system in this condition, and pharmacotherapy and psychotherapy appear to normalize the brain–mind of depressed individuals. On the other hand, the threshold problem and the causality problem both arise. First, depression seems to be on a continuum with normality; as per evolutionary theory, there are often good reasons for feeling bad. So there is debate about precisely where to draw the line between sadness and depression. Second, we continue to have surprisingly little insight into the precise pathogenesis of depression. The best we can currently say is that a range of different factors contribute to the onset and maintenance of this condition, including genetic and environmental factors.

Another group of conditions that involve the reward system are the addictions. Concepts of addiction have long focused on substances of abuse, but more recently the notion of behavioural addiction has received attention. We earlier discussed the seminal contributions of Isaac Marks, the South African born psychiatrist, to behavioural therapy: he was also one of the first to use the term behavioural addiction. The paradigmatic example of a behavioural addiction is pathological gambling, a well-studied condition. More recently, this construct has been extended to a range of other excessive behaviours, including exces-

17. See again, my volume on 'Philosophy of Psychopharmacology', which speaks to issues in '**cosmetic psychopharmacology**' [53].
18. See earlier footnote on Lycas.

sive gaming, sexual compulsivity, and even tango dancing, all of which have received less attention in psychiatric research.

There is a good deal of debate about the construct of behavioural addiction. An immediate question is whether behavioural addictions should be classified as addictions, rather than as, say, impulse control disorders. First, there are differences in the symptom profiles of substance use disorders and conditions such as gambling; for example, gamblers do not really develop withdrawal symptoms. Within the putative behavioural addictions, gaming disorder would seem to have a good deal in common with gambling disorder, tango dancing not so much. Second, while some research on the psychobiology of these conditions (e.g. neuroimaging studies) has suggested overlap, such data are far from conclusive. More research is emerging on gaming disorder, but very little is known about a number of other proposed behavioural addictions. Third, there is the question of whether it is clinically useful to classify conditions such as gambling disorder with substance use disorders. On the one hand, the treatments for these different conditions would seem to overlap only partially. On the other, there may be rather more overlap in the public health measures used to minimize harmful substance use and gambling.[19]

The American Psychiatric Association (APA) produces the Diagnostic and Statistical Manual of Mental Disorders. Editors of the 5th edition encouraged incorporation of neuroscientific data, and gambling was classified together with substance use disorders for the first time. The World Health Organization (WHO) produces the International Classification of Disease. ICD-11, like other WHO products, is particularly informed by issues of public health, and this is part of the reason that this classification system includes gambling and gaming disorder in the section on addictions. I was a member of a number of WHO ICD-11 committees, including those on mood and anxiety disorders, and on obsessive–compulsive and related disorders, and so was quite involved in this debate.

Initially, I was not particularly impressed by phenomenological and psychobiological evidence of an overlap of gambling and gaming disorder with substance use disorders: sure gaming and gambling involve reward systems, but substances have direct and dramatic deleterious effects on brain structure and function. Gaming, gambling, and a tipple can all be fun, but after several drinks the brain–mind is inebriated, and with enough alcohol intake, the brain–mind becomes comatose. Consistent with these distinctions, psychiatric medication has a crucially important role to play in treating the potentially life-threatening complications of severe alcohol withdrawal, but is a less central intervention in the treatment of gambling and gaming disorder.[20]

---

19. For more on the **behavioural addictions**, see Kenneth Rosenberg and Laura Feder [697] and Nancy Petry [698]. I was pleased to be able to contribute to both volumes.
20. The brain basis of substance use disorders and behavioural addictions will be discussed in more detail in the next chapter, when we discuss the nature of mental disorders.

However, the argument that public health approaches to these conditions overlap seemed to me a strong one. While there are certainly individual vulnerabilities to these conditions, they are much more likely to develop in some social contexts than in others. So, if one really wants to lower prevalence of smoking or of gaming, interventions such as increasing taxes (increasing punishment), decreasing advertising (lowering perceived rewards), and so forth, are needed. This sort of intervention requires significant political will, because governments benefit from taxes paid by the sin industries. Nevertheless, public health interventions that reduce smoking levels, alcohol use, and gambling stand to save any country millions of dollars.[21]

Positive psychology

Taken together, a consideration of different psychological approaches to happiness, and of the role of reward systems in different psychiatric disorders, suggests that psychology has made more of a contribution to understanding pathologies of happiness, and has had less to say about well-being. That said, recent decades have seen the robust emergence of the fields of positive psychology and positive psychiatry. The term 'positive psychology' was coined by the humanistic psychologist Abraham Maslow in the 1950s, and the field was given particular impetus by the work of Mihaly Csikszentmihalyi in the 1980s on flow, and by the subsequent writings of Martin Seligman on authentic happiness. These ideas have had significant influence in a range of broader forums, as exemplified by the annual 'World Happiness Report'.[22]

Positive psychology argues that psychologists ought to focus more on well-being, and that there is an emerging science of how to do so. It has contributed a literature on how to alter negative biases, and how to improve resilience. Thus, for example, where clinical psychology has focused on how trauma may lead to posttraumatic stress disorder, positive psychology has investigated posttraumatic growth. There's also a growing literature on the neuroscience of positive psychology constructs, as well as on positive psychology interventions, including those focused on mindfulness practices. Analogously, positive psychiatry has an explicit focus on patients' strengths and resilience, and on interventions to increase positive mental health outcomes such as well-being.[23]

---

21. See D.J. Stein and colleagues [699].
22. See **Martin Seligman** [671] and **Mihaly Csikszentmihalyi** [700]. For more on the politics and economics of happiness, see Derek Bok [701], Robert and Edward Skidelsky [702], Benjamin Radcliff [703], A.E. Clark and colleagues [704], M. Jain and colleagues [705], John Helliwell and colleagues [706]. Work on Gross National Happiness is particularly noteworthy, see Kent Schroeder [707].
23. The **scientific literature on happiness and well-being** is large and expanding; see, for example, Daniel Kahneman, Edward Diener, and Norbert Schwarz [708], P. Kesebir and E. Diener [709], and D.G. Myers and E. Diener [710]; and see later footnote listing positive psychology texts. Aristotle's ideas have significantly influenced work on the psychology of happiness and well-being [711], but see also next footnote. For the neuroscience of positive psychology, see Joshua Greene, India Morrison, and Martin Seligman [712]. For more on positive psychiatry, see Dilip Jeste and Barton Palmer [713], and G.A. Fava and J. Guidi [714].

Positive psychology has, however, been criticized on a number of grounds. In line with criticism of philosophical positions that focus on happiness as an individual's subjective state, it has been argued that positive psychology ignores important objective aspects of the world relevant to human flourishing. Relatedly, positive psychology often seems to take a scientistic stance that reduces happiness to a set of skills, ignoring the complexity of human flourishing.[24] From a practical standpoint, many have pointed out that focusing on happiness is the sort of thing that actually decreases happiness; Mill was one of the first to articulate this 'happiness paradox', saying "Ask yourself whether you are happy and you cease to be so".[25]

The focus of positive psychology on the value of meditation exemplifies some of these issues. As we've noted, mindful attention to thoughts and emotions seems sensible. However, such attention is surely only a first step; the idea that if everyone simply did meditation for a few hours a day, then the problems of both individuals and the world would disappear, not only lacks evidence, but also seems deeply reductionistic. From a practical standpoint, while meditation may be useful for some, it may be valueless or even harmful for others.[26] The term 'McMindfulness' is useful in emphasizing the commercialization of meditation practices, and the way in which these focus on allowing emotions to wash over oneself, so deflecting attention away from active engagement with the world to ensure that these emotions no longer arise.[27]

---

24. While Seligman has arguably attempted to move beyond positive thinking by emphasizing the virtues [715] and flourishing [716], **positive psychology has been criticized** for not sufficiently emphasizing practical wisdom and for a range of other conceptual and methodological shortcomings [717–719]. For a volume that criticizes an overly simplistic approach to 'flow', emphasizing that expertise requires thought and effort, see Barbara Montero [720]. See also later footnote on the value of negativity.

25. See Mills' 'Autobiography' where he wrote: "Those only are happy (I thought) who have their minds fixed on some object other than their own happiness; on the happiness of others, on the improvement of mankind, even on some art or pursuit, followed not as a means, but as itself an ideal end. Aiming thus at something else, they find happiness by the way. The enjoyments of life (such was now my theory) are sufficient to make it a pleasant thing, when they are taken en passant, without being made a principal object. Once make them so, and they are immediately felt to be insufficient. They will not bear a scrutinizing examination. Ask yourself whether you are happy, and you cease to be so. The only chance is to treat, not happiness, but some end external to it, as the purpose of life" [721]. For more on the **happiness paradox**, see B. Eggleston [722]. Many have eloquently described this paradox including Victor Frankl; for more recent perspectives, see J. Gruber and colleagues [723] and Ruth Whippman [724].

26. It's been suggested that "A wandering mind is an unhappy mind" [725], and there is a growing literature on the **neuroscience of mindfulness** [726]. However, the **limitations and side effects of mindfulness** also deserve emphasis [727–729]. See later footnote for similar sorts of references in relation to positive psychology.

27. For the term '**McMindfulness**', see Ronald Purser [730]. Relatedly see Ronald Dworkin [731], Barbara Ehrenreich [732], William Davies [733], Ashley Frawley [734], Jordan McKenzie [735], Peter Doran [736], Edgar Cabanas and Eval Illouz [737], and David Forbes [738].

While positive psychology has possibly made some useful contributions, **an integrative perspective suggests that a more negative pathway to happiness may be helpful**. Such an approach would aim to avoid the Panglossian aspects of positive psychology, to accept the importance of life's constraints and the value of negative experiences and emotions, and to emphasize the need to engage with the world to address the flourishing of self and others. In these sorts of deliberations, the particulars are important; we'll consider some specific exemplars of pleasure as this chapter proceeds, and of pain and suffering in the next.[28]

4.3 Neurophilosophy and neuropsychiatry

How does modern cognitive-affective science inform our understanding of happiness? A great deal of work has been done in recent decades. A number of findings seem to me to be key.

First, Kent Berridge and Terry Robinson's work has emphasized important psychobiological differences between what they term 'liking' and 'wanting'. 'Liking' circuitry reflects hedonia; this is objectively demonstrable when human infants or animals are given sugar, and they smack their lips with pleasure. In adult humans, this reward circuitry is engaged by a range of stimuli, from tasty food through to beautiful faces. 'Wanting' circuitry, on the other hand, is objectively demonstrable when specific learned cues trigger behaviour (e.g. an animal in a 'Skinner box' will press a lever to get a food reward, exemplifying incentive salience).[29]

The use of quotation marks to describe 'liking' and 'wanting' emphasizes that hedonic reactions and incentive salience are objectively defined and largely unconscious processes that differ from conscious phenomena such as subjective desire. In humans, incentive salience may manifest implicitly (as unconscious 'wanting') or explicitly (as conscious craving). Whereas dopamine was once thought to be key to hedonia, there is increasing evidence that 'liking' is mediated by the opioid system, while 'wanting' is mediated by the dopamine system. It turns out that the neural circuitry underlying 'wanting' is more easily activated than the neural circuitry underlying 'liking'; Berridge has laconically

28. For more on the '**negative pathway**', see earlier footnote; for more on 'Panglossian', see later footnote. Along these lines, from an economics perspective, while some work has integrated economic theory with work on flourishing [739], it has also been argued that rather than focusing on happiness measures, it is more useful to focus on removing sources of unhappiness [740]. Given the links between Stoicism and CBT, it's also relevant to note that while some see Stoicism as comprising social critique [98], others hold that Stoicism has been associated with acceptance of the status quo [741].

29. See T.E. Robinson and K.C. Berridge on the **incentive sensitization theory of addiction** [742] and K.C. Berridge and M.L. Kringelbach [568]. See also popular writing by Morten Kringlebach [743], Paul Bloom [744], David DiSalvo [745], David Linden [746], and Dean Burnett [747]. For a critical reading, see Foucault [748]. Although the reward circuitry involved in hedonic responses involves mid-anterior orbitofrontal cortex, nucleus accumbens, and ventral pallidum, hedonic hotspots are seen in areas ranging from brainstem through to cortex, and so hedonic responses can be seen even in anencephalic individuals without a cortex.

pointed out that this is consistent with an evolutionary strategy of consigning animals more often to a state of desire than to a state of pleasure.[30]

Second, it is noteworthy that animals may have to choose between obtaining small rewards now, or postponing this in order to achieve large rewards later on. Behavioural ecology has developed theories of optimal foraging, which explain when an animal might choose to move from one location (where there is some food) to a further location (where there may be more food). This phenomenon has also been called temporal discounting: there are times when it is useful to give into dopaminergically driven 'wanting', and there are times when it is useful to engage other neurocircuitry in order not to do so (a decision technically termed 'intertemporal choice'). The psychiatrist George Ainslie has argued that in our current environment, where so many rewards are readily available, we often choose such suboptimal immediate rewards over considerably better delayed rewards.[31]

It turns out that the medial portion of the orbitofrontal cortex (mOFC) plays a particularly important role in intertemporal choice. One can conceptualize the mOFC as integrating a range of cognitive and emotional information in order to do this. In some individuals and in some conditions, such as alcoholism, there is a kind of 'reward myopia', with too much temporal discounting: giving into short-term pleasures without regards for long-term consequences. In other individuals and conditions there is too little temporal discounting (e.g. anorexia nervosa): short-term pleasures are postponed for so long that it's unhealthy. Ainslie emphasizes, however, that all of us are continuously faced with intertemporal choice, and are continuously tempted by smaller/sooner rewards. Finding a balance between pleasure (immediate gratification) and purpose (gratification over the longer term) is therefore an ongoing struggle.[32]

Third, there is the issue of setpoints of happiness. Early empirical studies of happiness have suggested that each individual has a setpoint to which they rapidly return after being exposed to either rewards or punishments. Elation after winning the jackpot dissipates in the face of new stressors; despair after losing a limb in a

30. See K.C. Berridge on **wanting and liking** [749]. The work of the psychologists James Olds and Peter Milner on rats that repeatedly press a lever that delivers direct brain stimulation to reward centres seems to provide a neuroscientific analogue of Nozick's experience or pleasure machine (see earlier). However, on closer examination, this work models 'wanting' rather than 'liking'. Rats with electrodes implanted into 'wanting' circuitry behave like addicts: they repeatedly respond to incentive salience cues, preferring this to all other activities (eating, drinking, mating, suckling). Analogous phenomena have been seen in humans, who report no pleasure from this behaviour [568]. The point that pleasure is relatively fleeting was made earlier on by Hume; he had his protagonists discuss whether it is true that "Health is more common than sickness; pleasure than pain; happiness than misery. And for one vexation which we meet with, we attain, upon computation, a hundred enjoyments" [750]. Schopenhauer, who had a dismal view of nature and human nature made similar points (see later footnote).

31. For more on **foraging theory**, see D.W. Stephens and J.R. Krebs [751] and P.M. Todd, T.T. Hills, and T.W. Robbins [752]. For more on temporal discounting, see George Ainslie [753].

32. Ainslie sees this temptation as part of human nature, our 'original sin' [753]. See also earlier footnote on pleasure and purpose.

motor vehicle collision diminishes with adaptation to new circumstances. We are always, therefore, on an 'hedonic treadmill'. It has even been suggested that we each have 'a happiness pie' to which life circumstances make only a small contribution. The 'hedonic treadmill' and 'happiness pie' theories may not be entirely correct, though; animals including humans that have experienced early adversity are more likely to have adverse responses to future stressors, and humans who have experienced major negative life events such as death of a spouse or unemployment may have lasting changes in their subjective well-being.[33]

Complex traits such as well-being are typically influenced by both genetic and environmental factors. It seems, for example, that the heritable personality trait of extraversion is associated with greater subjective well-being, while introversion is associated with less subjective well-being. Preliminary genomic studies have suggested that subjective well-being (perhaps akin to the notion of hedonia or 'pleasure') and psychological well-being (perhaps somewhat akin to the notion of eudaimonia or 'flourishing') have a high genetic correlation, and are at least partly heritable. Further work is needed to explore the exact psychobiological mechanisms underlying subjective and psychological well-being.[34]

Fourth, there can be 'mistakes' in our reward system. While objective evidence of 'liking' and subjective reports of hedonia typically go hand in hand, experiments that use framing to alter subjective experience can produce a dissociation between evidence of liking and subjective reports. In addition, the neural circuitry underlying 'wanting' can be sensitized: administration of dopaminergic agents to an animal leads to increased incentive salience without an increased hedonic response. This is an important model of human addiction: the sensitized animal presses for food ('wanting'), even when the food provides no pleasure ('liking'). An evolutionary perspective on this phenomenon emphasizes that the brain–mind has systems which ensure exploration for rewards, but that an addictive substance (e.g. cocaine) exerts an impact that is many times greater than the impact of a natural award. Thus addictive substances effectively 'hijack' the brain–mind, which subsequently focus solely on exploring for this reward.[35]

---

33. See N.J.L. Brown and J.M. Rohrer [754] on the happiness pie and M. Luhmann and S. Intelisano [755] on hedonic adaptation. The notion of individuals being on an **hedonic treadmill** is paralleled by the **progress paradox**, whereby increases in national income and standards of living are not accompanied by increases in happiness [756–758]. This paradox has also been subject to critique [759].
34. For work on **personality and well-being**, see Brian Little [760]. For the genetics of well-being, see Michael Pluess [761] and B.M.L Baselmans and M. Bartels [762]. Conversely, it's possible that having purpose may have beneficial brain–mind effects [763].
35. See R.M. Nesse and K.C. Berridge [764]. The phenomenon of 'wanting' in the absence of 'liking' can also be seen as an adverse event after administration of dopaminergic medication. For example, patients with Parkinson's disease, where there is dopamine depletion, may be prescribed dopamine agonists, and consequently may develop compulsive gambling, compulsive sexual behaviour, or other addictive-like and nonpleasurable behaviours. It's also noteworthy that substance abuse has on occasion been described in the animal kingdom [765].

Barry Everitt and Trevor Robbins have emphasized that substance abuse has both impulsive and compulsive aspects. During adolescence, for example, there may be significant emotional dysregulation. As part of a pattern of impulsive behaviour, substance use may be initiated. Over time, however, ongoing administration of substances leads to changes in the brain–mind. People with chronic addictions display a range of compulsive behaviours. From a psychobiological perspective, Robbins and colleagues have demonstrated that the shift from impulsive to compulsive functioning is accompanied by a shift in control over drug-seeking behaviour from the ventral to the dorsal striatum.[36]

Fifth, the brain–mind demonstrates significant neuroplasticity and is able to learn and to change throughout the course of a lifetime. We have mentioned that happiness setpoints have a significant genetic component, but that they can be altered by the environment. This section has made the point that addictive substances negatively alter the brain–mind, but we earlier noted that psychotherapy can positively alter the brain–mind. Evolutionary theory notes that we are not wired to maximize our happiness; instead our perpetual hedonic mill ultimately aims to maximize our replication. Nevertheless, we can also develop habits that increase our flourishing.

There is a growing literature on the psychobiology of habits. Neuroplasticity is greater earlier in life, so it's best to follow Aristotle's advice and to develop healthy habits from an early age. Because of the way neuroplasticity works, learning new behaviours involves beginning with more conscious attention to what we are doing (i.e. involving more explicit cognitive-affective processes) before transitioning to more automatic executions of what we have learned (i.e. involving more implicit cognitive-affective processes). Fortunately, there is evidence that neuroplasticity can itself be increased by such factors as exercise and sleep.[37]

4.4 Sharpening ideas on happiness

With these kinds of consideration in mind, how might we improve our levels of happiness? To answer, we need to get further into the nitty-gritty. In this section we'll consider several different kinds of things that are rewarding and that may bring happiness: food and drink, play and music, exercise and running, physical and natural beauty, and attachment and love.

---

36. See B.J. Everitt and T.W. Robbins [766].

37. Although **neuroscientific work on habits** has often focused on simple stimulus–response associations, it has also begun to address more complex behaviours [767, 768]. For work on how to harness neuroplasticity, see J. Shaffer [769]. A range of books on habits draw on philosophical and neuroscientific work; see for example, Charles Duhigg [770], Jeremy Dean [771], Claire Carlisle [525], Vincent Deary [772], Gretch Rubin [773], James Clear [774], and Wendy Wood [775]. The idea that cognitive-affective processes can be more automatic versus more considered is consistent with dual process theory, mentioned earlier.

Food and drink

Even single cell organisms are designed to approach nutrients. And primates like ourselves thrive on a range of different foods. Furthermore, there's some evolutionary evidence for the old saw that we are what we eat. Thus it's been argued that during evolution access to essential fatty acids was crucial in allowing the expansion of the human cortex. Richard Wrangham, a student of Jane Goodall, noted that no human could survive on a chimpanzee diet of fibrous fruit, and put forward the hypothesis that the invention of cooking played a key role in human evolution, facilitating a change in human physiognomy.[38]

Some foods, such as carbohydrates and chocolates, are particularly rewarding for humans. The neural circuits involved in craving for these foods overlap with the 'wanting' circuitry that is sensitized by addictive substances. Access to carbohydrates was limited in the environment of early adaptation, but there is increasingly easy access to sugars in today's world. This mismatch has contributed to the current epidemic of obesity. There is a sense in which the concept of 'food addiction' seems reasonable: as in the case of other behavioural addictions, there is a spectrum from normal behaviour, through to behaviour that only some would regard as excessive, and on to behaviour that almost all would agree is pathological.[39]

Alcohol gives a lot of people pleasure. Unfortunately, however, it's also a highly toxic addictive substance that impacts negatively not only on the brain, but on just about every other organ in the body. True, there's a literature that suggests that a small amount of red wine each day is associated with increased heart health. However, the mechanisms underlying this association are unclear; it's possible, for example, that healthier people are more able to tolerate a small amount of alcohol. Furthermore, there's an argument that given all the other

38. See Wrangham [776], who dates this change to 1.8 million years ago. Based on his findings at Pinnacle Point on the South African coast, Curtis Marean hypothesizes that humans made another significant shift in their eating habits, moving from foraging inland to relying on shellfish, around 164,000 years ago [777]. University of Cape Town archaeologist John Parkington and his colleagues have emphasized the potentially important role that marine sources of **essential fatty acids** had for the evolution of human brain [778]. While there is agreement that a shift to reliance on African coastal resources was key for human evolution, there is ongoing debate about the details of this hypothesis [779]. Particularly controversial in the field of essential fatty acids is the work of David Horrobin, who also emphasized the role of omega-3 fatty acids in evolution, and who hypothesized that deficits in fatty acid metabolism were responsible for schizophrenia [780]. His hypothesis was not widely accepted in academia, so he decided to put it into practice in business, selling evening primrose oil (EPO) for a range of conditions. He became one of the wealthiest people in the United Kingdom, but product licenses for EPO were eventually withdrawn because of lack of efficacy. The journal he established, 'Medical Hypotheses', continues to publish offbeat ideas.

39. For more on **evolutionary aspects of obesity and evidence for food addiction**, see Michael Power and Jay Schulkin [781] and E.L. Gordon and colleagues [782].

negative effects of alcohol, there's no safe level of daily alcohol use. The psychiatrist David Nutt has therefore proposed developing psychotropic agents that have the same pleasant psychological effects as alcohol, but that don't harm the body. Louis Pasteur asserted that "A bottle of wine contains more philosophy than all the books in the world"; we can only hope that Nutt's invention will be equally extraordinary.[40]

The use of diet and nutraceuticals to improve mood and decrease anxiety is a relatively new but rapidly growing and exciting area of research. A modified Mediterranean diet, which is rich in vegetables, fruit, and whole grains, and which has a particular emphasis on oily fish, olive oil, legumes, and nuts, was found useful for depressive symptoms. For people with binge eating or substance use disorders, a range of evidence-based treatments are now available; these include pharmacotherapy as well as psychotherapy techniques such as mindfulness. As in the case of the behavioural addictions, a public health approach seems key: there is evidence that taxes on dangerous goods such as soda and alcohol diminish their consumption.[41]

Play and music

Play sometimes gets overlooked as a rewarding activity. But play is a ubiquitous phenomenon in mammalian species, where it seems to have a particularly important role in neurodevelopment. It's easy to see how much puppies enjoy playing: they have a play 'bow', they wag their tails, and they prefer play to most other activities. It's harder to see play in some other animals: rats, for example, don't seem to respond that obviously when you tickle them. However, Jaak Panskepp showed that although rats don't laugh out loud on being tickled, they emit a high-pitched vocalization which can be picked up on ultrasound.[42]

Once again evolutionary theory may provide useful insights. Play provides species that hunt an opportunity to learn how to do this. In species with clear hierarchies it provides practice at negotiating such hierarchies. Rather than thinking about jokes as revealing the unconscious, it may be more useful to see jokes in terms of mammalian play. The neurocircuitry of humour involves areas involved in detecting and resolving incongruity, as well as reward areas. For

---

40. For more on **alcohol risk-benefits and alcohol alternatives**, see G. Chiva-Blanch and colleagues [783] and D. Nutt [784]. David Nutt has long offered generous advice and a kind word to me.
41. For work on **nutritional psychiatry**, see W. Marx and colleagues [785], including Michael Berk—a creative South African born psychiatrist, whom I've long known and collaborated with. For work on **public health approaches to altering behaviour** in general and alcohol in particular, see Richard Thaler and Cass Sunstein [786] and T. Babor and colleagues [787].
42. For more on **animal play**, see M. Bekoff and J.A. Byers [788], Stuart Brown [789], Richard Restak [790], Jonathan Balcombe [791], and J. Panksepp [792].

humans, humour not only engages our relationships and our shared practices, but it can also challenge them, so making an important contribution to human flourishing.[43]

Play in humans takes an extraordinary range of different forms, including dance and music, and the visual arts. Aristotle noted the arts are a natural form of pleasure, and contemporary neuroscience again sheds light on the relevant psychobiology and on the value of the arts for neurodevelopment.[44] In the modern world, many kinds of work (including philosophy and psychiatry) have elements of play. As Daniel Dennett puts it, "Philosophy at its best is informed play of the highest order". While exploration of the environment and play are thought to involve somewhat different neurocircuitry, for many humans exploration of new environments and discovery of new knowledge would seem to have much in common with play; 'Aha!' seems closely linked to 'Ha-ha!'[45]

The term 'epistemic emotions' has been used to describe joys related to learning, curiosity, and astonishment. As Aristotle said, "to learn is a natural pleasure, not confined to philosophers, but common to all men". And very much in line with this thinking, Dewey argued that like the artist, the teacher needs to

---

43. For more about the neuroscience and psychology of **humour**, see P. Vrticka and colleagues [793] and Rod Martin and Thomas Ford [794]. Different neural circuits may be involved when humour is more voluntary, more aggressive, or more facilitative of social relationships [795–797]. The idea that humour engages but also challenges social relationships and practices is put forward by Simon Critchley [798]. For more on humour, flourishing, and the virtues, see Mordechai Gordon [799]. There is a considerable body of philosophical work on humour, going back at least to Aristotle, who was not only one of the first to expound the idea that humour is based on incongruity (e.g. creating an expectation, and then violating it), but who also advised moderation with regards to humour, steering clear of both the boorish (who avoided humour) and the buffoons (who took humour to excess in a vulgar way). The balanced approach is to be humorous at the right time and place, and to the right degree, as John Morreall puts it [800]. An early contribution—one of many made to philosophy—was from the Nobelist Henri Bergson [801]. For more recent work, see Ted Cohen [802], the wonderful set of volumes by Thomas Cathcart and Daniel Klein, beginning with "Plato and a Platypus Walk Into a Bar – Understanding Philosophy Through Jokes" [803], Jim Holt [804], Stephen Gimble [805], and Terry Eagleton [806]. See also Richard Wiseman for psychological research on the world's funniest joke [807], and Harold Mosak for a discussion of humour in psychotherapy [808].

44. The art historian Ernst Gombrich has been characterized as the first to apply **cognitive science to aesthetics**: he used the concept of schemas to explore how background knowledge influences aesthetic experience [809] (see also later footnote on Nelson Goodman). For more on the joy of dance, see Barbara Ehrenreich [810]. For more on neuroscience and the pleasure of music, see E.A. Stark and colleagues [811].

45. Many philosophers have contributed to our **thinking about work**, perhaps most famously Karl Marx. In more light-hearted vein, Alain De Botton concludes his volume on the 'Pleasures and Sorrows of Work', with the thought that "Our work will at least have distracted us, it will have provided a perfect bubble in which to invest our hopes for perfection, it will have focused our immeasurable anxieties on a few relatively small-scale and achievable goals, it will have given us a sense of mastery, it will have made us respectably tired, it will have put food on the table. It will have kept us out of greater trouble" [812]. See D. Dennett [813]. The link between 'Aha!' and 'Ha Ha' is explored by Matthew Hurley, Daniel Dennett, and Reginald Adams [814].

have a combination of playfulness and seriousness, in order to best instil new habits of mind in his or her students. Moritz Schlick, the leader of the Vienna Circle, unusually for a logical positivist, took this topic even further: he emphasized the joys of play and of creative work, arguing these activities carried their own purpose, and so bring meaning to life.[46]

Based on his pioneering work on play, Panksepp advocated that we need to give children much more time for play. The school day needs to include recess time, perhaps at more frequent times than is currently the case. Play in the digital world may have some benefits, but physical play has a number of important advantages: so the former should be limited and the latter encouraged. Indeed, ordinary digital play is on a spectrum with the most recently recognized behavioural addiction: gaming disorder. Also, given the growing professionalization of youth e-sports and sports, it's perhaps relevant to point out that once such play is professionalized it shifts into the realm of labour, with both the pros of work (e.g. payment) and its cons (e.g. work-related injuries).[47]

Exercise and running

Exercise is not usually thought of in terms of pleasure, but from the perspective of embodied cognition-affect, a key component of pleasure is its association with particular body movements. Humans have certainly learned to play in a whole range of different ways—from children playing tag to adults disco-dancing—and exercise does often involve an element of play. It also turns out that simple jogging, although perhaps not the most sophisticated form of play, is something that brings real pleasure and real health benefits to many.[48]

In the animal kingdom, unnecessary exercise is often avoided; it's important to conserve energy. It's possible, though, that long-distance running is something that humans are uniquely built to do. We are not the fastest species, but given sufficient time, we can outrun other species: aspects of our physiology, such as our sweat glands, allow us to keep going and keep going. Hunter–gatherers in the central Kalahari Desert are known to able to chase down antelope. Daniel Lieberman and Dennis Bramble have gone on to argue that human anatomy and physiology suggests that we evolved as specialized long-distance runners. It's

---

46. The term '**cognitive emotions**' has also been used by Israel Scheffler [815]. For more on Aristotle and the pleasure of learning, see James Warren [816]. For more on this aspect of Dewey, see A. Skilbeck [817]. Alison Gopnik and Andrew Meltzoff have described the joy of learning in infants, and argue that infant theory construction lies on a spectrum with adult scientific work [818]. For more on curiosity, see Mario Livio [819]. For more on play and games as meaningful, see M. Schlick [820], Bernard Suits [821], and Ian Bogost [822].

47. See Jaak Panksepp and Lucy Biven [358]. For more on the potential distress and impairment associated with gaming, see H.J. Rumpf and colleagues [823].

48. For more on **embodied cognition and exercise**, see Massimiliano Cappuccio [824].

only very recently that we learned to grow food, and that we had to deal with the associated problems of too little exercise and too many carbohydrates.[49]

There is a growing range of science demonstrating that exercise has a range of positive health effects, including for the brain–mind. Just as we should ensure that we get sufficient sleep and eat healthily, so we should ensure that we get enough exercise. Easier said than done, particularly for those who don't like jogging and gym. Perhaps people have to choose an exercise that they really enjoy: for me this has been participation in the martial arts, particularly karate. I'm so focused on the specific techniques or sequences that I'm learning that I'm not thinking about exercise at all. The martial arts are part play, part exercise, and part assertion of hierarchy: a formidable way of inducing pleasure.[50]

Physical and natural beauty

Aristotle held that beauty had to do with symmetry and order, and it turns out that faces that are more symmetrical are universally considered more attractive. Remarkably, the neurocircuitry that is activated by food and drink rewards is also activated by these kinds of faces.[51] Edward Wilson, the father of sociobiology, has put forward the 'biophilia hypothesis'; this states that humans are inherently attracted to certain kinds of landscapes, and to natural beauty. While this remains a speculative idea, it's long been hypothesized that being in nature is healthy for us, and there is growing empirical evidence for this.[52]

It's easy enough to understand why rodents quickly learn to press a lever for sugar (operant conditioning), or why Pavlov's dogs salivated when a bell indicated that it was feeding time (classical conditioning).[53] It's harder to understand why humans find symmetrical faces or natural beauty rewarding. One idea

---

49. See Louis Liebenberg [825], D.E. Lieberman and D.M. Bramble [826], Christopher McDougall [827], and Daniel Lieberman [828]. Complementing this work, the South African sports scientist Timothy Noakes has put forward a '**central governor hypothesis**', which emphasizes that exercise fatigue is determined by the brain, rather than by peripheral muscles. I've done some work in collaboration with him and others on this [829].

50. We earlier mentioned Freud's theory of dreams; for more contemporary perspectives on the value of **sleep**, see J. Allan Dobson [830] and Matthew Walker [831]. For advice on **exercise**, see Damon Young [832]. For more on the martial arts and philosophy, see Graham Priest and Damon Young [833], Barry Allen [834], and Jonathan Gottschall [835].

51. See A. Chatterjee and colleagues [836]. Paralleling the development of neurophilosophy and neuroethics, is the comparatively new field of **neuroaesthetics** (see S. Zeki and J. Nash [837], Denis Dutton [838], A. Chatterjee [839], A.P. Shimamura [809], G.G. Starr [840], and Eric Kandel [841]). Again, such work needs to avoid neuroreductionism and neuroscientism [842, 843].

52. The term '**biophilia**' was coined by Erich Fromm and popularized by Wilson. Many have emphasized the link between nature and well-being. For a review of the mental health impacts of being in nature, see G.N. Bratman and colleagues [844]. Some have gone so far as to talk of a 'nature-deficit disorder' [845]. The neurocircuitry of exposure to natural beauty has also been studied; speculatively this involves decreased activity in areas associated with self-focused withdrawal and rumination [846].

53. It turns out, however, that **Pavlov** never trained a dog to salivate to the sound of a bell [847].

drawn from evolutionary theory is that symmetry serves as a marker of genetic health; we are attracted to symmetrical faces because they indicate that the individual has high 'fitness'. Speculatively, our preference for particular kinds of natural beauty is again consistent with our evolutionary history.[54]

Whatever the reason for our preference for beauty, it has remarkable consequences for people's lives. Attractive people have a range of advantages: they are more likely to be viewed in a positive light (the 'halo effect'), they find partners and employment more easily, and they earn higher salaries (the 'beauty premium'). Conversely, people with skin blemishes are often viewed in a negative light and may suffer discrimination.

Unsurprisingly, many people are plagued by doubts about their body-image. The philosopher Alain de Botton reassuringly points out that perfect symmetry is uninteresting: we are each attractive in unique ways. There may be some universal commonalities, but subjective responses to beauty are variable.[55] As for natural beauty, there's every reason for everyone to get in touch with nature more often.[56]

Attachment and love

We typically think of attachment as occurring between mothers and infants. However, a range of interpersonal relationships provide pleasure, so in this section we'll briefly cover early attachment, romantic attachments, and community attachments.

Strong maternal–infant attachment is one of the defining features of mammalian species and is crucially important in primates. Current neuroscience sheds light on the underlying mechanisms; we now know quite a bit about the neurocircuits and the neurohormones involved. When mothers observe their infants, for example, particular parts of the brain are activated. From the time of Harry Harlow, who separated monkey infants from mothers, we have known that disruption of early attachment has hugely negative consequences on neurodevelopment. Subsequent work across species, including humans, has only further emphasized this point, and has also pinpointed the consequences of such disruption on specific brain circuitry and chemistry.[57]

Asexual reproduction seems a lot simpler than sexual reproduction, and trying to understand why sex arose has been a key focus of evolutionary theory.

---

54. **Evolutionary approaches to beauty** and the arts have typically focused on sexual selection (see Matt Ridley [848], Nancy Etcoff [849], Geoffrey Miller [850], David Rothenberg [851], David Buss [852], Richard Prum [853], and Michael Ryan [854]). Some evolutionary work has taken a broader approach that encompasses a range of other considerations [855].

55. Although not all data are consistent, the bulk of the data support the hypothesis that **attractiveness is advantageous** [856, 857]. For more on skin disease and stigma, see C. Walker and L. Papadopoulos [858]. For more on the beauty of imperfection, see Alain de Botton [859].

56. For more on getting in touch with **nature** and related philosophical issues, see Tristan Gooley [860], Esther Sternberg [861], Florence Williams [862], and David Cooper [863, 864].

57. For more on mother–infant attachment, see Sarah Hrdy [865]. For more on **Harry Harlow's** work, see Deborah Blum [866].

Behavioural neuroscience has begun to shed light on the brain circuits and chemicals that are involved in romantic attachment. Before Tom Insel became director of the National Institute of Mental Health he did some fascinating work on attachment in voles (a rodent species). He showed that differences in the prairie vole—which is monogamous, and the montane vole—which is polygamous, boil down to a small variation in a key neuropeptide, oxytocin. Oxytocin is able to facilitate social affiliation by modulating reward pathways. How applicable this specific finding is to primate species is not altogether clear, though.[58]

Primate societies are enormously complex, involving a range of hierarchies, allegiances, and attachments. Robin Dunbar has put forward the thesis of the social brain: that the primate brain evolved in order to allow better processing of all the cognitive and affective information needed to succeed in complex societies. The growing field of social neuroscience has shed light on the psychobiology of a range of these phenomena: we now have data, for example, on the neurochemical correlates of increased status within a primate hierarchy, perhaps relevant to a neurobiology of pride. Sport provides many people with pleasure, perhaps in part because of the passionate alliances formed. Spending time with friends, and giving to others, turn out to be an important source of happiness. Conversely, we have a growing appreciation that loneliness may be associated with a negative impact on our health and well-being.[59]

---

58. See T.R. Insel [867]. Several of the volumes in earlier footnotes on beauty also discuss the **psychobiology and evolution of human sex**, for example, David Buss [852]. A sceptical attitude is warranted for some writing in this area; see, for example, the contrasting views expressed in David Barash and Judith Lipton [868], Thomas Lewis, Fari Amini, and Richard Lannon [869], Christopher Ryan and Cacilda Jethá [870], Lynn Saxon [871], and Helen Fisher [872]. Robert Wright has pointed out that in contemporary life, serial monogamy has become a norm. He argues, however, that this is not an inevitable consequence of our psychobiology, and that it is up to us to decide whether enduring monogamous relationships are more valuable [873]. Alain de Botton argues wittily and poignantly that marriage necessarily entails tensions, that the best approach is to be realistically pessimistic about the possibility of resolving these, and to celebrate fidelity as a high point of the ethical imagination [874].

59. See C. Gamble, J. Gowlett, and R. Dunbar [875] and Brian Hare and Vanessa Wood [876]. See also Dorothy Cheney and Robert Seyfarth's "**Baboon Metaphysics: The Evolution of a Social Mind**" [877] (which the authors see as a response to Darwin's remark that "He who understands baboons would do more towards metaphysics than Locke") and Matthew Lieberman [878] (who discusses the pain of social rejection, referred to earlier). For more on the evolutionary psychology of sport, see Desmond Morris [879]. For more on loneliness, as well as the value of solitude, see Hannah Arendt [880], Anthony Storr [881], Anneli Rufus [882], John Cacioppo and William Patrick [883], Susan Cain [884], Sarah Maitland [885], and Stephen Batchelor [886]. For Jaspers, "Solitude in nature can indeed be a wonderful source of self-being; but whoever remains solitary in nature is liable to impoverish his self-being and to lose it in the end. To be near to nature in the beautiful world around me therefore became questionable when it did not lead back to community with humanity and serve this community as background and as language" [47]. The African-born philosopher, Anthony Grayling, puts it more pithily, "Life is all about relationships. By all means sit cross-legged on top of a mountain occasionally. But don't do it for very long" [887]. For textbooks of social neuroscience, see Jean Decety and John Cacioppo [888], Russell Schutt, Larry Seidman, and Matcheri Keshavan [889], and Stephanie Cacioppo and John Cacioppo [890].

In the African context, we have the important concept of ubuntu. This principle is based on the Nguni saying: 'Umuntu ngumuntu ngabantu': a person is a person through others. Great South Africans such as Nelson Mandela and Desmond Tutu have highlighted how important ubuntu is to their way of thinking and leading, and how it drove key aspects of the Truth and Reconciliation Commission in the aftermath of apartheid. An Ubuntu ethic overlaps with communitarian approaches that have been put forward across the world. Still, for those of us who live in Africa, the warmth of our communities is often truly remarkable.[60]

The implications of neuroscientific findings on attachment seem like common sense, but are no less the important for that. Infants thrive when they are provided nurturance, and early neurodevelopment is maximal when the environment provides lots of love and lots of cognitive stimulation. Individuals and cultures differ in the way they form bonds and fall in love, but the quality of our friendships and romantic relationships plays a major part in the overall quality of our lives, and we should therefore devote lots of time and attention to strengthening these bonds. Family and social life inevitably involve complexity, competition, and stress, but humans flourish more when they have good attachments to family and social networks.[61]

The big question of happiness

Are there any broad conclusions about the big question of happiness that we can draw from this brief discussion of food and drink, play and music, exercise and running, physical and natural beauty, and attachment and love? A number of general points seem worth highlighting.

First, although there is some universality to what makes us happy, there is also quite a bit of individuality. I'm reminded of a clip I once saw on TV about two athletes competing at the Olympics. One was a firm favourite for gold and had spent all his time aiming for gold. He came second and was hugely disappointed. Another was someone who was not expected to medal. She had trained hard, but not to the point of exhaustion. She came in 3rd, winning a bronze medal. She was absolutely delighted with this outcome. Given this kind of

---

60. For more on the **philosophy of ubuntu**, see T. Metz [891]. Communitarianism emphasizes the connection between the individual and the community, see Daniel Bell [892]. Once again, Aristotle is prescient, noting that "Man is a social animal, indeed a political animal, because he is not self-sufficient alone, and in an important sense is not self-sufficient outside a polis" (in his 'Politics'). There has been philosophical debate about how best to conceptualize ubuntu; in line with a focus on Aristotle, see G.M. Kayange [893]. Other thinkers of the Axial Age also spoke deeply to the importance of community.

61. Philosophical writing on **love and on friendship** goes beyond our scope. A number of our favourite authors have, however, made contributions to this literature, most notably Aristotle (who saw friendships as based on pleasure, utility, and virtue; and who described friends as 'mirrors' who are key to flourishing; see Damian Caluori [894]) and Spinoza [895], but with some points also made by Hume [896] and Dewey [116]. For additional texts, see Roger Scruton [897], Robert Solomon and Kathleen Higgins [898], Neera Badhwar [899], Harry Frankfurt [900], Troy Jollimore [901], Alexander Nehamas [902], Christopher Grau and Aaron Smuts [903], and Lydia Denworth [904].

variation in terms of what precisely will make any one person happy, we should be careful about drawing overly broad conclusions. As John Stuart Mill argued, "The same things which are helps to one person towards the cultivation of his higher nature, are hindrances to another".[62]

Second, we can be pretty lousy at predicting our own happiness (or 'affective forecasting'), and at doing what is good for us. This reflects in part the earlier point that we are more often in a state of 'wanting' than 'liking', that we are not evolved for happiness.[63] Earlier, we also talked about mistakes that the reward system is prone to, such as too little or too much temporal discounting. Further, we typically tend to overrate how happy we will be after a goal is fulfilled; soon after we return to our baseline setpoint of happiness, continuing to seek additional rewards. And we typically tend to underestimate our ability to cope when a goal is not fulfilled; again we return to our baseline setpoint of happiness quite quickly. One consequence is that happiness may well be more about the journey than the destination. Another consequence is that failure is not always as important as our perception of and approach to setbacks. Further, it seems reasonable for people to have a balanced portfolio of short- and long-term goals, allowing immediate as well as delayed gratification (Table 7).[64]

Third, perhaps the most important way in which nice things are not good for us, is when we aim for rewards that are not truly meaningful. The extreme

---

62. See John Stuart Mill's 'On Liberty', in 'Essays on Politics and Society' [905]. The same point regarding **individual variations in eudaimonia** has been made by Aristotle [516], Hume, and Dewey, amongst others. Sonja Lyubomirsky's 'The How of The Happiness: A New Approach to Getting the Life You Want' provides lots of individualized tips [906].

63. Evolutionary theorists have emphasized the importance of an evolutionary perspective on pleasure and pain, and a number have pointed out that **we are not evolved for happiness** [907, 908]. This more recent work contradicts Darwin, who wrote, "If all the individuals of any species were habitually to suffer to an extreme degree they would neglect to propagate their kind; but we have no reason to believe that this has ever or at least often occurred. Some other considerations, moreover, lead to the belief that all sentient beings have been formed so as to enjoy, as a general rule, happiness" [909]. Aspects of Schopenhauer's view of the 'will to life' anticipated both Darwin and Freud; he pithily pointed out that "The coming generation is provided for at the expense of the present" [910].

64. The idea that '**life is a journey, not a destination**' has been falsley attributed to Ralph Waldo Emerson (see quoteinvestigator.com). Julian Baggini has suggested that happiness is neither the journal nor the destination, but rather a consequence of travelling in the right way [911]. We'll return to this metaphor in the last chapter. For Seneca, "no prizefighter can go with high spirits into the strife if he has never been beaten black and blue" (in his 'Moral Letters to Lucilius'). Marcus Aurelius followed up with, "So remember this principle when something threatens to cause you pain: the thing itself was no misfortune at all; to endure it and prevail is great good fortune" (in his 'Meditations'). For Friedrich Nietzsche, "what does not kill me, makes me stronger" [912], and so "To those human beings who are of any concern to me I wish suffering, desolation, sickness, ill-treatment, indignities" [913]. More light-hearted is a quotation falsely attributed to Winston Churchill: "Success consists of going from failure to failure without loss of enthusiasm" (see quoteinvestigator.com). For books on **reframing failure**, see Carol Dweck on a growth mindset [914], Kathryn Schulz [915], Alina Tugend [916], Ryan Holiday on a Stoical approach [917], Sarah Lewis [918], and Ray Dallo [919].

TABLE 7 Pleasure and purpose.

	Pleasure	Purpose	Pleasure and purpose
Historical antecedents	Epicureans	Stoics	Aristotle[a]
Temporal discounting	Short-term reward may predominate	Long-term reward may predominate	Short- and long-term rewards are valued[b]
Theological perspectives	Some authorities celebrate this	Some authorities focus primarily here	Maimonides

[a]*H. Paul Grice has joked that* **Aristotle had a "Prussian view"** *of life. The point he was making is that Kant's deontological position allowed at least some pleasure. In contrast, a Stoic, such as Epictetus, asks that all behaviour is virtuous, and warns about watching over oneself as over an enemy lying in wait [925]. That said, for Aristotle, as noted earlier, flourishing includes both pleasurable and purposeful activity.*
[b]*Both pleasure and purpose may be experienced in states of low arousal and high arousal, see Robert Cloninger [926].*

example of this is drug addiction. Drug addiction 'hijacks' our reward system: we feel pleasure momentarily, but this is inevitably accompanied by the pain of withdrawal, as well by impairment in our ability to function. This kind of phenomenon also characterizes a range of other apparently pleasurable behaviours, that start to get into the vicinity of behavioural addictions, including compulsive gambling, compulsive gaming, compulsive buying, and compulsive sex. For a particular individual, though, apparently addictive behaviour may simply reflect high engagement, and may in fact be meaningful, again pointing to the fuzzy boundaries around concepts of disease. Still, the exemplar of addiction is a reminder that both pleasure and purpose are needed for a fulfilling life. Both the Epicurean focus on pleasure and the Stoical focus on obtaining a proper perspective seem reasonable, and again a balanced approach seems useful (Table 7). Importantly, although our reward systems can go horribly wrong, these systems are plastic and can be rectified.[65] The balanced kind of approach to happiness advocated for here seems to me to fit with some theological advice. In the Jewish Talmud there is significant debate between the school of Shammai and the school of Hillel; the former argued for a more ascetic approach, while the latter argued that God was served by appreciation of his bounty, providing the foundation for a later Hasidic movement which centralized the expression of joy. Maimonides, an important medieval commentator, who followed Aristotle

65. For boundaries of **pathological gaming**, see J. Billieux and colleagues [920]. For discussion of the controversy about pathologizing **compulsive sex**, see S.W. Kraus and colleagues [921]. For a classical evolutionary account of addiction, see R.M. Nesse and K.C. Berridge [764]. Evolutionary accounts of other behavioural addictions have also been put forward; see Gad Saad [922], Stephanie Preston, Morten Kringelbach and Brian Knutson [923], and Bruce Hood [924].

and influenced Spinoza, argued for a balanced life. In his own life, he was a physician engaged with the world and vested in the health of his patients, but at the same time he was a philosopher writing about the meaning of life. In biblical terms we need to work towards a balance, having something of Esau (more pleasure) and something of Jacob (more purpose).[66]

The balanced approach suggested here is also consistent with much in positive psychology, which has emphasized findings from the science of happiness. That said, as noted earlier, we need to be wary of Panglossian positive psychology. Apart from the fact that McMindfulness avoids consideration of social context, it's far too optimistic for my liking. We surely must accept that life has real constraints, and that aspects of the human condition suck. Pleasures are fleeting, and wants are enduring. Evolution did not aim to maximize happiness, but only ensured that we constantly strive for it. Some of our greatest pleasures, such as those of attachment and love, inevitably run the risk of loss. Communities are crucial for our contentment, but they are also invariably competitive and stressful. Some pleasures turn out to be based on disadvantage or hurt to others (winning, schadenfreude, and the sweetness of revenge). Taken together, then, unalloyed optimism seems irrational. We will return to this issue in the next chapter, which focuses on pain and suffering.[67]

4.5 Conclusion

Cognitive-affective science helps inform a number of age-old philosophical questions about the nature of happiness and how to achieve it. In particular, neuroscience sheds light on why particular activities and objects give us so much

66. Although **Maimonides**, in his 'Guide for the Perplexed', often follows Aristotle, he does not always. For example, on the issue of the golden mean, he argues that the person whose character traits are balanced is wise (hakham or sage), but that a person who goes beyond the mean when circumstances warrant is truly pious (hasid or saint). He recognized, however, that in order to build society, we needed sages rather than saints (see Jonathan Sacks [927], who argues that the saint seeks only his perfection, while the sage seeks the perfection of society). Adopting a more secular approach, Hume advised, "Indulge your passion for science … but let your science be human, and such as may have a direct reference to action and society. Be a philosopher; but amidst all your philosophy, be still a man" [133]. After all, as he pointed out, only a philosopher would be sufficiently biased to argue that a life of reflection is the only path to happiness [677]. See also Susan Wolf on 'moral saints' [928], and later discussion of 'derech eretz'.

67. For more on love as a source of wonder, but also danger, see Martha Nussbaum [507]. She writes, "Emotions should be understood as 'geological upheavals of thought': as judgments in which people acknowledge the great importance, for their own flourishing, of things that they do not fully control – and acknowledge thereby their neediness before the world and its events"; we'll cover this sort of point in the later discussion of virtue ethics. For more on **schadenfreude**, see Richard Smith [929], Wilco van Dijk and Jaap Ouwerkerk [930], and Tiffany Smith [931]. For Schopenhauer, schadenfreude is "the worst trait in human nature" [932], but this literature also points to its value. The psychiatrist Theodore Dalrymple admits that "The sight of harmless endeavor gone wrong always comforts me; it makes one's own failures seem so much the less dispiriting. I can think of few things worse than to be constantly surrounded by triumphant success" [933].

short-term pleasure, but also should not be elevated to the sole goal of our lives. An evolutionary understanding of our reward system indicates that evolution aims for propagation of the species, rather than for its happiness. Psychiatry, informed by neuroscience, sheds light on how certain pathologies of pleasure emerge, as well as on the possibility of treating these pathologies. Evidence-based positive psychology may provide some useful guidance as we strive for both short-term pleasure and long-term purpose.[68]

Returning to philosophy, a neurophilosophical perspective holds that human happiness entails a search for both pleasure and purpose. A range of human activities provide opportunities for achieving such pleasure and purpose, not the least being love and work.[69] At the same time, in order to flourish, we need a number of conditions in place, including family and community networks.[70] In addition, particular patterns of thought-feelings may help maximize our happiness setpoint and may help increase the likelihood of our experiencing happiness. Such patterns of thought-feelings may, for example, contribute to our developing and strengthening habits that improve our sense of appreciation, that encourage healthy behaviours, and that contribute to our well-being.

That said, we need a certain amount of skepticism in order to avoid an irrationally optimistic positive psychology. As William Davies so articulately puts it, "We are told to do things for each other, 'notice' the world around us, participate in social events. These touchy-feely tips and preaching smack of positive psychology, and get on people's nerves. Is there an ethical issue

---

68. Oliver Burkeman differentiates **positive thinking versus positive psychology**: "Positive psychology's insights have the inestimable advantage of being backed by real experimental research, but there's another reason why they feel so right in comparison with positive thinking. They're modest, varied, heterogeneous: they speak to our intuition that happiness has a mixture of causes; that it involves trial and error, and broadly chimes with common sense; that there isn't a single secret or quick fix, waiting to be uncovered, and that looking for one might make you miserable. The advice is straightforward. Remember to be grateful. Spend your money on experiences, not objects. Volunteer. Nurture your relationships. Spend time in nature. Make sure you encounter new people and places. And never assume that you know what will make you happy" [630]. Such enthusiasm is, however, tempered by key conceptual and methodological shortcomings of some of the positive psychological literature, see previous footnote.

69. As Freud said, "**Love and work** are the cornerstones of our humanness". Dewey was also a proponent of work, saying "An occupation is the only thing which balances the distinctive capacity of an individual with his social service. To find out what one is fitted to do and to secure an opportunity to do it is the key to happiness" [934]. That said, technological advances may mean that some are permanently without work, albeit still occupied [935]. These sorts of issues raise the question of how much control we have over our happiness; see earlier footnote on Robert Solomon's claim that emotion entails active choice, and later discussion of free will.

70. Bertrand Russell put it pithily, saying "The good life must be lived in a good society, and is not fully possible otherwise" [936]. More recently, Pedro Tabensky has emphasized the intertwining of **individual and community eudaimonia** [937]. Aristotle (see previous footnote on humans as political animals) and several of our favourite philosophers would likely agree. See also Lisa Tessman's work on virtues in the context of societal oppression [938].

regarding the socialisation and exchange of goods in our capitalist, increasingly privatised economy? Certainly there is. Is there a medical and economic issue regarding levels of depression and anxiety in our individualised, atomised age? Certainty there is. But neither of these requires optimisation, optimism or the abandonment of irony". Just as we emphasized the importance of avoiding reductionism and scientism in thinking about the brain–mind, so we ought to be wary of attempts to oversimplify and technologize happiness.[71]

With these points in mind, I would support those who have criticized overly optimistic positive thinking, and who have argued for a 'negative path to happiness'. First, a great deal of the advice given in self-help books about positive thinking is extraordinarily superficial or plain wrong.[72] Furthermore, as encapsulated by the concept of the 'happiness paradox', it turns that constantly striving to be happy is just the sort of thing that will make us miserable.[73] Conversely, constantly trying to avoid the negative aspects of life is an impossible task, and so may deepen unhappiness. Finally, an uncomfortable possibility is that some of what makes us happy may hurt others. We'll discuss the appropriate balance to take between being optimistic versus pessimistic in more detail in the next chapter.[74]

---

71. See William Davies [733], and earlier footnotes on technologizing happiness and on McMindfulness.

72. For some examples, see S. Lilienfeld and H. Arkowitz [939], N.J.L. Brown and colleagues [940], C.E. Cavanagh and K.T. Larking [941]. See also earlier footnotes on defensive pessimism and on the happiness pie.

73. Those who are sceptical of an overly positive psychology, and who advocate for the **value of negativity** include not only Oliver Burkeman, but also Alain de Botton [83], Barbara Held [942], Julie Norem [691], Paul Pearsall [943], Arthur Kleinman [944], Eric Wilson [945], Neel Burton [946], Pascal Bruckner [947], Nassim Nicholas Taleb [948], Simon Critchley [949], Christopher Hamilton [950], W. Gerrod Parrott [951], Michael and Sarah Bennett [952], Kelly McGonigal [953], Mark Manson [954], Svend Brinkmann [955], Blaine Fowers, Rank Richardson and Brent Slife [956], and Randy Nesse [575].

74. **Positive psychology** has at times attempted to distance itself from positive thinking; see Martin Seligman [671], Corey Keyes and Jonathan Haidt [957], Sonja Lyubomirsky [906], Susan David [958], and Caroline Webb [959]. The literature on happiness is immense and keeps growing. It's perhaps worth mentioning that a number of leading psychotherapists have contributed volumes in this area; see Arnold Lazarus and Allen Fay [960], Jeffrey Young [961], Albert Ellis [962], Mardi Horowitz [963], and Steven Hayes [964]. Also, a number of books have specifically attempted to base their approach on neuroscience evidence or evolutionary thinking; see Bjorn Grinde [965], C. Robert Cloninger [926], Stefan Klein [966], Desmond Morris [967], Daniel Nettle [968], Daniel Gilbert [969], Dacher Keltner [970], Rick Hanson [971], Peter Whybrow [972], and Glenn Geher and Nicole Wedberg [973]. Jonathan Haidt's work is particularly integrative, drawing on both philosophy and psychology [974].

Chapter 5

Pain and suffering

Some of the really big questions about life have to do with pain and suffering. All of us experience pain and suffering at one time or another, ranging from physical pain through to mental anguish, and encompassing fear, anxiety, sadness, and depression. Why is there so much suffering in the world? And how should we as humans best respond to the fact that life is so filled with suffering? Philosophers have grappled with these questions for aeons. And given that psychiatry focuses on people with mental suffering, any foundation for this field must provide a reasonable approach to addressing this kind of question.

As elsewhere in this volume, this chapter will take an integrative approach that is informed by cognitive-affective science. Multicellular animals have developed specialized neuronal systems which orchestrate approach (to reward) and avoidance (of punishment). The previous chapter focused on reward circuitry, here we'll focus on the fear and disgust circuitry that ensures avoidance. From an evolutionary perspective, it turns out that avoidance of threat and danger, and unpleasant responses such as pain, are mechanisms that have important survival value. From the perspective of evolution, any negative impact on our well-being isn't the point.

Given that pleasure and pain are opposites, this chapter will in some ways mirror the previous one. At the same time, pain and suffering need careful attention as unique phenomena. In order to sharpen our view of pain and suffering, this chapter will begin by looking at negative emotion in general, but will go on to explore a range of specific negative emotions, including fear and anxiety, disgust and repugnance, and jealousy and envy. Each has particular evolutionary value, but each may also be accompanied by a downside. So once again, we'll need to think through how best to balance things.

5.1 Philosophy and pain/suffering

The African-born Nobelist, Albert Camus, writing during the dark times of World War II, asserted that the single most important question was whether to suicide. While some philosophers have focused on the pleasures of life, others have emphasized the tragedy of life. Life is a sexually-transmitted disease, as the wag wrote. Schopenhauer is doyen of the downbeat philosophers, and his work influenced a range of subsequent writers. More recently, the South African philosopher David Benatar has clearly articulated an antinatalist position, which

Problems of Living. https://doi.org/10.1016/B978-0-323-90239-7.00008-0

rests on the argument that coming into existence entails serious harm. Relatedly, there are those who question whether there has been human progress.[1]

There is a range of constructs that are related to pain and suffering. At a higher level of abstraction, there is the problem of evil, and we'll tackle this in the next chapter. The issue of suicide raises the questions of alienation and anomie, which have been examined by a number of great thinkers, most notably Karl Marx and Émile Durkheim. In this chapter, however, we'll focus on the construct of disease, which has been a key issue for philosophy of medicine and philosophy of psychiatry. Answers to the question 'what is disease?' turn out to reflect perennial debates in philosophy of science and language (Table 8).[2]

A classical perspective views *disease* as an objective construct that can be defined by necessary and sufficient criteria. The classical tradition has played an important role in psychiatry by influencing the development of the diagnostic criteria of the 3rd edition of the Diagnostic and Statistical Manual of Mental Disorders (DSM-III). DSM-III was a key advance for the field insofar as it ensured that reliability of psychiatric diagnosis was as high as reliability in other areas of medicine. However, as discussed earlier, the DSM approach also raises several issues, including the threshold and reification problems (are DSM disorders really valid?), and the causality problem (how is classification related

---

1. "There is only one really serious philosophical problem, and that is suicide", Camus writes in 'The Myth of Sisyphus' [975]. Cicero's axiom, 'To philosophize is to learn how to die', represents another key idea that many have addressed [976]. One wag who equated life with sexually transmitted disease was the psychiatrist Ronald Laing, who stated that, "Life is a sexually transmitted disease and the mortality rate is one hundred percent". For **Arthur Schopenhauer**, "There is only one inborn error, and that is the notion that we exist to be happy … So long as we persist in this inborn error, and indeed even become convinced in it through optimistic dogmas, the world seems to us full of contradictions. For at every step, in great things and small, we are bound to experience that the world and life are certainly not arranged for the purpose of containing a happy existence" [910]. Much has been written on Schopenhauer; I might mention the work of Irving Yalom [977], Robert Wicks [978], and Frederick Beiser [979]. For more on antinatalism, see David Benatar [980]. We'll later return to the question of progress, but for now let's mention the work of John Gray [981]. Also worth citing is Karl Jaspers, who wrote "Man is not only finite but knows of his finiteness. He is not satisfied with himself as a finite creature. The clearer his knowledge and the deeper his experience, the more he gets to know finality and with this the radical deficiency in every mode of his being and doing. All other finite things as well – the embodiment of which we call the world – are not as such enough for him. Everything that is the world leaves him dissatisfied, no matter how deep his involvement and how absorbed may be his participation" [344].

2. **Émile Durkheim** is a particularly relevant scholar to consider given his contributions both to suicidology and to philosophy. Durkheim can be characterized as a social realist, although his views also overlapped in part with the American pragmatists [982, 983]. He argued that the European society of his time was without cohesion and solidarity, so that moral rules were no longer binding, and individuals acted according to their self-interest. In such a society suicide rates would increase: first individuals would no longer see society as providing purpose, and second, passions would be unregulated leading to unrealistic expectations and unhappiness. He understood the 'hedonic treadmill', writing "The more one has, the more one wants, since satisfactions received only stimulate instead of filling needs". Durkheim believed that people need constraints to flourish and that a cohesive society provides a regulative force that plays "the same role for moral needs which the organism plays for physical needs".

TABLE 8 Disease, illness, and integrated positions.

	Disease-focused position	Illness-focused position	Integrated position
Philosophy of science and language	Positivism, meaning requires verification	Hermeneuticism, meaning is based on validation	Mechanism and meaning are important; science is theory-bound and value-laden
Definition of disorder	Necessary and sufficient criteria can be used to operationalize a disease	Expression and experience of illness are context bound	Disease-illness conceptualized with **medical metaphors** that may/may not apply
Ontological status of physical and mental disease	Both physical and mental illnesses can be operationalized	Disease and 'problems in living' should be conceptualized differently	Wetware changes, involving a range of mechanisms, can lead to distress and dysfunction, or sickness
Principles vs particulars	Medicine and psychiatry are concerned with formulating principles regarding natural kinds	Medicine and psychiatry are concerned with appreciating the particulars of conventional kinds	Medicine and psychiatry reasonably weigh up principles and particulars, and different kinds of kind
Facts vs values	Diagnosis is about making a factual determination: are criteria met? (naturalism)	Diagnosis is about making a value judgement (normativism)	Diagnosis, although theory-bound and value-laden, reasonably weighs up facts and values

to underlying mechanisms?). DSM-III is therefore criticized by both neuroscientists (who want it to be more neuroscientific) and social constructionists (who regard it as a response to social deviance).[3]

From a critical point of view, a disease is more similar to a weed than to an element in the periodic table; it is defined within a particular time and space, and it involves the exertion of power or authority. The label of mental disorder then leads to specific experience: the lived experience of *illness*. The critical tradition includes work by Thomas Szasz, Erving Goffman, Ronald Laing, David

3. For references to the substantial literature on **defining medical and psychiatric disorder**, see my 'Philosophy of Psychopharmacology' [53].

Cooper, Michel Foucault, and others. My own first brush with psychiatry was reading about Szasz and other critics of the field in an undergraduate psychology class. This perspective was valuable in teaching me how much of a social activity psychiatric diagnosis is. However, this tradition often goes too far in ignoring the real structures, processes, and mechanisms that underlie the brain–mind, and in rejecting the possibility that our wetware can reasonably be judged to be disordered.[4]

From an integrative point of view, judgements of disease-illness involve both objective and subjective features, bringing together views of disease that have been termed 'naturalist' and 'normative'. Such judgements require a knowledge of underlying biological mechanisms but also an understanding of human practices, and so an integration of principles and particulars, and of facts and values (Table 8).[5] Science, including psychiatric science, is a social activity, but it discovers real structures, processes, and mechanisms in the world, and over time it therefore gains greater and greater explanatory power. With rigorous debate, there

---

4. See, for example, Thomas Szasz [984], Ronald Laing [985], David Cooper [986], Michel Foucault [987], Erving Goffman [988]. Cooper, a South African-born psychiatrist, coined the term 'antipsychiatry' [989]. See later footnote on the term 'reasonably'.

5. Christopher Boorse was key in advancing a naturalist position, arguing that disease can be defined in objective, value-free terms as a "deviation" from statically typical functions that contribute to survival and reproduction, with illness being a disease that is "serious enough to be incapacitating" [990]. His view has been criticized by those who have taken a normativist position, arguing that disease is essentially an evaluative construct [991]. See earlier footnotes for more on distinctions between biomedical explanation and illness experience, and on facts vs values. Many authors have written on facts and values, and on **bridging naturalism and normativism, in medical and psychiatric diagnosis** [53]. For examples, see Karl Jaspers [992], Georges Canguilhem [993], Edmund Pellegrino and David Thomasma [994], Bill Fulford [995], Lawrie Reznek [995a], K.F. Schaffner [996], William Stempsey [997], John Sadler [998], Tim Thornton [999], Jonathan Metzl and Anna Kirkland [1000], Richard McNally [1001], and George Graham [1002].

6. A key debate in this literature is that between **Jerome Wakefield**, who sees mental disorders as harmful dysfunctions [1003], and Lilienfeld and Marino, who see mental disorders in terms of Roschian categories [1004]. Note that Wakefield's position, which conceptualizes dysfunction in evolutionary terms, is close to that of Don Klein's earlier work [1005]. I have sympathy for both Wakefield's attempt to bridge facts (about dysfunction) and values (regarding harm), and for Lilienfeld and Marino's approach to categorization. In a series of articles Wakefield provides a sophisticated defence of his position, appropriately criticizing a view that disorder does not correspond to anything in reality [1006]. Nevertheless, it is not clear that evolutionary dysfunction is an essential natural kind that can be defined using necessary and sufficient criteria [53]. For one thing, for many traits there will be environments that select for, or select against, that trait; so from an evolutionary point of view, health is no more natural than disease [621], although admittedly this doesn't stop biologists from using the terms 'healthy' and 'disease' [1007]. Furthermore, as a later footnote points out, defining human function in terms of evolutionary theory is flawed insofar as natural selection aims for reproductive success rather than for flourishing. Perhaps tellingly, a recent review of the 'selected effect' account by P.E. Griffiths and J. Matthewson notes that this approach may suggest grounds for assigning an individual to a zone between two categories, that is 'objectively vague' [1008]. For more on homosexuality in the DSM, see Ronald Bayer's 'Homosexuality and American Psychiatry: The Politics of Diagnosis' [1009].

can also be a sharpening of its values; thus, for example, careful weighing up of the relevant arguments led to the exclusion of homosexuality from the DSM system.[6]

As in the case of 'flourishing', the construct of 'disorder' relies on a consideration of human nature, but is nevertheless complex and fuzzy; close argument, using rigorous reasoning-imagining, is needed to fill out our rough sketches in a reasonable way. An approach that carefully employs concepts of human function, and cautiously incorporates both relevant principles and particulars, as well as facts and values, arguably hearkens back to Aristotle. In more modern times, Karl Jaspers emphasized that our judgement of these matters in the case of psychiatry needs to rely on both erklären (and our explanations of brain biology) and verstehen (and our understanding of humans).[7]

Importantly, when we discuss the natural and conventional kinds, as well as the facts and values that are relevant to disease, we necessarily make use of metaphoric thinking. An often used metaphor of *disease-illness* is of a 'defence that is breached'. Thus, for example, a person is going about their life, an infectious agent attacks them, the body's defences are overwhelmed, and so a disorder occurs. In this metaphor what is key is that the person gets help from an expert, who has an armamentarium of antiinfectious agents. Their primary responsibility is to adhere to treatment. A second metaphor is focused on 'breakdown'. Thus, for example, a person is going about their life, they come across a barrier, they therefore have to come to a halt, and a disorder occurs. An intervention works to get them around or over the barrier, and to get better.

A third metaphor, and one that has a number of advantages, is one of 'imbalance'. Thus, for example, a person is going about their life, an intrinsic or an extrinsic factor or their interaction leads to some sort of imbalance, and so a disorder results. In this case, the intervention works to restore balance. The metaphor of 'defence' makes a good deal of sense for infectious diseases, and the metaphor of 'breakdown' might be useful for a stressor-related illness.[8]

---

7. See later footnote on Aristotle's discussion of sketching out healthiness and goodness. For approaches to mental disorder that explicitly emphasize Aristotle's account of human function (as bridging facts and values, and normativism and naturalism), see C. Megone [1010] and R. Hamilton [1011]. The terms **'reasonable'** (rather than, say, 'rationally') and **'reasoning-imagining'** are used in this section and elsewhere to indicate that the relevant scientific judgements are not merely algorithmic, and to emphasize the important role of metaphor in scientific decision-making: Susan Haack has used the phrase 'within reason' in a somewhat similar context [120]. See later footnotes on Martha Nussbaum's use of 'reasonably' and on moral reasoning-imagining. Similarly the focus on here is on **scientific particularism**; see later footnote for discussion of moral particularism. Jaspers wrote that "As scientists we want to know: what kind of phenomena are possible in the human psyche? As practitioners we want to know what are the means whereby we can advance the very diverse desirabilities of psychic life? For these purposes we do not need the concept of 'illness in general' at all and we now know that no such general and uniform concept exists" [992], cited by S. Nassir Ghaemi [242]. For more on essential natural kinds and conventional kinds, see earlier discussion of invasive species and related footnote.

8. **Militaristic metaphors** of infectious disease have, however, been criticized, with the suggestion made to replace them with **journey metaphors** [1012].

However, an 'imbalance' metaphor has a long history in both Western thought (e.g. Hippocrates' and Galen's humours) and Eastern thought (e.g. yin vs yang), and it is consistent with a good deal of modern physiology (i.e. control systems ordinarily ensure a stable equilibrium—homeostasis) and evolutionary theory (which highlights the importance of 'trade-offs').[9] The imbalance metaphor may encourage a search for both intrinsic and extrinsic factors that increase vulnerability to disease-illness and influence the sick role, as well as a search for multiple different treatment approaches.[10]

These kinds of metaphors work well in 'typical diseases'. In these typical cases, such as acute bacterial pneumonia, the main responsibility of the ill person is to adhere to treatment; they fulfil the 'sick role' articulated by the sociologist Talcott Parsons. Just as we can quickly classify a robin as a bird, there is ready agreement that pneumonia is a disease. Ordinarily people who get pneumonia are immediately given the sick role: they are not expected to work, we try to get them treatment as quickly as possible, and we support them as best we can in their efforts to recover. Using the 'imbalance' metaphor, the lungs have too much inflammation, and treatment will aim to reduce inflammatory processes. The same kind of thinking might hold for a person who develops posttraumatic stress disorder in warfare: they are supported to get back in balance.[11]

But these metaphors don't work as well in 'atypical diseases'. In atypical conditions, such as obesity, the condition may seem to be part of the person, and the person may be seen as having some responsibility for this. Just as we don't immediately recognize an ostrich as a bird, we are slower to recognize obesity as a disease. We are slower to give people the sick role, as our expectation is that they themselves must take responsibility for changing themselves and their diet. The same considerations would hold for alcoholism. Using the 'imbalance' metaphor, obesity and alcoholism may reflect some sort of imbalance in physical or mental systems, but the individual needs to change in order to get better.

When we think through whether condition x is a disease or not, then, we use this kind of metaphoric thinking. We easily see an infection as a disease because it fits our MEDICAL metaphors (i.e. 'breach in defences', 'breakdown', 'imbalance'). However, we are also inclined to see obesity and alcoholism in terms of good or bad behaviour, that is using a MORAL metaphor. There are those

---

9. In the Judaic tradition, Maimonides wrote, "It is known … that passions of the psyche produce changes in the body that are great, evident and manifest to all. On this account … the movements of the psyche … should be kept in **balance** … and no other regimen should be given precedence" [1013]. For more on medieval ideas of balance, see Joel Kaye [1014]. See later footnote on the Swedish notion of lagom.

10. The World Health Organization has defined health as "a state of complete physical, mental and social well-being", a definition that has often been criticized. A more negative pathway to thinking about happiness and healthiness leads to an alternative proposal that is consistent with the **imbalance** metaphor; health is the ability to adapt in the face of social, physical, and emotional challenges [1015].

11. For more on the **sick role**, see Talcott Parsons [1016]. For more on categorization, see E. Rosch [179] and Lakoff and Johnson [159].

who argue that alcoholism is brain disorder, i.e. a typical medical disorder. But there are others who argue that this metaphor of addiction is simply incorrect. The neuroscientific work on sensitization of the 'wanting' system in addictions seems consistent, however, with a nuanced position which acknowledges brain alterations (per the MEDICAL metaphor), but at the same time continues to hold individuals responsible (per the MORAL metaphor).[12]

The debate is particularly contentious in the case of behavioural addictions, such as gaming disorder, where it cannot be argued that an external toxin has changed the brain, but where in some individuals there appears to be sensitization of the 'wanting' system and loss of control. In order to make a decision about whether these conditions are disorders, we must weigh up these metaphors and their implications. Thus we can reframe the debate on behavioural addictions as one in which some use a MEDICAL metaphor, and others hold that a MORAL metaphor is more appropriate. In any particular case, however, it may also be useful to draw on both MEDICAL and MORAL metaphors: frameworks that see addiction as lying *between compulsion and choice*, or that focus on *responsibility without blame* usefully bring together aspects of these two metaphors.[13]

Again, the perspective taken here is one that encourages integration. Some have contrasted the objective construct (disease), the subjective experience (illness), and the social role (sickness). From an integrative perspective, all three aspects are important. Human thought is necessarily metaphoric: some disease-illnesses are able to serve as central exemplars but others are less typical. While there is understandably debate about the validity of atypical conditions, this does not mean that 'anything goes'.[14]

Finally, while different metaphors are useful for thinking through disease-illness, the 'imbalance' metaphor may have a number of advantages, including embracing the possibility of integrating a number of different explanations (explanatory pluralism). It's certainly a metaphor that has been widely employed across the geographic range of writers of the Axial Age, and it's also one that is in line with current scientific concepts of homeostasis and trade-offs. Of note,

---

12. For more on the **brain disease model of addiction**, see W. Hall and colleagues [1017]. The capitalization of 'MEDICAL' and 'MORAL' here follows the convention established by Lakoff and Johnson when emphasizing metaphoric cognition [159].

13. For more on whether gaming involves simply high engagement or rather loss of control, see again J. Billieux and colleagues [920]. For more on whether addiction is an involuntary compulsion or a voluntary choice, see R. Holton and K. Berridge [1018].

For more on the framework of *responsibility without blame*, see H. Pickard [1019]. For more on the **philosophy of addiction**, see Don Ross, Harold Kincaid, David Spurrett, and Peter Collins [1020], Adrian Carter, Wayne Hall, and Judy Illes [1021], Jeffrey Poland and George Graham [1022], Neil Levy [1023], Nick Heather and Gabriel Segal [1024], and Hanna Pickard and Serge Ahmed [1025]. See also later footnote on akrasia.

14. There has been some work on **health and disease from the perspectives of critical realism and embodied cognition**, which may also be informative. See, for example, Malcolm MacLachlan [1026], David Pilgrim [1027], Drew Leder [1028] (who argues for thinking of the body as a gift), and Michele Maiese and Robert Hanna [1029].

an imbalance metaphor has been applied in discussions of both medicine and morality, likely reflecting overlaps and intersections between our concepts of healthiness and goodness.[15]

5.2. Psychiatry and pain/suffering

In this section we will first discuss various theoretical positions within psychiatry on anxiety and depression, we'll then discuss in more detail anxiety and mood disorders—psychiatric conditions in which the characteristic symptoms are closely related to pain and suffering, before ending with a discussion of the pros and cons of medicalizing anxiety and depression.

Theoretical positions in psychiatry on anxiety and depression

Freud's view on anxiety changed over the course of his life. Early on he saw anxiety in terms of the transformation of sexual libido, but later on he saw anxiety as a threat that gives rise to repression. While the idea that anxiety is a strong motivator remains valid, other aspects of his view, such as giving fear of castration a key role in generating anxiety, haven't garnered much support. Similarly, Freud put forward different views of depression, including the idea that depression is self-directed aggression due to an excessive superego. Again, some aspects of his work remain influential, for example, Freud's view that depression can be due to loss, particularly when this leads to reminders of earlier loss.[16]

These specific hypotheses about anxiety and depression should be seen against Freud's overall perspective on human life. In 'Civilization and its Discontents', Freud wrote that "One feels inclined to say that the intention that man should be 'happy' is not included in the plan of 'Creation' ". There he also articulately mapped the sources of human suffering; this comes "from three directions: our own body, doomed to decay; from the external world that refuses to grant our wishes; and from our relations to other individuals". Freud

---

15. Aristotle was, as ever, prescient, emphasizing the **indeterminacy or fuzziness of complex constructs related to both goodness and healthiness,** noting "Let this be agreed on from the start, that every statement *(logos)* concerning matters of practice ought to be said in outline and not with precision, as we said in the beginning that statements should be demanded in a way appropriate to the matter at hand. And matters of practice and questions of what is advantageous never stand fixed, any more than do matters of health. If the universal definition is like this, the definition concerning particulars is even more lacking in precision. For such cases do not fall under any science *(technē)* nor under any precept, but the agents themselves must in each case look to what suits the occasion, as is also the case in medicine and navigation" (in his 'Nicomachean Ethics', cited by Nussbaum [527]). See also earlier footnote on medical practice as entailing phronesis, and later footnote on psychotherapy and virtue ethics. More recently, Georg von Wright pointed out similarities in how we think about goodness and healthiness [1030]. The question of whether this reflects similarities in their underlying causal properties is worth considering [1031], and see also earlier footnote on Boyd, and later footnote on moral naturalism and the Cornell school.

16. For more on **psychoanalytic theories of anxiety and depression**, see A. Compton [1032] and B. Robertson [1032a].

became increasingly convinced that the only way to allay human suffering was by psychoanalysis. While psychoanalysis has informed psychodynamic thinking, which may be useful for some patients in some contexts, this seems an overly confident conviction in the value of a therapeutic culture.[17]

Skinner, as we've pointed out, put the notions of reward and punishment at the centre of his thinking about behaviour. From the perspective of operant conditioning, depression occurs when the environment doesn't provide sufficient positive reinforcement. Depression may be triggered by a loss, which leads to loss of positive reinforcement, and then spirals into a vicious cycle when social withdrawal leads to further loss of positive reinforcement. Some aspects of this work again remain influential; behavioural activation, for example, continues to be an important intervention for depression. However, those who have tried to put Skinner's ideas about utopianism into practice seem not to have been able to diminish human suffering very much.[18]

We've noted that cognitive-behavioural therapy has much in common with the approach of Stoicism. Aaron Beck, a founder of cognitive-behavioural therapy, put forward the notion of a cognitive triad of negative automatic thinking about the self, the world, and the future. He also described negative self-schemas, which might develop in the context of early adversity, and could be triggered by subsequent stressors. Depression is characterized by a range of logical errors including drawing negative conclusions without supporting evidence (arbitrary inference), and focusing on the worst aspects of a particular situation (selective abstraction).[19]

From the biopsychosocial perspective of contemporary psychiatry, anxiety and depression may result from biological, psychological, or sociological mechanisms, as well as their interaction. We understand a fair bit about the proximal mechanisms underlying the anxiety and mood disorders—ranging from neurobiological mechanisms related to neurocircuitry and neurogenetics all the way through to the social determinants of these conditions. There has also been work on the relevant distal mechanisms, that is, why depression and anxiety might have evolved, and what's adaptive about them. At the same time, there is a lot we're not clear about, as we'll focus on in the next section.

---

17. See Freud [1033]. Whether or not we agree with Freud's views about how to respond to suffering, he clearly played a key role in the development of a contemporary **'therapy culture'**. As W.H. Auden wrote in his poem, 'In Memory of Sigmund Freud', "if often he was wrong and, at times, absurd, to us he is no more a person now but a whole climate of opinion". For more on Freud in particular, see Claude Lévi-Strauss on psychoanalysis as shamanic, Philip Rieff on the triumph of the therapeutic [1034, 1035], Richard Webster on Freud as Messianic [1036], and Anthony Storr on Freud as a guru [1037]. For more on psychotherapy in general, including work by both left-leaning and right-leaning authors, see Barbara Ehrenreich and Deirdre English [1038], Frank Furedi [1039], Eva Illouz [1040], Richard Rosen [1041], Steve Salerno [1042], Christina Sommers and Sally Satel [1043], and Tom Tiede [1044].
18. For more on **behavioural theories of depression**, see P.M. Lewinsohn and I.H. Gotlib [1045]. For more on the implementation of **Skinner's** ideas, see Hilke Kuhlmann [1046].
19. See Aaron Beck and Brad Alford [1047].

Mood and anxiety disorders

Mood and anxiety disorders are the most prevalent psychiatric conditions. The global burden of disease study has been a major contributor to our understanding of their negative impact. This study was initiated by Chris Murray and colleagues, who developed a rigorous framework for understanding disability across different disorders, for measuring such disability on a yearly basis, and for using these findings to impact policy. Bill Gates chose to support this work because it provides an important foundation for thinking how best to get the biggest bang for the buck, in terms of health spending. Data on the disability due to mental disorders were astonishing: depression and anxiety were associated with a massive burden of disease, with depression a leading contributor to disability worldwide.[20]

We earlier noted that psychiatric classification systems have a number of flaws. Let's consider these in relationship to mood and anxiety disorders.

First, where do we draw the line? We earlier talked about the threshold problem in psychiatry. This is particularly acute for mental disorders which are characterized by symptoms that can be normal. Sadness is entirely expectable after the loss of a loved one, and fear is an entirely understandable response to threat. Depression and anxiety disorders are, however, characterized by sadness and fear that is judged to be excessive, in terms of their severity and duration and associated dysfunction. The DSM uses specific cut points to guide the clinician, for example, requiring that the person feels down every day, most of the day, for at least 2 weeks. Critics argue that this amounts to pseudo-specificity, leading to both underdiagnosis and overdiagnosis. There is also heated argument about whether depression and anxiety in the context of loss and threat are really disorders.[21]

Second, there is some debate about what exactly the symptoms of depression are? The DSM system is widely used, but so is the ICD. After all, a particular set of diagnostic criteria may be useful in a highly resourced setting where the service is run by a clinician–scientist (the DSM is often employed in such settings), while a different set of diagnostic criteria may be more useful in a poorly resourced primary care setting where the service is run by a nonspecialized health care worker (the ICD aims to be clinically useful in this context). This point again suggests that psychiatric diagnoses do not carve nature at her joints: they are not essential natural kinds, but rather are best **conceptualized as fuzzy natural kinds (characterized by particular networks of causal**

20. For more about the **global burden of disease**, see Jeremy Smith [311] and A.J. Ferrari and colleagues [1048].

21. For more on the DSM approach to the **threshold issue**, see M.B. First and J.C. Wakefield [1049]. For more on depression and anxiety in particular, see A.V. Horwitz and J.C. Wakefield [1050, 1051]. See, however, the different view of R.C. Kessler and colleagues [310].

mechanisms).[22] Furthermore, examination of different criteria sets over time shows that some symptoms that were historically thought useful for diagnosis have been overlooked by modern criteria. In addition, there may well be some cross-cultural variation in the experience and expression of these conditions; different groups have different 'idioms of distress'.[23]

Third, what are the causes, and best treatments of mood and anxiety disorders? As noted we've learned a good deal about the psychobiological mechanisms involved in these conditions. We've also developed a range of different treatments which work. And they work as well as any interventions in general medicine. At the same time, we have to admit that our understanding of these mechanisms, and our ability to manage mood and anxiety disorders, is limited. We don't fully understand the brain circuitry or genetics underlying depression and anxiety, we can't predict who will respond to current treatments and who won't, and while the majority of people with mood and anxiety disorders will respond to current interventions there is a sizable minority who will not.

Earlier we also considered a few principles for thinking about psychiatry (Table 3). Let's review these, as applied to anxiety and mood disorders.

First, psychiatry necessarily involves an appreciation of multiple levels of reality, from molecular structures through to social determinants. While very specific hypotheses about one or other neurochemical, or indeed one or other environmental factor, are useful to explore in mood and anxiety disorders, clinicians need integrative models that cut across the different levels of

---

22. See earlier footnotes on natural kinds and on essentialism. The idea that mental disorders are **fuzzy natural kinds**, characterized by networks of causal mechanisms, is consistent with the critical realist focus on structures, processes, and mechanisms. Such an approach has been well articulated by Kenneth Kendler and colleagues [1052], who in turn draw on Richard Boyd's concept of species, which suggests that these are 'homeostatic cluster property' natural kinds (i.e. they are characterized by mutually reinforcing networks of causal mechanisms) [624]. An alternative approach argues that mental disorders are **practical kinds**, a view that may be consistent with a pragmatic realist perspective, see Peter Zachar [1053]. For more on natural kinds and psychiatric disorders, see H. Kincaid and J. Sullivan [1054], which includes a chapter by me on psychopharmacology. Such work again supports a view of psychiatry as theory-bound and value-laden. For more on fuzziness in psychiatry, see Geert Keil, Lara Keuck, and Rico Hauswald [1055], which includes a chapter by me on typical versus atypical mental disorders. A third kind of natural kind also bears mention; Marc Lange has argued that diseases are **medical natural kinds**, pointing to biological explanations of incapacity [1056]. Of note, however, he suggests that in the future, as the various 'omics' (i.e. genomics, proteomics, metabolomics, etc.) enter the clinic, these sorts of natural kinds will no longer lie at the core of diagnosis and prognosis. However, at this stage it seems difficult to predict how advances in omics will impact nosology [1057].

23. For an example of how DSM criteria may not align with historical criteria, see again K.S. Kendler [309]. A view that mental disorders are not essential natural kinds may help address the issue of apparently high comorbidity of mental disorders (i.e. patients using mental health care services are often diagnosed with more than one condition). While in some cases one mental disorder may lead to another, in other cases Ockham's razor suggests that comorbidity is an artefact of the system: there may be one underlying condition leading to a range of different symptoms [1058]. For more on **Ockham's razor** see Elliott Sober [1059].

psychobiological reality. Stress-diathesis models of mood and anxiety disorders, which focus on how predispositions to these conditions interact with stressors to precipitate symptoms, for example, are useful in that they integrate aspects of both nature and nurture. Neurocircuitry models of mood and anxiety disorders can also be fairly integrative, accounting for how neurocircuitry is related to symptoms, how such neurocircuitry can be negatively impacted by a range of vulnerability factors, and how this can be positively impacted by different therapeutic interventions.

Second, to achieve appropriately complex and integrative accounts of psychopathology, psychiatry must pay attention to a number of different kinds of explanations. While it's important to understand the specific neurocircuitry underlying mood and anxiety disorders, it's also relevant to understand the evolutionary value of mood and anxiety symptoms. This is not to say that mood and anxiety disorders are themselves adaptive: they are not. However, a framework that explains why humans (and likely other primates) are vulnerable to pathological depression and anxiety helps build an integrated understanding. An evolutionary perspective, for example, helps us in appreciating why different fears emerge at different ages, with separation anxiety seen in infants (given how important the parental–infant bond is), specific phobias emerging in childhood (when things like snakes and spiders can be dangerous), and social anxiety emerging in adolescents (when other humans are an increasingly important threat).

Third, psychiatry necessarily involves an appreciation not only of mechanism but also of meaning. Evolutionary perspectives help us to understand that in many instances depression and anxiety are expectable and understandable responses to particular situations. Arthur Kleinman, who pioneered the field of anthropological psychiatry, emphasized that any individual with a disease will have an explanatory model of why the condition developed; this will in turn contribute to the expression and experience of the illness. Even when a decision is made to prescribe pharmacotherapy for treatment of a mental disorder, the medication will have specific meaning for the individual who is using it.[24]

Fourth, psychiatry ought to be open about how much we don't know. Genetic studies of tens of thousands of people, for example, are very impressive, but ultimately conclude, perhaps not particularly helpfully, that a broad range of genes, each with a small effect, contribute to specific states or traits. We have no biomarkers for differentiating useful sadness from dysfunctional depression, nor do we have a clear understanding of underlying mechanisms that differ between normal sadness and pathological depression. This has a number of key implications. First, we are far better at saying what doesn't cause depression and anxiety than at saying what does. Be wary of any monocausal model! Second, it's tricky to persuade donors to get excited about these conditions. In the case of polio, you just have to get the vaccine to everyone. In the case of malaria, bed

24. See again **Arthur Kleinman** [347].

nets are a highly effective and cost-efficient solution. In the case of depression and anxiety, there is no silver bullet: from a public health perspective a range of different interventions are needed, a hard sell.[25]

Fifth, in working with patients with anxiety and depression, psychiatry needs to employ practical judgement. Diagnostic manuals provide useful operational criteria, but deciding where to draw the line between normal defence and abnormal disorder requires a large measure of 'clinical judgement'. Case formulation and prognostication require a range of different risk and resilience factors to be considered; to date predictive algorithms have not proven particularly useful in psychiatric practice, so that clinicians must weigh up the relevant mechanisms and meanings, with as much science (episteme) and craft (techne) as they can muster. Similarly, in deciding on whether and how to intervene, clinicians need to consider a range of factors, including the evidence base, and patient preferences. Good clinical practice entails practical wisdom and requires successfully balancing perennial tensions between the objective and the subjective, explanation and understanding, and conflicting values.[26]

Medicalization of pain and suffering

Psychiatry has focused more on mood and anxiety disorders than on the broader construct of suffering. On the one hand, the field has certainly moved forwards in terms of more knowledge and more efficacious interventions for specific mood and anxiety disorders. On the other hand, its knowledge of depression and anxiety remains limited in key respects, and when it comes to the broader arena of human suffering its knowledge is particularly thin. In a later section of the chapter, we will go into more detail about a number of specific phenomena that may be useful in sharpening our understanding. For now, I want to briefly consider the question of medicalization of depression and anxiety, given the current status of psychiatric science.

There are those who argue that mood and anxiety disorders are brain disorders, and that given the burden of disease associated with them, they have been relatively neglected by health care systems and by research funders. There is significant underdiagnosis of these conditions across the globe, and many structural and attitudinal barriers to their treatment; given our evidence base of efficacious and cost-effective interventions, these need to be scaled up dramatically. On the other hand, there are those who argue that the offer of 'a pill for every ill' leads to overmedicalization. From this perspective we need to be very careful about equating suffering with medical illness: we need, for example, to instead address the social determinants which contribute to such suffering.

---

25. See again D.J. Stein [486].

26. See earlier footnote for more on medical decision-making as requiring practical judgement or medical phronesis, as well as on evidence-based and values-based practice and a flourishing-oriented medicine, and later footnote on epistemic virtues in science and medicine.

There are those who argue that if it is understood that people with mental symptoms are suffering from an illness this destigmatizes and empowers them, and there are those who argue that a brain model reduces a individuals' sense of responsibility and agency.[27]

Once more I want to acknowledge that in arguments of this sort, both sides have much to say that is entirely valid. It strikes me, for example, that there is both overdiagnosis and underdiagnosis of mood and anxiety disorders. In the high-income world, there is every reason to emphasize that mental suffering is not a disease, and that there are a range of useful ways to address it. In the low-income world, on the other hand, there is significant underdiagnosis of these conditions, symptoms may be seen as evidence of moral failing, and there are both structural and attitudinal barriers that prevent evidence-based medical practice. Once again, in any particular situation we need to see whether a MEDICAL metaphor or a MORAL metaphor is more appropriate. While a reductionistic medical model may be harmful, a more nuanced model that provides space for personal responsibility and agency may be helpful. Perhaps findings from modern cognitive-affective science can facilitate a more complex and comprehensive model, a point we'll explore more in the next section.[28]

5.3. Neurophilosophy and neuropsychiatry

How does modern cognitive-affective science inform our understanding of pain and suffering? A number of findings seem to me to be key.

First, there are similarities in the neurocircuitry of physical and mental pain. This may be useful in emphasizing that mental pain is not merely a figment of the imagination, but is as 'real' an experience as suffering from a physical condition. A fascinating neuroimaging study by Naomi Eisenberger and colleagues examined brain–mind activity during social exclusion. They found that during such exclusion there was activation of both the anterior cingulate cortex and the right ventral prefrontal cortex: areas that are also engaged by physical pain. Social exclusion was experienced, almost literally, as painful. Cingulate activation was associated with increased distress, while prefrontal activation was associated with decreased distress; this is consistent with the hypothesis that prefrontal cortex–cingulate connectivity is involved in emotion regulation.[29]

At the same time, however, there are also a range of different kinds of suffering; overlaps and distinctions in the psychobiology of fear and sadness are consistent with phenomenological evidence that these are partially related constructs. In many of the anxiety disorders, there are disruptions in fronto-amygdala neurocircuitry. In obsessive–compulsive disorder, in contrast, there is

---

27. From the time of Aristotle it has been suggested that **depression and anxiety provide philosophical insights** [500, 1060].

28. For more on the **debate about medicalization**, see E. Kaczmarek [1061], E. Parens [1062], and D.J. Stein and colleagues [1063].

29. See N.I. Eisenberger and colleagues [1064].

excessive activation of cortico-striatal circuitry. These kinds of findings do not resolve philosophical or psychiatric questions about the appropriate boundaries of disorder. However, they emphasize that brain–mind disorders are embodied in quite specific ways.

Second, it is notable on the one hand that trauma is ubiquitous, but also that it can have enduring consequences. Trauma ordinarily elicits a normal stress response, and in the majority of cases this gradually subsides. However, in a minority of cases there are enduring symptoms, including intrusive memories, avoidance, and hyperarousal: in such cases a diagnosis of posttraumatic stress disorder (PTSD) is appropriate. Whereas the field once believed that PTSD was an understandable response to an abnormal event, the current view is therefore that PTSD represents an abnormal failure to recover in the aftermath of exposure to trauma. Thus the field focuses intensively on what wetware mechanisms predispose people to develop PTSD, in the hope that this will ultimately lead to better interventions.[30]

A 'scarring hypothesis' asserts that childhood adversity can have enormously negative consequences. This is supported by neuroscientific data, which demonstrate that early adversity can lead to disruptions in brain circuitry and chemistry, and accompanying changes in cognitive-affective processes. Importantly, though, some degree of exposure to early adversity is entirely normal, and there is great individual variation in response to such adversity. An 'inoculation hypothesis' argues that some exposure to stress is necessary in order for individuals to develop resilience. Individual variation is emphasized in the orchid versus dandelion hypothesis, which holds that some children (dandelions) thrive no matter what the conditions are, while others (orchids) thrive only in nurturing environments. From an evolutionary perspective, early adversity may trigger adaptive strategies that enhance survival under harsh conditions.[31]

Third, there are setpoints in unhappiness and fear. Some people are born with behavioural inhibition; they show less responsivity to external stimuli while still in the womb, they have increased vulnerability to anxiety symptoms during childhood and adolescence, and in early adulthood they have amygdalas that are more readily activated by threatening stimuli. There is growing information available about the 'genetic architecture' of anxiety: multiple gene variants increase risk for anxiety symptoms, and multiple features of the environment may also do so. At the other end of the spectrum are those who are born with too little fear, and sometimes too much impulsivity.[32]

---

30. See D.J. Stein and colleagues [1065]. Work on the **embodiment of pain** was given particular impetus by Elaine Scarry's work [1066, 1067].
31. In a previous footnote we noted how the Stoics and others emphasize the **value of adversity** (e.g. for Seneca, "Everlasting misfortune does have one blessing, that it ends up by toughening those whom it constantly afflicts"). For a more modern and evolutionary perspective on this point, see B.J. Ellis and colleagues [1068]. The dandelion vs orchid hypothesis is put forward by W. Thomas Boyce [1069].
32. See Jerome Kagan [1070].

As noted earlier, there may unfortunately be decreases in our setpoints over time; animals including humans that have experienced early adversity are more likely to have adverse responses to later stressors, and humans who have experienced major negative life events such as death of a spouse or unemployment may have lasting changes in their subjective well-being. At the same time it is notable that over the course of life there seems to be an increase in general life satisfaction. Studies of the impact of genes (the genome), of environments (the envirome or exposome), and of their interactions on well-being are helping us understand the different mechanisms involved.[33]

Fourth, there can be mistakes in our fear and attachment systems. Alarm thresholds may be set inappropriately or we may fear the wrong sorts of things. For example, while small animal phobias make sense from an evolutionary perspective, in our current environments it makes no sense for people to be petrified of spiders, and it's too bad that we are not more fearful of things like dangerous driving. Just as we differentiated short- and long-term rewards, so it's useful to distinguish between short-term pain and enduring suffering. Short-term anxiety in the context of a threatening environment may well be a useful signal that we need to take very seriously. In contrast, persistent overwhelming anxiety in the absence of threat may well point to dysfunction.

Judgements as to whether there is inappropriate anxiety and how best to understand it may, however, be contentious. In some ways, PTSD would seem to be a typical disorder: the disruption can be conceptualized as a breach of defences, a breakdown, or an imbalance. However, it has been argued that PTSD is a construct that has been developed within particular sociocultural contexts, and that the 'medicalization of trauma' is best understood as a political decision. My view is that an integrated position is strongest (Table 8); certainly judgements of disease are a social activity and they reflect particular values, but PTSD is also an entity that is characterized by alterations in psychobiological structures, processes, and mechanisms. Thus we can make reasonable judgements about whether or not a diagnosis of PTSD is appropriate.[34]

Fifth, there is neuroplasticity, and setpoints can be altered and mistakes rectified. We can learn to be less fearful, as Aristotle opined in ancient Greece and as Joseph Wolpe showed in his early work in Johannesburg. A range of other negative cognitive-affective experiences, including sadness, disgust, and jealousy, are evolved responses that make a good deal of sense within particular contexts, but we can learn to modulate them. Neuroimaging has demonstrated that the increased activation of fronto-amgydala circuitry in anxiety disorders, and of cortico-striatal circuitry in OCD, can be normalized by both pharmacotherapy and psychotherapy. This sort of work started to emerge during the time of my own postdoctoral fellowship, and I've continued to find it inspirational;

33. See Michael Pluess [761]. See later footnote on quality of life and ageing.
34. See D.J. Stein and colleagues [1065].

it's a wonderful demonstration that the wetware of the brain–mind can be altered by targeting a range of mechanisms in a range of ways.[35]

That we can decrease psychiatric symptoms with targeted interventions allows us some optimism, but I want to tread with caution. The first self-help book was perhaps physician Samuel Smiles' volume in 1859, which encouraged readers to persevere, "with determined resolution to surmount difficulties". Similar views are expressed in current popular volumes on grit and resilience. Indeed, following Aristotle we certainly ought to be developing habits that lead to a virtuous cycle of increased inner strength and courage. The Stoic idea of reframing adversity in a positive way remains core to contemporary cognitive-behavioural therapy, and there is some evidence that we are able to change our cognitive-affective styles for the better. At the same time, as Smiles' contemporary, Charles Dickens, made clear, a narrow focus on perseverance and success may well be unduly optimistic and overly simplistic. Certainly, research that aims to strengthen resilience is at a very early stage, and to date the evidence that short-term psychological interventions can increase resilience in healthy people is limited. Furthermore, we need to be careful about a Panglossian positive psychology, which ignores the point that sometimes we need to address the structural causes of suffering, rather than learn to cope with them better.[36]

5.4 Sharpening our views of suffering

With these kinds of consideration in mind, how might we cope with the suffering that life brings? Again, it seems to me that in approaching this question, some of the devil is in the detail. In this section we'll consider several different phenomena that are associated with pain or suffering: sadness and despair, fear and anxiety, disgust and repugnance, jealousy and envy, and self-injury and self-harm. We're focused, then, on the sorts of suffering about which cognitive-affective science may have something to say, putting to the side, but not forgetting, the many other kinds of suffering that life can bring, including physical disorders, famine and pestilence, and torture and abuse.

Sadness and despair

Depression is a highly prevalent phenomenon across the globe. It ranges from mild sadness to abject despair. It has been said that just prior to suicide, an individual may be in extreme mental pain. The University of Cape Town went

---

35. See S. Brooks and D.J. Stein [1071].

36. For more on **resilience**, see A. Chmitorz and colleagues [1072] and Angela Duckworth [1073]. For more on **Smiles and Dickens**, who did not sell as well as his contemporary, see J. Meckier [1074]. Dickens' work parodied Smiles' focus on perseverance and success, and demonstrated the tragicomic nature of life. See also Ralph Waldo Emerson's slightly earlier essay on 'Self-Reliance' [1075], as well as earlier footnote on McMindfulness, and later footnote on Pangloss. The phrase 'cruel optimism' comes to mind here [1076].

through a difficult period around 2015–2018, with multiple student protests about fees and other key issues. My colleague, Dean Bongani Mayosi, a brilliant clinician–scientist, took his life. I have often asked myself, what was he thinking and feeling just before he did this? How could this brilliant man, who had accomplished so much, and who had so much more to do, have been in such pain that he saw suicide as his best option?[37]

Much remains to be learned about the psychobiology of depression. Clearly, though, primates are concerned about their place in the social hierarchy and about the extent of their social connections. When we humans lose our place in the social hierarchy, or when we lose our social connections, we experience psychic pain. Contemporary neuroscience has made progress in understanding the neuroscience of social defeat, of maternal separation, and of loneliness; these events lead to significant changes in our brain circuitry and chemistry. From an evolutionary perspective, however, the psychobiological system that underpins sadness is a conserved one, present in a range of primate species, and therefore clearly adaptive. One key potential benefit of this system is that it may help us to reset our priorities after defeats and losses.[38]

What are the implications of this overly brief and simplistic account of the neuroscience and psychology of depression for understanding suffering? Clearly, sadness is part of life: sadness is a marker of what we value. Randy Nesse argues, in line with the stance taken here on avoiding Panglossian positive psychology, that at times an appropriate response to depression is for us to give up on unobtainable strategic goals. On the other hand, in many situations it may be possible for us to make tactical changes, and to achieve our aims. We can be better warriors (when it comes to reversing our defeats in social hierarchies) and we can bond more effectively (when it comes to maintaining our social connections). Furthermore, we need the serenity to accept what we cannot change, the courage to change the things we can, and the wisdom to know the difference.[39]

Fear and anxiety

Anxiety disorders are the most common of the mental disorders in all parts of the globe. Furthermore, they are the disorders which often begin earliest,

---

37. See earlier footnotes on Camus, Durkheim, and **suicide**: a topic of a huge philosophical and psychiatric literature [1077]. See also D.J. Stein [1078].

38. See again Randy Nesse [575].

39. The wording here is from the **Serenity Prayer**, written by the American theologist Reinhold Niebuhr. Earlier on the medieval Jewish philosopher Solomon ibn Gabirol wrote: "At the head of all understanding – is realizing what is and what cannot be, and the consoling of what is not in our power to change". And even earlier, the Stoic Epictetus began his 'Encheiridion' with the words, "Some things in the world are up to us, while others are not. Up to us are our faculties of judgment, motivation, desire, and aversion – in short, everything that is our own doing. Not up to us are our body and property, our reputations, and our official positions – in short, everything that is not our own doing".

creating risk for a range of other later onset conditions, including depression and substance use disorders. They comprise a number of different conditions, including separation anxiety disorder (anxiety about being separated from a parent is normal in infants, but problematic later in life), specific phobias (e.g. fear of small animals), social anxiety disorder (where there is anxiety about being negatively evaluated in social situations), and generalized anxiety disorder (a condition characterized by excessive worry).

The psychobiology of fear and anxiety has been mapped out in fine detail. There are perhaps different systems for the fear response, and for the subjective experience of anxiety. Furthermore, the evolutionary value of anxiety seems clear. At different stages of our lives we need to avoid different threats, hence, at birth we experience separation anxiety, as children we may develop animal phobia, as adolescents social anxiety can be a major phenomenon, and as adults some of us become worriers. Just as electronic smoke alarm systems have to be fairly sensitive in order to be effective, so do brain–mind threat alarm systems. Thus, in healthy people, there will be many false alarms, accounting for the ubiquity of anxiety symptoms and disorders.[40]

This brief summary doesn't do justice to the complexity of the phenomenology and psychobiology of anxiety and its disorders. That admission made, what are the implications of our sketch of the neuroscience and psychology of anxiety for understanding suffering? From one point of view, we need to listen to our anxiety: at times there really is danger, and we need to flee. From another point of view, a key lesson is that given how often our wetware triggers false alarms, it's crucial that we face our fears. Paradoxically, facing our fear lowers the threshold for future alarms, while avoiding our fears only strengthens the alarm system, in a vicious cycle. This idea is one that is emphasized in cognitive-behavioural theories of anxiety, and that is supported by evidence for the efficacy of fear exposure in treating anxiety disorders.[41]

---

40. For more on the **neuroscience of anxiety**, see J.E. LeDoux and R. Brown [567] and D.J. Stein and R.M. Nesse [1079].

41. The **philosophy of anxiety** has been tackled by a range of authors, perhaps most famously the existentialist Kierkegaard [1080]. The work of the Stoics remains a particularly valuable resource. Seneca notes "There are more things ... likely to frighten us than there are to crush us; we suffer more often in imagination than in reality". He therefore advises, "How often has the unexpected happened! How often has the expected never come to pass! And even though it is ordained to be, what does it avail to run out to meet your suffering? You will suffer soon enough, when it arrives; so look forward meanwhile to better things" (in his 'Moral Letters to Lucilius'). Similarly, Epictetus pointed out that "Man is not worried by real problems so much as by his imagined anxieties about real problems". And Marcus Aurelius noted that "Many of the anxieties that harass you are superfluous ... Expand into an ampler region, letting your thought sweep over the entire universe". For overviews, see M.-A. Crocq [1081], Martha Nussbaum [1082], and Sergio Starkstein [1083]. See also earlier footnote on Darwin's point that expression of emotion can lead to it spiralling out of control.

Disgust and repugnance

Early in life, disgust is typically a response that involves rejection of foods; or gustatory disgust, characterized by avoidance of particular food or vomiting if some has been ingested. There are analogous avoidance mechanisms in other animals. However, during development humans are unique in also developing more abstract disgust reactions; in this sort of moral disgust there is a rejection of particular ideas. Anthropologists have noted how in all cultures the metaphor of 'clean' or 'pure' versus 'unclean' or 'polluted' is extremely important. A focus on the 'unclean' is a matter then of repugnance (a term that comes from the Latin for 'to oppose').[42]

Neuroscience research indicates that there may be some overlap in the neurocircuitry of physical disgust and moral disgust. The phenomenon of disgust can also be framed in terms of evolutionary value; disgust helps a range of animals to avoid and expel noxious matter. In humans, where our social lives are so very complex, the disgust system has been appropriated in the service of responding to complex social stimuli. The disgust system may begin as a simple reflex, but can be elaborated in complex ways over time, serving to embody cognitive-affective decision-making that is relevant to social situations.[43]

So, what do we learn from this? Again, I'd like to take a position that emphasizes balance. On the one hand, feelings of disgust are important and deserve a hearing. On the other hand, we need to have some skepticism about our values. Those brought up with particular views of sexuality, for example, may experience real disgust at the thought of homosexual relationships. We are not necessarily wired to be tolerant to different views; it takes an effort of will, and ongoing reflection on our thoughts and feelings. However, over time it seems that we humans have expanded our circle of altruism, allowing us to move away from a 'politics of disgust' and to be more considerate of a range of other humans.[44]

Jealousy and envy

While a good deal of psychiatric literature focuses on the emotions of sadness, fear, and disgust, a range of other affects play a key role in psychopathology.

---

42. For more on disgust in infants, rodents, and primates, see K.C. Berridge and M.L. Kringelbach [568]. For the shift from **physical to moral disgust**, see P. Rozin and colleagues [1084]. A classical anthropological text in this area is Mary Douglas's 'Purity and Danger' [1085].

43. For a review of the psychobiology of moral disgust, including brain imaging findings, see H.A. Chapman and A.K. Anderson [1086].

44. There is ongoing philosophical debate, however, about the limits of such expansion (for liberal views, see Martha Nussbaum [1087]; for conservative views, see Roger Scruton [897, 1088]). The term **'politics of disgust'** is from Nussbaum's volume, while the phrase 'expanding circle' is from Peter Singer's 'The Expanding Circle: Ethics, Evolution, and Moral Progress' [1089]. Notably, Singer and Scruton were amongst the first philosophers to engage with Edward Wilson's sociobiology, taking opposing stances. A number of philosophers have written about disgust, some ignoring neuroscience (e.g. Colin McGinn [1090]), and others engaging with it (e.g. Nina Strohminger and Victor Kumar [1091]).

Pathologies of jealousy include having jealous traits, obsessive jealousy, and delusional jealousy. An individual with jealous traits may have an inflexible approach to their relationships, insisting on rules designed to avoid infidelity. In obsessional jealousy, concerns about investigating and maintaining fidelity are sufficiently severe that they lead to marked distress or impairment. In delusional jealousy the individual loses touch with reality and is absolutely convinced of the infidelity of a faithful partner.[45]

From a psychobiological perspective, however, there is now some understanding of the relevant brain circuitry and chemistry that underlies jealousy. And from an evolutionary perspective, jealousy makes a great deal of sense. It's been pointed out that males are much more concerned with issues of paternity than are females, and that this leads to a range of sex differences in cognitive-affective processes related to partner relationships, including a much higher prevalence of concerns about partner infidelity in men. Envy, although one of the seven sins, may also be useful in motivating us to do more to emulate those whom we are envious of.[46]

What lessons can we glean from this? Once again, I am going to advocate for the Aristotelian golden mean, suggesting the value of an appropriate level of jealousy and envy. A monogamous relationship clearly requires each partner to display some degree of possessiveness. On the other hand, excessive suspiciousness is unlikely to be experienced by a partner as loving, instead he or she will experience such a phenomenon as overly restrictive and even attacking. It turns out that envy, too, is a common emotion, and that rather than increasing our suffering by feeling ashamed of our envy, it makes more sense to employ our envy in a constructive way.[47]

Self-injury and self-harm

Self-injurious behaviour ranges from nail-biting—which is extraordinarily common—to much rarer forms of severe self-injury. Technically severe nail-biting is a 'body-focused repetitive behavioural disorder', a category that includes people who pull out their hair excessively or who pick at their skin excessively. It's remarkable how common these conditions are; it's been said that 99% of people have rhinotillexomania (nose-picking), and the other 1% are liars. Quite remarkably Aristotle described hair-pulling and nail-biting as pathological habits involving positive reinforcement, that is, they are initially pleasurable. Much rarer are conditions like Lesch–Nyhan syndrome or other neurodevelopmental

45. See Gregory White and Paul Mullen [1092].

46. For more on the **psychobiology of jealousy**, see David Buss [1093].

47. For more on the **philosophy of envy**, see J. D'Arms and A. Kerr [1094]. Note that Aristotle provided one of the first definitions of envy, noting that "Envy is pain at the good fortune of others" (in his 'Rhetoric'). However, Nietzsche is the philosopher who has perhaps been most concerned with the importance of envy [1095]). For Nietzsche, "Envy and jealousy are the private parts of the human soul. Perhaps the comparison can be extended".

disorders with severe self-injury, such as biting off one's fingers. Another form of self-injury is seen in patients with so-called 'nonsuicidal self-injurious behaviour'; here there is damage to the surface of the body, of the sort that is likely to induce bleeding or bruising or pain.[48]

The neurobiology of self-injury is increasingly understood. It likely involves pathways related to grooming and to habit formation, including cortico-striatal neuronal circuitry and the opioid neurochemical system. Self-injury is often accompanied by a sense of relief, suggesting that arousal or emotional self-regulation (or self-soothing) play a role. In hair-pulling disorder and skin-picking disorder, the individual does not aim to hurt themselves, rather hair-pulling and skin-picking has become habitual, almost as if the grooming system has gone awry. In nonsuicidal self-injurious behaviour, in contrast, there is a specific intention to harm oneself, and this is done in order to obtain relief from a negative feeling or cognitive state, to resolve an interpersonal difficulty, or to induce a positive feeling state.[49]

What are the implications of this brief account of the neuroscience and psychology of self-injury and self-harm for understanding suffering? It's remarkable that we humans can so easily develop habits that are bad for us. In some cases the reward obtained from harmful habits is obvious; for example, in the case of alcohol abuse. In other cases, such as the body-focused repetitive behavioural disorders, it's harder to understand how habits develop. However, it turns out that such behaviours may have some value for the individual; they may be a form of emotional regulation. And it turns out that fairly simple cognitive-behavioural therapy techniques, known as 'habit reversal therapy', which focus on learning new and healthier habits, are useful in diminishing these symptoms. Notably, while habit reversal therapy requires some degree of self-awareness, it does not rely on new self-insights.[50]

The big question of suffering

Are there any broad conclusions about the big question of suffering that we can draw from this brief discussion of sadness and despair, fear and anxiety, disgust and repugnance, jealousy and envy, and self-injury and self-harm? A number of general points seem worth highlighting.

First, there is the philosophical question of whether ours is the best of all possible worlds (as Leibniz argued) or the worst (as Schopenhauer averred)? **I would argue for taking a middle ground, balancing optimism and**

48. For more on **Aristotle's views of psychopathology**, see G. Pearson [1096]. For more on **self-injury**, see Daphne Simeon and Eric Hollander [1097].

49. See M. Zetterqvist [1098].

50. People with hair-pulling disorder and skin-picking disorder may experience intense shame at being unable to control their symptoms. One of the privileges of my professional life has been to be involved in the TLC Foundation for **Body-Focused Repetitive Behaviors**, and I was delighted when skin-picking disorder was included in DSM-5 and ICD-11. For more on habit reversal therapy, see N.J. Keuthen and D.J. Stein [1099].

pessimism. Aristotle advised that life had its ups and downs, and that "a magnanimous person will neither be excessively pleased by good fortune nor excessively distressed by ill fortune". In a similar vein, both the Stoics and Spinoza argued that if we are clear-eyed about how the world works, instead of being disappointed by what it delivers, we can appreciate how we are part of its ebbs and flows. Similarly, the early American pragmatists argued for *meliorism*: the world is neither the worst nor the best possible, but it is capable of improvement, and rather than passively accepting fate with optimism or pessimism, we should focus on how practically to improve our lives.[51]

From a psychological perspective, approaches in positive psychology may border on saying that with the correct view, all will be well. This is Panglossian; there are aspects of the real world that veritably suck; depressive realism is a position with real validity.[52] Unfortunately, for example, the just world hypothesis, the idea that if we behave well we will be justly rewarded, is simply not true. The question of why good things happen to bad people has no reassuring

---

51. Variations on the **debate between Leibniz and Schopenhauer** include questions about the value of life (is it better or worse to have been born?), about moral 'forces' (does good outweigh evil?), and about human progress (have humans progressed?). Sophocles wrote, "Not to be born is the most to be desired; but having seen the light, the next best is to go whence one came as soon as may be". In the first century BC, two rabbinical schools, those of Shammai and Hillel, debated the question whether life was worth living or not not—"ṭov le-adam shenibra mishelo nibra". For more on antinatalism, see David Benatar [980], and for more on misanthropy see Andrew Gibons [1100]. Importantly, Leibniz fully acknowledged the existence of evil, while Schopenhauer wrote on how to diminish suffering. Schopenhauer movingly writes, for example, "In fact, the conviction that the world and man is something that had better not have been, is of a kind to fill us with indulgence towards one another. Nay, from this point of view, we might well consider the proper form of address to be, not Monsieur, Sir, mein Herr, but my fellow-sufferer, Socî malorum, compagnon de miseres! This may perhaps sound strange, but it is in keeping with the facts; it puts others in a right light; and it reminds us of that which is after all the most necessary thing in life—the tolerance, patience, regard, and love of neighbour, of which everyone stands in need, and which, therefore, every man owes to his fellow" [1101]. As regards the position of the Stoics, this is encapsulated by Seneca's comment: "What need is there to weep over parts of life? The whole of it calls for tears" (although in a passage more reminiscent of Aristotle's view, he also states that "the man who is not puffed up in good times does not collapse either when they change. His fortitude is already tested and he maintains a mind unconquered in the face of either condition: for in the midst of prosperity he has tried his own strength against adversity", see his 'On the Shortness of Life'). For more on the Stoics and Spinoza, see Firmin DeBrabander [1102] as well as Jon Miller [1103]. Enlightenment thinking is also relevant here: for a discussion of Hume' position, see Paul Russell [1104] and see also the debate between Hume and Adam Smith as to whether capitalism entailed real progress or a massive deception (see Dennis Rasmussen [24]). For a discussion of Dewey's tragic view, see R.A. Jacques [635] (and see earlier footnotes on Aristotle on tragedy, and on Dickens on tragedy). Finally, see also Claudia Bloeser and Titus Stahl [1105]; they include discussion of Aristotle, Spinoza, Hume, and the American pragmatists.

52. The word **'optimism'** derives form Leibniz's work [1106]. In the aftermath of the Lisbon earthquake in the 18th century, the French Enlightenment thinker, Voltaire, used satire in his novella 'Candide' to criticize the views of Leibniz, which are expressed by Candide's teacher Professor Pangloss. In the last line he advocates a practical way forwards, with Candide saying "we must cultivate our garden". See previous footnote on depressive realism.

answer.[53] At the same time, unremitting pessimism about this world and the future goes too far. There is often reason to be extraordinarily grateful that we are alive. Further, there is clear evidence of improvements in human quality of life over historical time. To balance optimism and pessimism, we need to develop habits of contributing to the world ('learned optimism'), and to find purpose in any adversity we have to overcome.[54]

Second, we use medical metaphors to think about a range of conditions that bring pain and suffering. **Such medicalization has pros (it encourages an evidence-based and humane set of interventions) and cons** (it may encourage the wrong sorts of intervention for a particular situation). Psychiatry runs the risk of both underdiagnosis and overdiagnosis, and we need to avoid both. Oftentimes the best treatment is no treatment.[55] We should welcome certain kinds of pain and suffering, and avoid some treatments.[56] Further, as we learn more and reason more, weighing up principles and particulars as well as facts and values, we can make progress in addressing these issues. This idea suggests that there may even be such a thing as moral progress, a possibility to which we will return.

While medical metaphors are useful in understanding and responding to certain kinds of pain and suffering, other metaphors are needed to help frame

---

53. For more on the **just world hypothesis**, see Melvin Lerner [1107]. For a related construct, see Baumeister's **'the myth of higher meaning'**, the idea that everything makes sense [1108]. Michael Puett and Christine Gross-Loh note that Confucian teaching emphasizes that the world is capricious rather than coherent [542]. The title of Kate Bowler's memoir, 'Everything Happens for a Reason: And Other Lies I've Loved' puts things pithily [1109].

54. Seligman's notion of 'learned optimism' draws on the work of Albert Ellis, who in turn drew on the Stoics. For more **optimistic books**, see Anthony Giddens [1110], who uses the phrase 'utopian realism'; John Tierney and Roy Baumeister [694], Noam Chomsky [1111], Peter Diamandis and Steven Kotler [1112], Charles Kenny [1113], Johan Norberg [1114], Steven Pinker [1115], Matt Ridley [1116], Hans Rosling [1117], Michael Shermer [1118], Raymond Tallis [1119], Chris Thomas [1120]. For more **pessimistic writers**, see John Gray [981], Emil Cioran [1121], Joshua Dienstag [1122], Max Horkheimer and Theodor Adorno [1123], Joel Kotkin [1124], Christopher Ryan [1125], Roger Scruton [1126], Justin Smith [1127], Oswald Spengler [1128], Lisa Tessman [1129], Eugene Thacker [1130], and Georg von Wright [1131]. For a discussion of optimism versus pessimism with regards to the environment, see Charles Mann's 'The Prophet and the Wizard' [1132]; the prophet is William Vogt, who emphasized our destruction of the environment, while the wizard is Norman Borlaug, who emphasized innovation to save the environment.

55. Many authors have emphasized the **limits of medicine** and iatrogenic illness. For early work, see Ivan Illich [1133]. For a more recent work that targets wellness interventions, in particular, see Barbara Ehrenreich) [1134]. Allen Frances, an architect of DSM-IV, and critic of DSM-5, has pointed out that **no treatment is sometimes the optimal intervention** [1135].

56. For the philosopher **Georges Canguilhem**, for example, to be healthy is not simply to live according to the norms of society, but rather to be able to transcend them. Thus, "To be in good health means being able to fall sick and recover" [993]. For an historical perspective on how pain has been seen as useful, and on metaphors of pain, see Joanne Bourke [1136]. Such views are partially consistent with Nesse's perspective on evolutionary medicine [575]. While I don't want to adopt a position that is overly restrictive regarding treatment ('pharmacological Calvinism'), I do want to recognize the need to carefully weigh up the benefits and risks of any proposed intervention, including both pharmacotherapy and psychotherapy (the adverse effects of which are commonly overlooked, see later footnote).

the broad range of pain and suffering that life brings. Religious texts may be helpful to some; the story of Job, for example, is often used in discussions of suffering.[57] Faced with an incomplete picture of why the world is as it is, we must maintain our hope. Camus, in his 'Myth of Sisyphus', suggests the possibility that Sisyphus, who has been condemned by the Gods to a never-ending, repetitive, and frustrating task, is able to recognize the futility of his situation, and imagines that Sisyphus is happy as he pushes his rock back up the hill. We don't have to solve all the world's problems, but nor are we free to desist from contributing while we can. These sorts of moral stories of suffering may help us to make sense of the world and to endure our fate.[58]

Third, **there are a range of practices that we can adopt to help minimize our suffering and build resilience**. These include accepting loss (of status, of networks) when appropriate, facing our fears and anxieties, and reflecting carefully about how valid our moral disgust is (Table 9). They include replacing habits such as nail-biting, which so many of us have, with better ways of self-soothing. We may wish to do some of the things discussed in the previous chapter to help increase our happiness. We may also want to consider evidence-based

---

57. **Charles Darwin** himself wrote in a letter, "It has always appeared to me more satisfactory to look at the immense amount of pain & suffering in this world, as the inevitable result of the natural sequence of events, i.e. General laws, rather than from the direct intervention of God though I am aware this is not logical with reference to an omniscient Deity". We'll return to the Book of Job later.

58. See 'The Myth of Sisyphus' [975]. Sisyphus may express a number of other emotions, though, including defiance and scorn (see Thomas Nagel [1137]). The idea here of not being free to desist is that of Rabbi Tarfon, a member of the school of Shammai, who stated, "You are not obliged to complete the work, but neither are you free to desist from it". The Jewish notion of 'tikkun olam' (literally 'repair of the world') has resonances with pragmatic meliorism. For more on a Judaic perspective on hope, see Jonathan Sacks, who argues for hope, rather than mere optimism [1138]. A number of others have expressed similar ideas from a secular perspective. For John Dewey, **meliorism** "attacks optimism on the ground that it encourages the fatalistic contentment with things as they are; what is needed is the frank recognition of evils, not for the sake of accepting them as final, but for the sake of arousing energy to remedy them" [1139]. He also stated that, "The end is no longer a terminus or limit to be reached. It is the active process of transforming the existing situation. Not perfection as a final goal, but the ever-enduring process of perfecting, maturing, refining is the aim of living" [1140]. Erich Fromm later said that "optimism is an alienated form of faith, pessimism an alienated form of despair", while Gramsci's motto was: 'Pessimism of the Intellect, Optimism of the Will'. More recently, Rebecca Solnit wrote, "Hope is an embrace of the unknown and the unknowable, an alternative to the certainty of both optimists and pessimists. Optimists think it will all be fine without our involvement; pessimists take the opposite position; both excuse themselves from acting. It's the belief that what we do matters even though how and when it may matter, who and what it may impact, are not things we can know beforehand" [1141]. Maria Popova pithily says, "Critical thinking without hope is cynicism. Hope without critical thinking is naïveté" [1142]. Derek Parfit writes, "Life can be wonderful as well as terrible, and we shall increasingly have the power to make life good. Since human history may be only just beginning, we can expect that future humans, or supra-humans, may achieve some great goods that we cannot now even imagine" [1143]. For more on the **philosophy of hope**, see Claudia Bloeser and Titus Stahl [1105] and Ronald Aronson [1144]; and for a neo-Aristotelian perspective, see Stan van Hooft [1145].

TABLE 9 Optimism and pessimism.

	Panglossian optimism	Unrelenting pessimism	Pragmatic meliorism
Historical antecedents	Leibniz	Schopenhauer	Dewey
Theological roots	Evangelic and Hasidic joy	Asceticism	Maimonides
Pessimism vs optimism	This the best of all possible worlds, just world hypothesis, myth of higher meaning	This the worst of all possible words, just world delusion, depressive realism	This is far from the best of all possible worlds, so we attempt to improve it (meliorism)
Disease and suffering	There is a pill for every ill	Illness provides opportunities for growth	Carefully weigh up facts and values: some things should be medicalized, others not
Concepts of resilience	The rules of resilience can be discovered and employed	Notions of resilience reflect and reproduce social values	Resilience is built by developing appropriate habits[a]

[a] *See earlier footnote on resilience training. For more on* **habits that may increase resilience**, *see Steven Southwick and Dennis Charney [1146].*

actions like getting enough sleep (very important for mental health) and wearing seatbelts (this will prevent trauma in vehicle collisions, and so minimize the chances that we get PTSD). Unfortunately there is no single silver bullet; given that a wide range of factors contribute to suffering, we need a broad approach.[59]

5.5 Conclusion

Cognitive-affective science and evolutionary theory provide a number of useful insights into the big question of pain and suffering. In particular, they help explain why pain and suffering are not part of life simply because Eve ate from the fruit of the tree of knowledge: rather pain is an ancient evolved response which is crucially important in a range of contexts. In social animals, the pain system has been co-opted for use in social decision-making and interactions, hence, for example, the pain of social rejection.

59. See J. Allan Hobson [830] and Matthew Walker on **sleep** [831]. See also D.J. Stein [486].

A psychiatry that is informed by cognitive-affective science and evolutionary theory will take a number of useful stances. First, the aim of treatment is not to entirely rid the individual of negative experiences; not all pain and suffering should be medicalized, and we need to avoid both underdiagnosis and overdiagnosis in psychiatry. Second, dealing with depression may involve behavioural activities like exercise, cognitive changes like rethinking one's approach, and social responses like becoming more connected. Third, one of the best ways of dealing with anxiety is facing one's fear (it's not as important to understand the origins of one's anxiety). As we've emphasized, there are few single causes of mental illness and no silver bullets in mental health care.

These resources may then help us address the big questions regarding pain and suffering. The questions of whether life is best seen as a comedy or a tragedy, and whether we should be optimistic versus pessimistic about the future, seem to me to have no absolute answer, and our responses on any particular day may well be context-bound. Some people simply seem to be born with good humour and a sanguine predisposition, others less so. Some people are born into relatively stable and well-resourced environments, others less so. Some days go well, some days go terribly.

As we face or embrace each day, the problem with taking a Panglossian positive psychology approach is that this papers over real pain and suffering and doesn't allow the possibility that important structural changes in ourselves and in our societies do need to take place. On the other hand, the problem with unrelenting pessimism is that it doesn't allow for the possibility there can be more to life than the 'nasty, brutish, and short'. We need to balance optimism and pessimism, we need to view our suffering as having meaning. At the n'th degree of removal, the second law of thermodynamics tells us that the entire world will eventually disappear. But in the interim, there is much to be grateful for, and much we can do to continue to make the world a better place.[60]

60. This is Thomas Hobbes's famous phrase, from his 'Leviathan' [1147]. Similarly, evolution is not about progress; Darwin himself rarely used the term 'evolution', instead speaking of 'descent with variation', and emphasizing that variations are random [620, 1148]. For a critique of Gould, see J. Li on **evolutionary progress** [1149]. The dismal long-term prospect for mankind was pointed out by Alvy Singer in Woody Allen's 'Annie Hall'; although as his mother responds "What has the universe got to do with it! You're here in Brooklyn! Brooklyn is not expanding!" and as his psychiatrist Dr. Flicker points out: "It won't be expanding for billions of years, Alvy, and we've got to enjoy ourselves while we're here, eh? Ha ha ha …" Consistent with Alvy's position, Bertrand Russell wrote "Optimism and pessimism, as cosmic philosophies, show the same naïve humanism; the great world, so far as we know it from the philosophy of nature, is neither good nor bad, and is not concerned to make us happy or unhappy. All such philosophies spring from self-importance, and are best corrected by a little astronomy". Still, as Winnie the Pooh, pointed out, "How lucky I am to have something that makes saying goodbye so hard".

Chapter 6

The good and the bad

The last two chapters have pointed towards the idea that while approach and avoidance mechanisms are present in many life forms, when it comes to *Homo sapiens*, meaning (or, more technically, 'embodied and embedded cognitive-affective processing') is key. We find activities that are both pleasurable and purposeful particularly rewarding, and we are able to endure various forms of suffering if we find meaning in doing so. We are homo symbolicus: unlike simple organisms that approach nutrients and avoid threat mechanistically and reflexively, we do so not only to increase pleasure and diminish pain, but also with purpose, with specific thoughts, feelings, and motives that are in accordance with our values.[1]

With this framework in mind, we can perhaps get to the big question of morality, the nature of good and evil. Good and evil are large constructs, and we are going to need some examples of each, in order to sharpen our approach. Childhood incidents have a way of enduring vividly, as I write this I recall two. A first was in primary school. There was a gang of boys that used to wait for one of our fellow students every day, and then try to hurt him, punching and pummelling him. The second was more reflective of apartheid South Africa. I recall watching a police van pull up alongside a man walking in our road, yelling at him for his 'pass', and then roughly pushing him into the back of their vehicle. Think about these incidents as an adult; however, perhaps the real evil lies in the administrations and bureaucracies that facilitate such behaviour.

I can also recall many times in my childhood and adolescence where I was bowled over by examples of good. Still in my teens, I decided to use a vacation to hitch around the country. Complete strangers would offer me a lift, provide me food, and show extraordinary kindness. Again in my teens, I remember being extremely nervous about an examination, and pacing up and down the corridors. A student, whom I didn't know well, happened to be in the vicinity, and I mentioned my nervousness. He sat me down, heard me out, and somehow managed to calm me down completely. Again such

1. As Nietzsche said, **"If we have our own 'why' of life we shall get along with almost any 'how'"** [912]. Or as Mary Shelley earlier put it "Nothing contributes so much to tranquillize the mind as a steady purpose, – a point on which the soul may fix its intellectual eye" [1150]. See also earlier footnote on homo symbolicus.

Problems of Living. https://doi.org/10.1016/B978-0-323-90239-7.00009-2

day-to-day benevolence is perhaps a pale reflection of the altruism of which humans are capable. In my first years at university, for example, I met people who had made incredible personal sacrifices in order to contribute to the struggle against apartheid. Thinking back on those times, it remains remarkable to me how South Africans avoided a full-scale civil war, and instead chose democracy.

These sorts of exemplars are perhaps overly simplistic. The moral issues of apartheid were, after all, often stark. My years in university management have put forward many situations where the issues appear much less clear. It's been said that "in any dispute the intensity of feeling is inversely proportional to the value of the stakes at issue – that is why academic politics are so bitter". Still, university decision-making has a major impact on careers and livelihoods. Sitting on selection committees, for example, has at times raised enormous quandaries for me: how precisely does one fairly determine which of two applicants, each of whom has entirely different strengths and weaknesses, is more deserving of the post? My hypothesis is that the spoils often go to the most likeable candidate rather than to the strongest: in this way university decision-making systems can at times be iniquitous.[2]

How do we think about these sorts of phenomena? In the age of neuroethics, there is a growing literature that attempts to think through questions of good and evil in terms of biology in general and neuroscience in particular, drawing on observations of altruism and cruelty across the animal kingdom. And in the age of neuropsychiatry, there is a growing literature that attempts to think through questions of good and evil in terms of psychopathology, and its discoveries about disorders of empathy. We'll attempt to synthesize some of this literature, and to outline an approach to thinking about good and evil. In the previous chapter we argued that our constructs of disease entail the use of both a medical metaphor and a moral metaphor; in this chapter we turn in more detail to the notion of moral metaphors.

6.1 Philosophy and morality

In her thought-provoking volume on 'Evil in Modern Thought', Susan Neiman argues that the problem of evil sits at the border of metaphysics and ethics, and that a focus on evil is a useful way of explaining the history of philosophy. She suggests that the earthquake that destroyed Lisbon in the 18th century marks the emergence of modern views of good and evil. Traditional philosophers took God as a given and were concerned with explanations for why his world including the sort of suffering seen in Lisbon. But over time this kind of perspective was subject to more and more criticism, until Nietzsche eventually claimed that

2. This aphorism has been called **Sarye's Law** [1151].

God was dead. Just as progress in science brought about the division of natural philosophy into philosophy and science, so with God's death there emerged the modern distinction between natural evil (as exemplified by Lisbon) and moral evil (as exemplified by Auschwitz).[3]

Modern moral philosophy has been particularly concerned with how best to conceptualize good actions. Consequentialism argues that the morality of one's actions should be judged on the basis of their outcomes; one form of this is utilitarianism, which argues that what is morally right is that which leads to maximal utility, often defined in terms of happiness (e.g. by Bentham and Mill, as discussed earlier). Deontology, on the other hand, argues that what is important for morality is that our actions stick to a set of universalizable rules regarding obligations and prohibitions (such as Kant's 'categorical imperative'); the means do not necessarily justify the ends. These sorts of rigorous attempts to define the good, and to live accordingly, are consistent with a naturalistic view that we ought to see evil as intelligible, and so fight against it.[4]

Both utilitarianism and Kantianism aim to provide necessary and sufficient characteristics of what is moral, arguably attempting to resolve moral questions using a kind of moral arithmetic. William Ross, a Scottish-born philosopher who taught at Oxford and who was the General Editor of the Oxford Aristotle translation series, argued that the view from either of these positions 'oversimplifies the moral life'. Instead he put forward multiple 'prima facie duties', so laying the foundation for a principle-based ethics. The Georgetown University philosophers Thomas Beauchamp and James Childress applied this framework to biomedical ethics, emphasizing the principles of autonomy, nonmaleficence, beneficence, and justice. They argued that these principles are part of a 'common morality' that is shared by members of society, and that is in turn based on common sense and tradition.

---

3. While the problem of evil dates back to the ancients, Neiman focuses on more modern thinkers [1152]. For related work on responses to Auschwitz, see Richard Bernstein [1153]. For more on the **philosophy of evil,** see amongst others Mary Midgley [1154], John Kekes [1155], Raimond Gaita [1156], Ruth Grant [1157], Terry Eagleton [1158], Luke Russell [1159], Chad Meister and Paul Moser [1160], and Thomas Nys and Stephen de Wijze [1161].

4. More modern utilitarians include Henry Sidgwick, George Moore, Richard Hare, and Peter Singer. Kant formulated his 'categorical imperative' in different ways, stating "Act only according to that maxim whereby you can, at the same time, will that it should become a universal law", and "So act as to treat humanity, whether in your own person or in another, always as an end and never as only a means" [1162]. Although redolent of the golden rule of "doing unto others as you would have them do unto you", proposed by a range of Axial Age authors, Kant and subsequent authors have noted key differences between these principles [1163]. See later footnote for more on the golden rule and reciprocity. The contrast between **evil as intelligible and evil as inexplicable** is Neiman's; she lists Leibniz, Pope, Rousseau, Kant, Hegel, and Marx in the former category, and she lists Bayle, Voltaire, Hume, Sade, and Schopenhauer in the latter one.

The term 'principlism' was introduced by two critics of their work, who argued that these principles lack theoretical unity, and so lead to conflicting conclusions.[5]

In opposition to the classical position that there are universal moral standards that apply to every situation (moral absolutism), a critical position may hold that nothing is morally right or wrong (moral nihilism). Moral disagreements seem to be commonplace, and certainly the conceptual structures of what is considered moral differ from time to time and place to place, suggesting that determinations of morality cannot be universalized (moral relativism). There are, however, divergent views of how to explain and respond to such disagreements. Indeed, the critical approach encompasses different views in meta-ethics, including cognitive subjectivism (morality is that which is approved of), noncognitive stances such as emotivism (moral terms express an attitude of approval), and moral skepticism (moral judgements cannot be justified). It may also be consistent with nonnaturalism; this again encompasses different views, including that moral philosophy is autonomous from the natural sciences, and that moral properties are not reducible to any natural property.[6]

Nietzsche is a particularly key author in the critical tradition; he denied a universal morality, and asserted that we need to go 'beyond good and evil'. Like Nietzsche, Martin Heidegger emphasized the textual nature of truth, seeing history not so much as an attempt to increase freedom or decrease misery, but rather as a poem—as a series of narratives with changes in humans' understanding of themselves and of what matters. A poststructuralist perspective may focus, therefore, on how different kinds of power exert their effects over history

---

5. Roger Scruton, in his critique of Derek Parfit's monumental synthesis of aspects of utilitarianism and Kantianism, goes so far as to say "One way of being a bad person is to think that [moral dilemmas] can be resolved by moral arithmetic" [1164]. While this may seem like a gratuitous insult, both conservatives (like Scruton and John Kekes) and liberals (like Isaiah Berlin) have argued that the implementation of overly confident and **overly precise moral rules** lies behind many of the tragedies of the modern era [672, 1165]. As Heinrich Heine wrote a century before Hitler came to power, "Do not take my advice lightly, the advice of a dreamer who warns you about Kantians, Fichteans, and *Naturphilosophen* ... A play will be enacted in Germany which will make the French Revolution look like a harmless idyll" [1166]. Of note, Ross was influenced by George Moore (both were nonnaturalists), but also opposed him (disagreeing with his ideal utilitarianism) [1167]. For more on **principlism**, see Tom Beauchamp and James Childress [1168], as well as K.D. Clouser and B. Gert [1169]. For a critique of overly precise moral rules in bioethics, see Savulescu [1170].

6. For more on moral conflict, see Walter Sinnott-Armstrong [1171]. There are a range of different views within **moral skepticism** [1172]—Sinnott-Armstrong himself argues for a *moderate moral skepticism*, holding that moral beliefs can be justified, but not extremely justified. For more on the range of work on **moral error theory,** the view that all moral judgements are mistaken, in particular, see Richard Joyce and Richard Garner [1173] and Jonas Olson [1174]. In addition, the meta-ethical distinction between naturalist and nonnaturalist approaches has been formulated in different ways, and so nonnaturalism is not included in Table 10.

and across geography. In this view, naturalistic explanations of evil smack of reductionism and scientism.[7]

Although debates about moral absolutism and moral nihilism have likely gone on for aeons, there is something peculiarly modern about their current incarnations. John Stuart Mill was the first to refer to the 'moral sciences', and an attempt to find the laws of utility or of duty seems redolent of efforts to discover laws in the sciences. In attempting to develop an integrative approach, circumventing the possible reductionism and scientism that a focus on moral laws entails, but also sidestepping the nihilism of modern Nietzschean declarations that 'God is dead', it is perhaps worth returning to the thinkers of the Axial Age. What is notable about many writings of ancient Greek, Hebrew, Indian, and Chinese philosophers is their focus on specific exemplars of moral behaviour. In addressing the big question of how to live, these thinkers often emphasized the practices and the particulars of virtuous behaviour. In our earlier discussion of reason and emotion we spoke of Aristotle's focus on the 'priority of the particular'—for Aristotle "Practical wisdom is not concerned with universals only; it must also recognize particulars, for it is practical, and practice concerns particulars".[8]

For Aristotle and other ancient Greek philosophers, then, ethics necessarily entailed 'virtue ethics': moral decision-making aimed at excellence or virtue (arete). Consider the 'sick role' discussed in the previous chapter. A consequentialist may argue that we should absolve those who are ill from their everyday obligations, and assist them, because this will ultimately maximize the good of all. A deontologist may argue that we rightly look after the ill, because it is our duty to do so. From the perspective of virtue ethics, when we decide to take care of someone, we are involved in a process of moral reasoning: one that ought to consider universal principles as well as particular situations, exhibit practical wisdom, and demonstrate the key virtue of nurturance.[9]

An immediate question regards the justification of the virtues. How do we know what the virtues are? A neoAristotelian perspective argues that the virtues contribute to flourishing. An immediate objection to this view is that our

---

7. See Allan Megill [1175]. For **Friedrich Nietzsche**, "There are no such things as moral facts" [912]. This point was later echoed by John Mackie in the opening sentence of his volume, 'Ethics: Inventing Right and Wrong' [1176]. Mackie writes about Hume; see his 'Hume's Moral Theory', oddly enough drawing some important morals from him [1177, 1178]. For a charitable reading of Nietzsche see Robert Solomon [1179], who argues that where the Stoics aimed for peace of mind (ataraxia), Nietzsche instead usefully emphasized vibrance, enthusiasm, and engagement with the world. See also Jacob Golomb and Robert Wistrich on the debate about the relationship between Nietzsche and fascism [1180]. Notably, Nietzsche has also been read as a naturalist and as a virtue ethicist [1181, 1182].

8. See John Stuart Mill [1183]. The phrase '**priority of the particular**' is from Nussbaum [527], who cites Aristotle's words form his 'Nicomachean Ethics'. See also previous footnotes on moral particularism in Aristotle and in Mencius.

9. For more on **Aristotle and contemporary virtue ethics**, see Timothy Chappell [1184]. For more on the array of virtues and vices, see Kevin Timpe and Craig Boyd [1185].

concepts of what counts as 'flourishing' are clearly value-laden, with people having entirely different views, and with individuals having different views at different times in their lives. Indeed we cannot ignore the complexity of constructs such as 'flourishing', and the different values that are reasonably brought to be bear on expounding it. Earlier we emphasized that nature is 'dappled', so that scientific reasoning-imagining requires conceptual and explanatory pluralism. Similarly, a neoAristotelian approach to morality acknowledges the complexity and fuzziness of nature, and proposes **value pluralism.**[10]

Nevertheless, as in the case of science, this is an 'objective pluralism' that eschews both moral absolutism and the nihilistic conclusion that 'anything goes'.[11] There is good evidence, for example, that human beings are social animals; virtues like nurturance do contribute to our flourishing because they help create a society with high levels of attachment and reciprocity. This perspective, then, entails **moral naturalism**; human ethics is entirely derived from our knowledge and experience of being human, of being creatures that have reasons and passions, with physical bodies and social networks.[12]

Margaret Mead, the well-known anthropologist, is reported to have said that evidence of a healed femur was evidence of a caring society; only by giving

---

10. Nussbaum, amongst others, reads **Aristotle as a value pluralist**, citing his point that there is agreement that eudaimonia is a key goal of life, "But concerning, eudaimonia, as to what it is, they are in disagreement" [300]. Aristotle also points out that this disagreement is not only between individuals, but also within any one individual. George Moore's claim that the concept of goodness cannot be further decomposed (foundational monism) has been criticized by authors who emphasize that such a category necessarily entails more than one value (foundational pluralism), perhaps just as the Aristotelian construct of 'flourishing' is not a monistic one [1186, 1187]. Note that although I have linked value pluralism with virtue ethics, value pluralism is not inconsistent with utilitarianism and consequentialism [1186].

11. The phrase '**objective pluralism**' is from Isaiah Berlin, who consistent with the view of moral objectivism that moral judgements can be true or false, held that there is a minimal moral universalism [58]. For more on value pluralism as an alternative to moral absolutism and moral nihilism, see Owen Flanagan [169], Ann Alexandrova [231], Elizabeth Anderson [1188], Isaiah Berlin [58, 1189], Gilbert Harman and Jennifer Jarvis Thomson [1190], John Kekes [1191], Joseph Raz [1192], Michael Slote [1193], and Susan Wolf [1194].

12. For different definitions of and approaches to current **moral naturalism**, see Matthew Lutz and James Lenman [1195]. For related work on natural law, see Tom Angier [1196] and Mark Cherry [1197]. Of note are distinctions between **neoAristotelians** (e.g. Phillipa Foot, Martha Nussbaum, Rosalind Hursthouse, Judith Jarvis Thomson, William Casebeer) who have focused on human functions, and members of the **Cornell School** (e.g. Richard Boyd) who have argued that 'goodness' and 'healthiness' are best specified by their underlying causal properties, in the same way that other natural kinds can be characterized in terms of their 'homeostatic cluster properties' [1198]. Either way, the sort of moral naturalism put forward here does not entail an objective moral authority existing outside of human life, an idea that has appropriately been critiqued by moral sceptics [1199], and has some overlap with naturalistic positions focused on 'pluralistic relativism' [1200, 1201]. For more on how **defining human function in terms of evolutionary theory is flawed** insofar as natural selection aims for reproductive success rather than for flourishing, see William FitzPatrick [1202]. For more on **Aristotle's view that human function involves a life of practical judgement**, and what this entails, see J.M. Alexander [173]. See also later footnote on practical judgement as an essentially contested construct.

someone with a broken femur the sick role, it is possible for this fracture to heal.[13] Moral decision-making in this context may have been reflexive, but to the extent that it entailed a deliberative and imaginative process, it can be considered reasonable. Along these lines, the philosophers Annette Baier and Alasdair MacIntyre have developed positions that are grounded in biology, and which emphasize vulnerabilities that are characteristic of our species (embodied dependencies). Baier drew inspiration from Hume's view that at their best humans put the response to human need above creed or doctrine, and his emphasis that "persons are born of earlier persons and learn to be persons from other persons". MacIntrye, who played a key role in the revival of Aristotelian virtue ethics, argues against the idea of a disembodied thinker who can reason about morality, and against the 'illusion of self-sufficiency', instead emphasizing the importance of virtues that are expressed in social practices, and that simultaneously draw on our understanding of human nature, and contribute to human flourishing.[14]

I'd suggest that an approach to moral naturalism that emphasizes that moral decision-making draws on our views of human nature, and contributes to human flourishing, is dubbed *Oxford moral naturalism*, in deference to the foundational influence of Elizabeth Anscombe, Phillipa Foot, Iris Murdoch, and Mary Midgley, who were contemporaries at Oxford. Together they moved the field of moral philosophy away from a narrow focus on abstruse technical points, and towards addressing the bigger question of how best to live. Rather than getting stuck in the linguistic and cognitive nature of alleged moral claims, they focused on concrete human problems and made actual moral claims about what is right and what is wrong, albeit noting the fallibility of such claims. Mary Midgley has argued that they were able to do this precisely because with men involved in the war effort, they had greater freedom and creativity. These authors specifically opposed scientism (e.g. human nature cannot be reduced to issues of survival and reproduction), saw facts and values as intersecting, and were interested in 'thick' moral concepts (Table 10). The American philosopher Martha Nussbaum has further strengthened this perspective, drawing together a close

---

13. For this story about **Margaret Mead**, see Ira Byock [1203].

14. As Julian Baggini puts it in his article on **Hume the humane**, "Hume was a great believer in paying attention to evidence and I think experience supports his model of morality better than the main competitors. The best human beings have not been driven by ideology, moral philosophy, and certainly not logic. They have always been people who have put the response to human need above creed or doctrine" [1204]. For these specific citations, see A. Baier [1205] and A. MacIntyre [1206]. The argument here is not intended to be that the virtue of nurturance is the sole virtue expressed in society, nor that it is applicable universally to all situations. Indeed, MacIntyre emphasizes that human practices are multiple (so that there are many virtues) and incompatible (so that conflict does not necessarily imply a flaw in character), and that an ethical approach which accepts particular means and ends as already given or necessarily worthwhile, therefore leaves out "those ongoing modes of human activity within which ends have to be discovered and rediscovered, and means devised to pursue them; and it thereby obscures the importance of the ways in which those modes of activity generate new ends and new conceptions of ends" [1207].

TABLE 10 Positions in moral philosophy.

	Classical position	Critical position	Integrative position
Philosophical roots	Kant, Bentham, Mill, principlism	Nietzsche, Wittgenstein, cognitive subjectivism, emotivism, moral skepticism	Aristotle, virtue ethics, pragmatic ethics, value pluralism[a]
Role of reason and emotion in moral decision-making	Reason is key for moral decision-making	Reason by itself is insufficient for moral thinking	Cognition and emotion needed for optimal moral decision-making[b]
How to ensure the good	Follow clear objective specifications to ensure particular values (moral absolutism)	Common subjective values and universal objective standards are not possible (moral nihilism)	Weigh up relevant facts and values, choose best metaphors, priority of the particular (moral naturalism)
Approach to evil	Bad decisions can be clarified (evil is intelligible)	Clarification is not always possible (evil is inexplicable)	This is not the best of possible worlds, we aim to improve it (meliorism)

Psychotherapy	Psychotherapy is objective; any ethical issues can be addressed by using rules	Psychotherapy is value-laden; may allow exploration of subjectivity	Psychotherapy is theory-bound and value-laden; moral decision-making can be reasonable (moral realism)[c]
Evolutionary ethics	Moral rules emerge from evolutionary principles	Naturalism and normativism must not be confused	Proto-morality emerges from evolutionary principles; humans can improve their morality (moral progress)[d]

[a] *Kant's categorical imperative, Bentham and Mill's utilitarianism, and modern principlism all provide precise specifications of morality (in this sense, ethics based on duty or utility can be considered 'principlist'). For Wittgenstein and Anscombe, such a concept of morality is a pseudo-concept [1208, 1209].* **Although closely associated with Aristotle, ideas consistent with virtue ethics have been forwarded by many others, including the Stoics, Spinoza, Hume, and the American pragmatists.** *For more on Stoic virtue ethics, see M. Sharpe [1210]. For more on Spinoza's ethics, see Michael LeBuffe [378], Don Garrett [1211], C. Marshall [1212], and Heidi Ravven [1213]. As noted earlier, although Hume has been characterized as a moral emotivist, other readings emphasize his virtue ethics and value pluralism [221, 677, 1214, 1215, 1182]. See also earlier footnote on Annette Baier's use of Hume in contributing to an ethics of care.*

[b] *Nussbaum has also emphasized that there are many different* **positions within virtue ethics**. *She helpfully distinguishes between virtue theorists who would like more emphasis on reason in ethics and in this sense are neoKantian (e.g. herself, Marcia Homiak, John McDowell, Iris Murdoch, Henry Richardson, Nancy Sherman, and David Wiggins) and those who would like more emphasis on passions such as those about care and in this sense are neoHumean (e.g. Annette Baier, Simon Blackburn, Philippa Foot, Alasdair MacIntyre, and Bernard Williams) [925]. On this score, see earlier footnote on Baier's ethics of care and Hume. Rebecca Goldstein adapts an adage of Kant, arguing "Reason without moral emotions is empty, moral emotions without reason are blind" [1216].*

[c] *For an application of Dewey's pragmatic ethics to* **moral decision-making in psychotherapy**, *see Svend Brinkmann [210]. He argues cogently that morality is real in the sense that we engage with it as part of life, and that is exists in historically evolved social practices.*

[d] *The literature on* **evolutionary ethics** *begins with Darwin. There are those who would reduce moral constructs to evolutionary ones [1217], and those who are more sceptical, emphasizing that naturalism and normativity are entirely different realms [1218, 1219]. Others see science and philosophy as reaching out to one another [1220]. For Lakoff and Thompson what is key are the metaphors we use to depict evolution; seeing evolution as a strict parent (encouraging survival of the fittest) or a nurturant one (what survives is that which is best nurtured) [159]. From an integrative perspective, moral knowledge is embodied, and with deliberation can improve. Virtue ethicists may, however, take different positions on moral progress. MacIntyre's thinking, for example, is influenced by Thomas Kuhn's notion of paradigms and their incommensurability [1221]. Nussbaum, however, reads Aristotle as seeing progress in morality as moving from a thin specification of the virtues to a thick one, in a process that is open to continuous revision, and agrees with this view [1222], and see later for her citation of Aristotle on scientific and moral progress).*

reading of the ancient Greeks as well as a concern with issues of well-being in contemporary society.[15]

The virtue of nurturance brings to mind a number of other key contemporary strands of work in ethics. First, an *ethics of care*, which focuses on the moral importance of caring for others, on the relevance of embodied dependencies to ethics, and on the importance of reasons and emotions regarding particulars in moral decision-making. Ethics of care is often viewed as feminist, with key contributors including Carol Gilligan, who critiqued models of moral development as gender biased, and Baier, who drew on Hume as the 'women's moral theorist'. Second, an *ethics of responsibility*, which may emphasize the moral importance of caring for others and for nature, or of acknowledging and actualizing the freedom of others. Emmanuel Levinas, Hans Jonas, and Jonathan Sacks, all intimately concerned with Auschwitz, have each contributed different strands of thinking to this work. Third, an *ethical cosmopolitanism* which emphasizes our moral ties to all humans, based on our shared humanity. This notion dates back to at least the Cynic Diogenes, who spoke of being a 'citizen of the world' (kosmopolitês), and has been given recent impetus by a range of authors including Kwame Appiah, who was raised in Ghana and the United Kingdom, and the South African-born Gillian Brock. Some contributions to each of these areas are explicitly neoAristotelian.[16]

---

15. As noted earlier, William Ross made an important early contribution at Oxford, contributing to work on Aristotle, and emphasizing the complexity of moral reasoning. The **revival of virtue ethics** is, however, typically attributed to the work of Elizabeth Anscombe, Bernard Williams, and Alasdair MacIntyre, each of whom also graduated from Oxford (in 1941, 1951, and 1961, respectively), and also criticized the focus of ethics on rules of utility or duty [1223]. Anscombe, who coined the term 'consequentialism', was a student, translator, and friend of Wittgenstein. She and her Oxford contemporaries Philippa Foot, Mary Midgley, and Iris Murdoch emphasized classical virtue ethics and/or moral naturalism. Foot followed Aristotle in linking virtue to human nature [1224], Midgley also argued for a return to Aristotle [1208, 1225], and Murdoch argued for a moral naturalism in which we are able to cultivate better ways of thinking and behaving [1226]. For more on opposition to scientism, see Midgley [580, 1227] as well as R. Hamilton [1228]. For more on facts and values as intersecting, see previous discussion and footnote, as well as a related literature in meta-ethics on integration of belief-like and desire-like normative judgements [1229]. For more on thick moral concepts, see earlier footnote on 'thin' versus 'thick' descriptions, and Bernard Williams's 'Ethics and the Limits of Philosophy' [299]. For more on Midgley's views of her Oxford colleagues, see 'The Owl of Minnerva: A Memoir' [1230]. In South African Nobelist John Coetzee's work, a key protagonist is redolent of Mary Midgley [1231]. Annette Baier had contact with Anscombe and her colleagues when she was a philosophy student at Oxford in the 1950s, and her work is aligned in key ways with Oxford moral naturalism. We return repeatedly to Nussbaum's reading of Aristotle and her capability approach, but note that she has argued that both utilitarianism and deontology may focus on virtues so that it is a mistake to contrast them with virtue ethics [925].

16. For more on Annette Baier's views and the **ethics of care**, see A. Baier [1232]. Her work has paid particular attention to the issue of trust. For a volume on ethics of care and embodiment, see Maurice Hamington [1233]. For more on the intersection of ethics of care and virtue ethics, see Michael Slote [1234] and A. Thomas [1235]. Emmanuel Levinas emphasized that philosophy was the 'wisdom of love' rather than the 'love of wisdom' (as mentioned later), and his **ethics of responsibility** held that responsibility towards the Other precedes any 'objective searching after truth', with 'ethics as first philosophy' [1232a]. Levinas's encounter with the Other in turn has commonalities with Martin Buber's I-Thou relationship [1233a]. For a perspective from embodied cognition on Levinas, see S. Gallagher [1236]. For more on the intersection of ethics of responsibility with virtue ethics, see A.A. van Niekerk and N. Nortjé [1237]. For more on **cosmopolitan ethics**, see Kwame Appiah [1238] and Gillian Brock [1239]. For more on cosmopolitan ethics and virtue ethics, see Martha Nussbaum [1240] and S. van Hooft [1241].

Dewey's earlier approach to ethics also has parallels with Aristotelian virtue ethics, insofar as he eschewed the possibility of finding ultimate ethical principles, and emphasized the relevance of habits or practices to virtue. In addition, Dewey emphasized that, like other forms of problem solving, moral problem solving was a deliberative process involving imaginative exploration of different courses of action (consistent with moral naturalism). For Dewey, although this process can go awry (we must be aware of our moral fallibility), it may ultimately lead to better practices, so allowing moral progress (consistent with meliorism). He argued that "the need for constant revision and expansion of moral knowledge is one great reason why there is no gulf between non-moral knowledge from that which is truly moral. At any moment concepts which seemed to belong exclusively to the biological or psychological realm may assume moral import".[17] That said, pragmatic ethics is also open to the possibility that some habits or practices appear to be equally satisfactory (consistent with value pluralism). Dewey's pragmatic position is not merely an intellectual stance; it can be considered a 'way of life'.[18]

Lakoff and Johnson echo Dewey in emphasizing that like other embodied cognitive-affective processes, moral decision-making often employs metaphors. They outline two particular models of parenting in ethical theory: the strict parent one (adopted by Kant) and the nurturing parent (seen in Hume's ideas and in Aristotelian virtue ethics). Moral decision-making is not a disembodied moral arithmetic, but rather entails engagement with these metaphors and models. More recently, Jonathan Haidt and colleagues have pointed out that key domains of moral reasoning include not only care, but also fairness, loyalty, authority, sanctity, and liberty. Thus moral reasoning must address a range of things that are valuable, as well as a range of different values.[19]

The perspectives taken by Dewey, Lakoff, Johnson, and Haidt are consistent with the focus in previous chapters on how cognitive-affective processes are embodied in the brain–mind, and embedded in society. An appreciation of moral cognitive-affective processes in particular may help us to navigate between a classical position that emphasizes moral absolutism and a critical position

17. It's noteworthy that the term 'ethics' comes from the Greek **êthos, or character,** while the term 'morality' derives from Latin **moralis, or habits**. For recent work that emphasizes how human reason involves future planning, see Martin Seligman, Peter Railton, Roy Baumeister, and Chandra Sripada [1242]. For this citation see Dewey's 'Theory of the Moral Life' [1243]; it is indicative of his moral realism. The prescience of Dewey's early neurophilosophical ideas is seen in his influence on current neurophilosophical pragmatism or **'neuropragmatism'** [1244, 1245].

18. For more on **pragmatic ethics** and its relationship to virtue ethics, see Steven Fesmire [1246], Jozef Keulartz and colleagues [1247], Henry Richardson [1248], and J.P. Serra [1249]. See also Glenn McGee [1250] on pragmatic bioethics, and G. Pavarini and I. Singh [1251] on pragmatic neuroethics. The iconoclastic value pluralist Elizabeth Anderson chose to call her professorship in philosophy the John Dewey Distinguished University Professor. Note, however, that not all agree that pragmatic ethics entails pluralism [1252]. For more on pragmatism as a 'way of life', see Hilary and Ruth Putnam [1253].

19. For more on **Jonathan Haidt**, see a range of his work [1254–1256]. Thaddeus Metz has suggested that in the African context, vitalism is also an important virtue [1257]. For more on pluralism about a range of things versus a range of different values, see M. Tucker [1258].

focused on moral nihilism, and towards an integrative moral naturalism. Still, critics of moral naturalism may object to such an approach by emphasizing the presence and depth of moral disagreement between individuals and societies, or by arguing that the emphasis of moral naturalism on practical judgement leaves moral decision-making as an entirely mysterious process.[20]

These objections are not, however, entirely persuasive. While there may well be significant moral disagreement this does not necessarily imply complete incommensurability of moral frameworks. The very existence of conceptual frameworks that allow moral disagreement to be aired and appreciated indicates at least some degree of commonality across views.[21] In both scientific and moral discussions, we are concerned with the 'dappled', 'promiscuous', and 'plural' nature of the world; we cannot expect that there will be straightforward agreement on the nature of 'thick vague' constructs such as 'flourishing', indeed we can expect a great deal of complexity and fuzziness in deliberations about virtues and vices.[22]

Furthermore, while rigorous and creative scientific and moral reasoning-imagining can certainly be awe-inspiring, such reasoning-imagining is no more mysterious in the moral realm than in the scientific realm. On the account put forward here, **both scientific and moral judgements are made on the basis of deliberative reasoning-imagining that considers the structures, processes, and mechanisms underlying our world, weighing up both principles and particulars.**[23] Both scientific and moral judgements involve a great deal of messy 'muddling through' rather than pristine algorithmic calculation, both are fallible and deserve to be conducted with considerable epistemic humility, but

---

20. For more on **virtue ethics and embodiment**, see J.H. Davis [1259]. For more on the argument that practical judgement in moral decision-making is mysterious, see Elinor Mason [1186].
21. For a useful discussion of this and related points, see Geoffrey Sayre-McCord [1260].
22. See previous footnote on varieties of 'pluralistic realism'. The phrase 'thick vague' is used by Nussbaum [319]. Perhaps relatedly, but more humorously, while the notion of 'doing unto others as you would have them do unto you' appears to be a useful general principle, the particularities of human life seem to entail that **reciprocity requires not only this 'golden rule', but also a certain amount of gossip** (see Michael Shermer [1261], J. van Niekerk [1262], Jeffrey Wattles [1263], and Emry Westacott [1264]). On a different note, the term '**disgusting happiness**' articulates the concern that for some people 'flourishing' may entail social practices that are harmful [667]; while such a view may take an overly dismal view of human nature (see Paul Bloomfield [1265], and see later discussion of psychopathy and altruism), it does seem that human flourishing is a complex and fuzzy construct. Animal flourishing is also a contested notion [1266, 1267], although the cruelty of humankind to animals is not [1268].
23. A view that moral decision-making can be deliberative and imaginative, and therefore reasonable, is consistent with **moral realism**. Moral realism encompasses the moral naturalism referred to earlier, but there are a range of different positions within moral realism [1269], with some moral constructivists also falling under this rubric [1270]. As is the case with the approaches of pragmatic, critical, and embodied realism to the human sciences, for moral realism there may be no 'view from nowhere' [1271]; nevertheless, across these forms of realism there is an appreciation of the structures, processes, and mechanisms of the natural and human worlds [314]. For more on natural kinds versus moral kinds, see G. Sayre-McCord [1272].

both can potentially grow richer and deeper and so achieve better insights and outcomes over time.[24]

6.2 Psychiatry and morality

What does psychiatry have to say about morality? The most relevant literature is likely that on psychopathy, although psychology and neuroscience also have some interest in altruism. The existence of psychopathy and altruism raises the question of human nature in general: are these two phenomena aberrations, or do all humans inherently have the capacity to be extraordinarily evil or good? We'll cover each of these phenomena in turn, mentioning some related issues such as narcissism. We'll then briefly discuss moral issues that emerge in psychotherapy.

Psychopathy and pathological altruism

In 1941 the psychiatrist Hervey Cleckley published his volume, 'The Mask of Sanity', including his case reports of people with psychopathy. The title of the book reflects Cleckley's view that psychopaths may appear completely normal to the outside world; indeed, they are often charming and engaging. However, in fact the 'mask' conceals an underlying mental disorder characterized by lack of a conscience. Psychologist Robert Hare, another seminal researcher in the field, titles one of his volumes 'Without Conscience', again pointing to this central characteristic of psychopathy.[25]

Once again issues around thresholds, reification, and causality arise. We'll discuss each in turn, beginning with the threshold issue. An early view was that psychopathy is a taxon: a discrete class of behaviours that has no overlap with normal behaviour. However, more recent evidence suggests that psychopathy—like many psychiatric phenotypes and personal traits—lies on a dimension, with individuals falling along a spectrum that ranges from few psychopathic traits to many. Kessler has pointed out that symptom dimensions and categories are in fact interchangeable: those with dimensional psychopathy scores above a particular threshold, for example, can be categorically diagnosed as psychopaths.[26]

---

24. The terms '**reasonably**' (rather than, say, 'rationally') and '**reasoning-imagining**' are used in this section and elsewhere to indicate that the relevant judgements do not simply entail moral arithmetic, and to emphasize the important role of metaphor in moral decision-making. Nussbaum has also relied on the term 'reasonably' (as opposed to 'rationally') in a somewhat similar context [1273], but note the criticism of Musa al-Gharbi that she has not fully recognized how value-laden such judgements are [1274]. For more on imagination in moral decision-making, see John Kekes [1155], Steven Fesmire [1246], Mark Johnson [1275], and John Kaag [1276]. See also earlier footnote on scientific 'reasoning-imagining'.

25. For more detail, see Hervey Cleckley [1277] and Robert Hare [1278].

26. For more detail, see J.F. Edens and colleagues [1279], G.T. Harris and colleagues [1280], and R.C. Kessler [1281] on the **categorical versus dimensional** assessment controversy.

The taxon versus dimension debate has implications for how we answer the question about whether humans are inherently evil. In the aftermath of Auschwitz, experiments by the psychologists Stanley Milgram and Philip Zimbardo suggested that in particular contexts, anyone can become a perpetrator; Zimbardo calls this the 'Lucifer effect'. Hannah Arendt's concept of the 'banality of evil', used in the context of Adolf Eichmann's trial, seems to complement this view. Further, based on evidence of our evolutionary history, it has been suggested that humans are 'Demonic Apes' and that aggression is a key part of our evolved make-up.[27] There's no denying that many primates species are 'Machiavellian' and that we engage in deception, manipulation, and a range of other machinations in our social engagements. Still, given interindividual differences on dimensions of psychopathy, as well as the intricacies of different social contexts, I would be wary of oversimplifications about human evil and aggression.[28] Indeed, it turns out that Milgram's and Zimbardo's experiments had key flaws; it turns out that perpetration of harm against humans is a complex phenomenon that is not merely the banal following of orders, and while each of us may have the capacity for aggression only a small number of us relish inflicting harm to the degree of an Eichmann.[29]

With regard to the issue of reification, it's useful to raise the question of the validity of current operationalizations of psychopathy. In particular environments, breaking social norms is itself normative: for example, children raised in the ganglands of the Cape flats in South Africa understandably see joining criminal gangs as a potential way to survive. It's notable that Hare's concept of psychopathy differs in key respects from the DSM-5 construct of antisocial personality disorder: Hare emphasizes personality traits like being callous and unemotional, while the DSM-5 focuses more on particular behaviours such as breaking the law. There are also important overlaps of psychopathy and Machiavellianism with narcissism; the three constructs have been called 'the dark triad'.[30]

---

27. Key proponents of the idea that humans are **innately aggressive** include Raymond Dart (see earlier footnote), the Nobelist Konrad Lorenz, and his popularizer Robert Ardrey. For more on the evolutionary roots of human aggression and deception, see David Barash and Judith Lipton [1282], David Buss [1283], Martin Daly and Margo Wilson [1284], Dario Maestripieri [1285], David Livingstone Smith [1286, 1287], and Richard Wrangham and Dale Peterson [1288].
28. For **critiques of innate aggression**, see Donald Pfaff [1289], Alfie Kohn [1290], and Ruter Bregman [1291].
29. For primatological and archaeological perspectives that emphasize both **our aggression and our hypersociability**, see Frans de Waal [1292] and W.M. Curtis [1293]. For more on Milgram's and Zimbardo's work and its flaws, see Stanley Milgram [1294], Philip Zimbardo [1295], A. Brannigan and colleagues [1296], and T. Le Texier [1297]. For more on the banality of evil, as well as critique of Arendt's views [1298], see her "Eichmann in Jerusalem: A Report on the Banality of Evil" and Deborah Lipstadt's "The Eichmann Trial" [1298, 1299]. For more on Arendt's broader views, including her work on "The Human Condition" [1300], which is dedicated to her former supervisor Karl Jaspers, see Selya Benhabib [1301].
30. For more on the gangs of Cape Town, see Don Pinnock 'Gang Town' [1302]. For more on the **dark triad,** see D.L. Paulhus and colleagues [1303]. Some have extended the dark triad to the dark tetrad, with the inclusion of sadism [1304]. For more on the philosophy of narcissism, see the entertaining work of Aaron James [1305] and Eric Schwitzgebel [1306].

These issues again bring us back to question of human nature and the extent to which psychopathic traits are universal. First, as the example of Cape Town's gangs demonstrates, DSM-5's focus on breaking of the law in order to differentiate those with and without antisocial personality traits seems problematic, in that under particular circumstances, healthy people may choose lives of crime. Second, and more contentiously, it may even be that some dark triad traits are healthy and adaptive. Certainly after a number of decades in academia I'm under no illusion that narcissism is a rare trait or that it necessarily hinders professional success. From an evolutionary perspective, under particular circumstances, impulsivity, deceit, and charisma can each be advantageous. That said, we should again be wary of oversimplification here; it seems that the minority of us with very high levels of dark triad traits may well behave in ways that lead to major disruptions in living, including inability to maintain lasting relationships.[31]

How good is our understanding of the aetiology of everyday perpetration and of extraordinary psychopathy? The idea that banal perpetration is entirely due to situational factors has given way to more complex models which emphasize that a range of causal mechanisms play a contributory role. Research on perpetrators in Nazi Germany, apartheid South Africa, and other contexts underscores how perpetration ranges in scope from banal through to sadistic, and how both individual and societal factors contribute to such perpetration.[32] Similarly, a range of explanations, including both proximal and distal factors, may contribute to psychopathy. Psychology and neuroscience research has pointed to significant alterations in the brain-minds of people with psychopathy, including alterations in executive functions, prefrontal cortex, and neurochemistry: findings that are consistent with increased impulsivity. There is also an increasingly sophisticated evolutionary literature which considers how psychopathic traits

31. For more on the ubiquity of dark triad traits, see Christopher Lasch [1307], Arthur Bohart, Barbara Held, Edward Mendelowitz, and Kirk Schneider [1308], and Will Storr [1309]. For more on how psychopathic traits can be advantageous, see Paul Babiak and Robert Hare [1310] and Kevin Dutton [1311]. It is also interesting to note how easily the rest of us can be taken in by psychopaths, see Maria Konnikova [1312]. For more on bullshit in academia, see G.A. Cohen [1313].

32. It has been suggested that Milgram's and Zimbardo's work complements Arendt's emphasis on the banality of evil in that these views emphasize **situationist obedience**. However, it is perhaps not fair to criticize Arendt as representing the perspective of situationist obedience [1314]. An alternative view, exemplified by Daniel Goldhagen in his controversial volume on 'Hitler's Willing Executioners' [1315], is that perpetrators in Nazi Germany were not simply following orders, but contributed to the Holocaust with relish [1316]. For models focusing on **multiple causal mechanisms**, see G.R. Mastroianni [1317] and C. Daniel Batson [1318]. For more on perpetrators in Nazi Germany, see Joel Dimsdale [1319], E. Staub [1320], and David Cesarani and Paul Levine [1321]. Dimsdale discusses the views of two key clinicians involved in assessments at Nuremburg, and concludes that "Kelley found some darkness in every person. Gilbert found a unique darkness in some. They were both right" [1321]. For more on perpetrators in the South African context, see D. Kaminer and D.J. Stein [1322].

may be adaptive in some contexts, perhaps contributing to explaining interindividual variations in dark triad traits.[33]

A discussion of causes of psychopathy leads on to the question of how best to manage psychopathy from a clinical perspective. It turns out that there is little evidence that psychotherapy and pharmacotherapy have any impact on psychopathic behaviour. On the other hand, the management of a family who are burdened with a psychopathic relative may well involve helping them to set limits, including bringing in the justice system where appropriate. While some have argued that psychopathy is a medical illness, and so is exculpatory, almost all legal judgements to date have insisted that people with psychopathic traits are responsible and accountable. We'll return to the broader issues of free will and willpower in more detail later.[34]

While psychopathy is characterized by inordinate disregard for others, altruism refers to behaviour that puts others first. The psychologist Abigail Marsh has differentiated three kinds of kinds of altruism: kin altruism, reciprocity-based altruism, and care-based altruism. Kin altruism is seen throughout the animal kingdom and refers to behaviour that benefits one's biological relatives, and so that ultimately allows reproduction of one's own genes. Reciprocity-based altruism refers to behaviour between allied conspecifics; participating in such reciprocity-based altruism may lead to increased rewards. Marsh has researched people with hyper-altruism or extraordinary altruism, such as those who donate kidneys to strangers: a fascinating group of individuals who display high levels of care-based altruism focused on assisting distressed individuals. Once again, the issues of threshold, of reification, and of causality can be considered.[35]

It seems clear that altruism is a dimensional trait, with individuals falling on a range from selfish to selfless. We noted earlier that dark triad traits often lead to significant disruption in a person's life, but it's not immediately apparent that altruism may also have negative impacts. There are, nevertheless, a number of ways in which excessive altruism may lead to harm: such behaviour may be unhealthy or harmful for the altruist, and it may at times cause more harm than good for others (e.g. the kidney donor's own family). People with

---

33. For more on the neurobiology of psychopathy and altruism, see A.A. Marsh [1323]. For more on evolutionary theory and psychopathy, see A.L. Glenn and colleagues [1324].

34. **Psychopathy** raises a number of philosophical issues, including the question of whether a psychopath bears moral responsibility [1325]. There seems to be a range of cognitive and affective deficits in psychopathy [1326], but it is not clear that these are sufficiently severe to entirely impair moral decision-making to the extent that psychopaths are not culpable [1327]. For more on philosophy of personality and character more generally, see Christian Miller, R. Michael Furr, Angela Knobel, and William Fleeson [1328], Peter Goldie [1329], and Christian Miller [1330].

35. For more on **hyper-altruism**, see Abigail Marsh [1331]. The term altruism was coined by Auguste Comte [1332], who developed a 'Religion of Humanity'. Darwin was the first to discuss **kin altruism**, although the term was coined by John Maynard Smith [1333]. William Hamilton pioneered work on such altruism [1334] and Robert Trivers pioneered evolutionary understanding of reciprocity and conflict more broadly [1335]. For more on evolutionary theory and altruism, see Elliot Sober and David Sloan Wilson [1336], Nigel Barber [1337], Stephan Klein [1338], and R. Kurzban and colleagues [1339].

hyper-altruism may be convinced that they are helping others, without necessarily considering the practical results of their 'help'.[36]

Exactly how we define hyper-altruism remains a matter for debate. Maternal behaviour can, for example, involve extreme self-sacrifice, and yet such behaviour would not necessarily be considered hyper-altruistic. In the case of psychopathy, we raised the questions of the extent to which nature and nurture influence psychopathy, and of whether adaptive and successful psychopathy are possible. Parallel issues arise in the case of altruism. Care-based altruism seems to be part of our make-up, but extraordinary altruism seems to be present in only a small minority of us. While altruism may ordinarily be adaptive, extraordinary altruism may at times be associated with significant harm. DSM and ICD have not considered whether hyper-altruism is a mental illness, perhaps reflecting how much altruistic behaviour is valued by society. However, a pattern of behaviour that reflects an underlying dysfunction and that is associated with distress and dysfunction would meet the definition of mental disorder put forward in DSM-5.

This point in turn raises the question of the psychobiology of hyper-altruism. Marsh has provided evidence that individuals with extraordinary altruism have increased right amygdala volume and increased responsivity in right amygdala to fearful expressions. So while psychopathy may be associated with decreased sensitivity to the distress of others, extraordinary altruism may be associated with increased sensitivity to such distress. Neuroscience research has demonstrated that specific neural circuitry is activated when mothers see their infants; it's possible that care-based altruism employs similar circuitry. While evolutionary theory predicts several forms of altruism, hyper-altruism does not necessarily seem adaptive.[37]

That human behaviour ranges so broadly, from psychopathic to altruistic, both across individuals, and within individuals, suggests that **any characterization of humans as essentially demonic or angelic is overly simplistic**. Adding further complexity is evidence that psychopaths do not have difficulties in demonstrating empathy—rather they may be better able to turn their empathy on and off, and that altruists are not merely motivated by prosocial behaviour—rather their behaviour reflects in part a response to their own distress. The vast majority of us are not psychopaths or hyper-altruists, and our social decision-making entails a range of cognitive-affective processes, including both prosocial empathy and self-oriented distress.[38] To return to the

36. See B. Oakley and colleagues [1340] and David Barash [1341].
37. The **amygdala** plays a key role in social decision-making; my colleague Jack van Honk has led research on South African patients with a rare genetic disease that leads to amygdala damage, so allowing focused study of this issue [1342].
38. For more on empathy switches in psychopathy, see Christian Keysers [1343], and for more on self-focused distress in altruism, see Abigail Marsh [1331]. The perspective taken here differs from those which focus simplistically on humans as good or bad, as well as from those who focus primarily on empathy as explaining psychopathy and altruism. The human propensity for **black and white thinking** gets us into all sorts of trouble [1344]. While empathy may be one key mechanism to ensure cooperation, punishment is another; psychopaths may be less sensitive to punishments for selfish behaviour, and hypothetically, hyper-altruists may be more sensitive to such punishment.

Aristotelian golden mean, it may be argued that what's important is to care for the right person and to the right degree and at the right time and for the right purpose, and in the right way.[39]

Moral issues in psychotherapy

It would seem that when health professionals attempt to change the behaviour of a person with psychopathy or with hyper-altruism they are not only providing a clinical intervention, but are also effecting a moral action. Do clinical interventions for other mental conditions also involve moral activity? Very broadly we might argue that health interventions that diminish the suffering associated with depression and anxiety are for the good. But people may embark on psychotherapy not so much because of specific symptoms, but rather with questions about how to live their lives.

One approach within psychotherapy theory is to ignore the value-laden nature of the enterprise. From this perspective, psychotherapy primarily involves assessment of specific symptoms and providing an intervention that targets these. Although a person may embark on psychotherapy to resolve issues rather than symptoms, a comprehensive evaluation will yield clear targets for treatment, such as unresolved conflicts or pathological personality traits. Thus psychotherapy involves objective scientific goals and is a clinical rather than a value-laden exercise. Should any ethical issues arise during the course of psychotherapy, principlism can be used to resolve these.[40]

A contrasting approach has been to see psychotherapy as an entirely value-laden enterprise. This approach is consistent with that of a broader critique of psychiatric science (Table 2). Thus, for example, a critical or postmodern view might see psychotherapy as merely another way in which the medical-industrial complex ensures that individuals reflect and reproduce its social norms. The rampant rise of a 'therapy culture' in modern life is certainly cause for pause: what particular values does this embody? The critical position covers a range of

39. Like psychopathy, **altruism** raises a number of philosophical issues, including work on effective altruism and on supererogation; see Thomas Nagel [1345], Stephen Post, Lynn Underwood, Jeffrey Schloss, and William Hurlbut [1346], Larissa MacFarquhar [1347], Andrew Flescher and Daniel Worthen [1348], Philip Kitcher [1349], Marin Nowak and Sarah Coakley [1350], Peter Singer [1351], and William MacAskill [1352]. **Empathy**, too, raises a range of philosophical issues [1353, 1354]. Our favourite philosophers have contributed relevant resources; this includes ideas put forward by Aristotle [1355], Spinoza [1211], Hume [1356], and Dewey [1357]. Arguably, each views selfish motives as intersecting with social cooperation, but this is perhaps most apparent in Hume's work, given that he was responding to specific debates of his time regarding human self-interest and sociability. Hume's view of human nature as expanding progressively to encompass others arguably predates current ideas that we have expanded our circle of morality [1089].

40. See, for example, S.J. Knapp and colleagues [1358].

different perspectives; some who take this position may regard psychotherapy as useful insofar as it allows the subjectivity of the individual to be creatively explored.[41]

Consider the patient who embarks on schema therapy. This approach assesses the extent to which individuals have early maladaptive schemas, which subsequently lead to inappropriate or suboptimal cognitive-affective responses to the world. The role of the therapist is to help the person to appreciate their cognitive-affective style and to develop more functional ways of being in the world. This therapy has both subjective and objective elements. On the one hand, the therapist is understanding the individual's particular cognitive-affective style and way of responding to the world. On the other hand, the therapist is challenging this style and way of being, and encouraging a more productive way forwards.

With this example in mind, I would argue that psychotherapy, like other health interventions, is both theory-bound and value-laden. The schema therapist is encouraging the person to adopt new habits; perhaps based on seeing themselves as more worthy (rather than as deeply flawed), or perhaps based on seeing others as allies (rather than as entirely untrustworthy). At the same, the schema therapist is fully aware that what matters differs from person to person; they have agency over what habits they choose to adopt. Put differently, normative decision-making in psychotherapy can have a reasonable basis, potentially avoiding both moral reductionism (which posits a disembodied scientistic moral decision-maker) and moral nihilism (which eschews the possibility of reasonable moral decision-making).[42]

There is an overlap between the position put forward here and the earlier focus on virtue ethics. The schema therapist's focus on the benefit of adopting and practising particular cognitive-affective styles in order to increase flourishing is consistent with Aristotle's emphasis on developing practical judgement through repetition, and with his emphasis on the virtue of the golden mean.[43] Although each individual undergoing schema therapy will need to ascertain their own particular aims (figuring out what is needed for them to flourish), given human knowledge and understanding of humans, certain cognitive-affective styles may be more likely to lead to flourishing (it's not the case that

---

41. Notably, **Alasdair MacIntrye**, in his 'After Virtue' describes the character of The Therapist, noting the dangers of therapy focusing too narrowly on symptom reduction (with the pretence of ethical neutrality), or of invading too broadly (with the pretence of universal applicability) [1359]. See earlier footnote on the triumph of the therapeutic.

42. See also William Doherty [1360] and Svend Brinkmann [210]. See also earlier footnote on narrative therapy, and later footnote on developing new narratives, for example, the work of Timothy Wilson [1361].

43. As noted in an earlier footnote, Aristotle's tables of virtue and vice, provided in his Eudemian Ethics, overlap to some extent with maladaptive and adaptive schema-driven cognitive-affective styles. This in turn raises the broader question of philosophy as therapy in general, and of philosophical counselling in particular. See also earlier footnote for more on philosophy as therapy. For more on **philosophical counselling**, see Elliot Cohen and Samuel Zinaich [501], K. Banicki [1362], E. Fischer [1363], and Lou Marinoff [1364].

'anything goes'). At the same time, the therapist must address each issue that emerges during psychotherapy with practical judgement.[44]

There is less literature on the moral aspects of pharmacotherapy, but this is also worth considering. The physician-philosopher Julian Savulescu has argued that given our major moral shortcomings and the imperilled state of the world, moral bioenhancement is crucially needed as a supplement to ongoing moral education. Less contentiously, the psychiatrist Sean Spence has pointed out that in many clinical situations, pharmacotherapy serves as a moral intervention. A man who suffers from a psychotic disorder may be violent when ill, but adherence to his medication ensures that the likelihood of him harming others is substantially reduced. Pharmacotherapy, like psychotherapy, is theory-bound and value-laden, and reasonable decisions can be made about psychopharmacological intervention.[45]

6.3 Neurophilosophy and neuropsychiatry

As humans we have brain–minds that allow us to engage in a complex dance of competitive, collaborative, and caring social behaviour. The cognitive-affective neuroscience literature refers to the 'social brain', and there is a growing understanding of the neurocircuitry involved in social and moral decision-making. Analogues of this neurocircuitry are seen in a range of species where social-decision making is important, and there are prototypes of morality in these species. Nonhuman primate social interactions, like ours, seem inevitably to involve some manipulation and deception, but also some cooperation and even altruism. There are a number of features of human social and moral decision-making that are particularly noteworthy.[46]

---

44. There is a long literature on **virtue ethics in medical practice**; see Edmund Pellegrino and David Thomasma [1365]. For more on **virtue ethics in mental health treatment and enhancement**, see H. Berg [409], Blaine Fowers [1366], Barbro Fröding [1367], R.P. Hamilton [1368], Antonia Macaro [1369], Mike Martin [1370], C. Proctor [1371], Jennifer Radden and John Sadler [1372], and Duff Waring [1373]. See also earlier footnotes on medical practice as entailing phronesis, and later footnote on epistemic virtues in medical practice.

45. For more on **moral bioenhancement**, see my volume 'Philosophy of Psychopharmacology' [53], Ingmar Persson and Julian Savulescu [1374], S.A. Spence [1375], and Harris Wiseman [1376].

46. For more on social decision-making across species, see S. Tremblay and colleagues [1377]. Darwin was the first to posit that moral judgements are rooted in our evolutionary past; see previous footnote. Early on Huxley, Nietzsche, Dewey, and others addressed the ethical implications of his hypothesis [92, 1378]. Edward Wilson renewed interest in this point with his volume on 'Sociobiology: The New Synthesis' [1379]. The subsequent growth of the literature addressing **evolutionary components of human morality** has been remarkable; see Robert Wright [873], Peter Singer [1089], Jonathan Haidt [1256], Phillip Kitcher [1349], Paul Bloom [1380], Christopher Boehm [1381], Jean-Pierre Changeux and Paul Ricoer [1382], Jean-Pierre Changeux [1383], Patricia Churchland [1384], Frans de Waal [1385, 1386], Jean Decety and Thalia Wheatley [1387], Daniel Dennett [1388], Blaine Fowers [1389], Michael Gazzaniga [1390], Marc Hauser [1391], Robert Hinde [1392], Richard Joyce [1393], Steven Mascaro, Kevin Korb, Ann Nicholson, and Owen Woodberry [1394], Mary Midgley [1395], Hilary Putnam, Susan Neiman, and Jeffrey Schloss [1396], Matt Ridley [1397], Michael Ruse and Robert Richards [1398], and Walter Sinnott-Armstrong and Christian Miller [1399].

First, social and moral decision-making is typically nonconscious and automatic, and involves a range of embodied cognitive-affective processes. Earlier we noted, for example, that decision-making about moral issues may involve disgust-related neurocircuitry. Other key cognitive-affective processes involved in complex social decision-making include positive regard for others (e.g. empathy) and negativity towards others (e.g. anger). Although there is proto-moral decision-making in other species, in humans cognitive-affective reasoning, including social and moral decision-making, is more complex; humans use metaphoric reasoning, for example, seeing a particular act as unnatural, as evil, or as out of balance. Ordinarily, our moral decision-making is rapid, with posthoc rationalization, but it can also be deliberative. Either way, our moral decision-making is prone to going awry; one of the most remarkable lessons of Auschwitz is the extent to which perpetrators, including highly educated Nazi scientists and doctors, were convinced of the morality of their actions.[47]

The trolley problem is useful in showing how social and moral decision-making engages different brain–mind cognitive-affective processes that are often far from algorithmic. In the trolley problem a train will kill five people but there is an option to press a switch that will lead to it killing only one. A variation is the bridge problem; a train will kill five people but there is an option to push a man in front of it so that only he is killed. Despite the outcome equivalence of the trolley and the bridge dilemmas, people are often happy to press the switch, but at a gut level are often reluctant to push a man. Joshua Greene notes that the trolley problem is a more impersonal and cognitive decision, while the bridge problem is a more personal and social–emotional decision. In a fascinating neuroimaging paper, he shows that the trolley problem engages more cognitive parts of the brain, while the bridge problem engages brain areas involved in social and cognitive processing.[48]

---

47. We earlier noted Dewey's emphasis on deliberative **moral decision-making**; this focus dates back to Aristotle, who argued that we are not born good or bad, but that by applying our minds deliberatively we could choose how to live [516]. For more on the proximal and distal mechanisms underlying empathy, and it's role in social decision-making, see J. Decety and colleagues [1400] and F.B.M. de Waal and S.D. Preston [1401]. For more on moral decision-making as automatic, with posthoc rationalization; see J. Haidt [1402] and D.A. Pizarro and P. Bloom [1403]. See also earlier footnote on dual process theories. For more on moral decision-making going awry, see Max Bazerman and Ann Tenbrunsel [1404]. For one trenchant, but perhaps unsurprising, example of moral fallibility, see E. Schwitzgebel and J. Rust J on the behaviour of ethicists [1405]. For more on the moral decision-making of Nazi doctors, see E.D. Pellegrino and D.C. Thomasma [1406].

48. For more on '**trolleyology**', see David Edmonds [1406a]. The trolley and bridge (or fat man) dilemmas were introduced by Phillipa Foot and further developed by Judith Jarvis Thomson amongst others. For more on their neuroscientific underpinnings, see J. Greene [1407]. Greene concludes that all nonconsequentialist moral views, including those of virtue ethics, involve more emotion and greater rationalization, and that moral realism is untenable [1408]. However, Greene's deductions have been rigorously criticized [1409]. Indeed, I would suggest that his neuroimaging findings are consistent with the idea that practical judgement entails engagement with both the universal and the particular (see previous footnote on moral particularism). For more on cross-cultural aspects of these dilemmas, now particularly relevant to autonomous vehicles, see E. Awad and colleagues [1410].

Second, social and moral decision-making is extraordinarily varied and complex; we are continuously engaging with different hierarchies, competing with and caring for others, reciprocating with kindness or with revenge. Such decision-making is not only embodied in a range of different brain–minds, but it is also embedded in a range of different social contexts. A contrast can perhaps be drawn between the harm we do when we eat animals (doing harm in an unthinking way), and the harm we do when we take pleasure in hurting another (causing harm in an intentional way); social decision-making in these quite different contexts is likely to involve quite different cognitive-affective processes. Between these extremes many of our social decisions may involve some part virtue and some part vice.[49]

Discussions of 'evil' may benefit from taking such distinctions into account. There seem to be a large number of different kinds of evil, and once again the devil is likely in the detail. In the next section, we'll say more about the psychobiology of a number of different kinds of potentially evil phenomena, ranging from aggression through to self-deception. The extent to which we engage in criminal versus caring behaviour is dependent not only on our brain–minds but also on context; social structures can do a great deal to nudge us towards good behaviour. Reassuringly, only a small proportion of people display very high levels of psychopathy and dysfunction, consistent with presence of a mental disorder.

Conversely, when we discuss the 'good', neuroscience and psychology indicate that people differ enormously in their individual make-up and social circumstances, so that different decisions and behaviours may well be appropriate for different people. The idea that different people may have quite different but entirely reasonable ways of achieving health, or balance, or purpose seems consistent with the position of value pluralism. Value pluralism does not however mean that 'anything goes'; in a particular case cognitive-affective reasoning may be more or less sophisticated and more or less compelling.[50]

Third, cognitive-affective processing changes during the course of an individual lifetime; during childhood our cognitive-affective processing becomes

---

49. For Aristotle 'The good is said in many ways', and some have drawn on his work to also consider different ways of talking about evil [1411]. Drawing on Aristotelian ideas, Mary Midgley sees evil as a failure to live as virtuously as we are capable of [1412]. For more on **'ordinary vices'**, see Judith Shklar [1413] and Emry Westacott [1264]. For Onora O'Neill, "The problematic area, where we dispute the dirtiness of hands, is when we turn to those who become the executors, minions and other acolytes, or even the partly willing victims, of bloodstained and other power" [1414].

50. Isaiah Berlin attempted to draw on the strengths of both the **Enlightenment and Counter-Enlightenment**, writing, "My point is that some values clash: the ends pursued by human beings are all generated by our common nature, but their pursuit has to be to some degree controlled – liberty and the pursuit of happiness, I repeat, may not be fully compatible with each other, nor are liberty, equality, and fraternity. So we must weigh and measure, bargain, compromise, and prevent the crushing of one form of life by its rivals" [1165]. For a more recent perspective on weighing different goods, see Thomas Hurka [1415].

more sophisticated and is able to make more compelling judgements, a change that is accompanied by alterations in neurocircuitry. Perhaps this is also true of human communities; the philosopher Peter Singer has argued that humans have expanded the circle of our moral thinking to include a range of 'others'. There are some who have suggested that an evolutionary understanding of morality suggests that our moral reasoning has no objectivity. Darwin himself provides an articulate argument along these lines.[51] However, while social and moral decision-making have adaptive value, there is no reason not to employ moral reasoning-imagining with greater and greater sophistication, learning from past mistakes in such reasoning-imagining, and developing more compelling practices.[52]

The argument has much in common with the views of scientific explanation expressed earlier. Clearly science is a social activity. At the same time, science advances through a process of deliberative reasoning; in doing so it provides greater and greater explanatory power, and it works better. Similarly moral decision-making necessarily involves human reasoning-imagining and interaction. There is no other yardstick of what is moral other than the ones we humans fashion, there is no 'view from nowhere'. At the same time it us up to us to develop our moral decision-making in more and more sophisticated ways, providing more reasonable and imaginative justifications, overcoming

---

51. For more on changes in cognitive-affective processing over the course of neurodevelopment, see G. Leisman and colleagues [1416]. See Peter Singer [1089]. For a more recent volume on evidence for **moral progress**, see Allen Buchanan and Russell Powell [1417]. Darwin writes, "It may be well first to premise that I do not wish to maintain that any strictly social animal, if its intellectual faculties were to become as active and as highly developed as in man, would acquire exactly the same moral sense as ours. In the same manner as various animals have some sense of beauty, though they admire widely different objects, so they might have a sense of right and wrong, though led by it to follow widely different lines of conduct. If, for instance, to take an extreme case, men were reared under precisely the same conditions as hive-bees, there can hardly be a doubt that our unmarried females would, like the worker-bees, think it a sacred duty to kill their brothers, and mothers would strive to kill their fertile daughters; and no one would think of interfering. Nevertheless the bee, or any other social animal, would in our supposed case gain, as it appears to me, some feeling of right and wrong, or a conscience" [1418]. This point is redolent of Aristotle's point that "The good is not single for all animals, but is different in the case of each" (see his 'Nichomachean Ethics'). At the same time, there is evidence that Darwin was deeply influenced by the romanticism of authors such as Alexander von Humboldt, and so viewed the evolution of human morality in altruistic terms; the underlying selection mechanism was the human effort to achieve the 'greatest good' or the 'vigour and health' of community members, allowing him to conclude that "the reproach of laying the foundation of the most noble part of our nature in the base principle of selfishness is removed" [1419].

52. Aristotle emphasized that we can make **mistakes about virtues** [516], a point also made by Spinoza [378], Hume [1420], and Dewey [1246]. More recently, Robert Nozick has pointed out that just as neurobiology accounts for perception but not for the physics of light, so sociobiology fails to debunk an objective basis for morality [1421]. For a more recent discussion, see W.J. FitzPatrick [1422]. For more on the neurocircuitry of moral decision-making, see William Casebeer [1423].

our moral blindspots, and developing practices that are imbued with greater humanity and that work better.[53]

Fourth, it may be useful to pay close attention to metaphoric thought in moral decision-making. We've noted the view that Kant's strict parent metaphor contrasts with the nurturing parent metaphor of virtue ethics. We've emphasized that psychotherapy is necessarily value-laden, and it's noteworthy that a great deal of psychotherapy focuses on the importance of early maternal attachment, of clarifying family roles and responsibilities, and of establishing family boundaries and rules. It's notable that both the strict parent and the nurturant metaphors rely on a relatively individualistic family framework; African and Chinese thought might rather emphasize that parenting is a collective effort, as articulated in the proverb 'it takes a village to raise a child'. Another African proverb pithily emphasizes the complexity of dependency relationships; "The child who is not embraced by the village will burn it down to feel the warmth".[54]

The management literature has many conceptual frameworks for approaching leadership. However, this literature does not often explicitly think of organizations and institutions in terms of family structures. Still, every family and every business has a hierarchy, with roles and rules more or less explicitly spelled out. Such roles and rules may be more or less helpful; family and business functioning can go awry in any number of ways. Fortunately, it is also possible for families and businesses to improve the way they function. It's ordinarily helpful, however, for those in authority to work to ensure that the needs of others are robustly addressed; and it's ordinarily helpful for those who are lower in the hierarchy to act in a way that facilitates the general goals. The metaphor of balance might again be useful; strictness is useful under some circumstances, nurturance under others, and collective responsibility is also particularly key at times.

6.4 Sharpening our thinking about morality

The question of evil is undoubtedly a wicked problem: a problem that is really complex. In some ways, as Neiman points out, evil represents a metaphor that refers to that which is simply inexplicable, the worst of ourselves and our world. Confronted with examples of evil, the focus of this metaphor is not so much on understanding, as on countering evil, on locking it away for good. In the age

53. We'll later provide a quotation from Aristotle that compares scientific progress with moral progress. The phrase '**view from nowhere**' is from Thomas Nagel [1271]. The notions of explanatory power and reasonable justifications are consistent with ideas in critical realism and in virtue ethics, while the notion of working better is used in pragmatism realism and pragmatic ethics. See also the footnotes in the previous chapter on optimism, pessimism, and meliorism.

54. Some have argued that there is an underlying psychobiology of choices that favour conservatism versus liberalism [1424]. For more on African thought, see earlier footnote on ubuntu and communitarianism. For more on the focus on families and communities in Chinese thought, see David Wong [1201].

of neuroethics and neuropsychiatry, however, we must also take a naturalistic perspective that considers how to make sense of a range of evils, while at the same time avoiding scientistic reductionism or a dismissal of such phenomena as simply inexplicable. In this section, we'll discuss in turn aggression and perpetration, hierarchy and status, cronyism and other 'isms', forgiveness and reconciliation, and self-deception.[55]

Aggression and perpetration

I started out this chapter with an early memory of bullying. It's extraordinary how aggressive humans are. Child abuse and partner violence are commonplace. When it comes to partner violence, males inflict significantly more physical damage, but it turns out that women are more likely to be perpetrators of verbal and physical abuse than men. And then there is our proclivity for war: an activity that no matter its justification is inevitably accompanied by massive suffering. The phrase 'Demonic Ape' seems an appropriate one; no other species rivals *Homo sapiens* when it comes to inflicting wanton damage on conspecifics.[56]

There is a growing literature on the psychobiology of aggression. Those with high levels of psychopathy seem to have specific brain signatures, which may make them more liable to be callous and impulsive. At the same time, dark triad traits are distributed throughout the population, and many of us have the potential to perpetrate aggression, particularly more indirect and subtle forms of aggression. The phrase 'sweet revenge' reflects the psychobiological reality that when the punishment of an individual is seen as appropriate, reward neurocircuitry is engaged. Brain signatures underlying aggression may well have a genetic component, but as always both nature and nurture play a role. From an evolutionary perspective it seems hard to argue that ours is a particularly non-violent species: we hunt for food, we fight for status, we battle for resources.[57]

It is notable that psychiatry has been relatively quick to pathologize excessive sadness and anxiety, but much slower to pathologize excessive aggression. DSM-5 does include a condition called 'intermittent explosive disorder', and epidemiological data emphasize its high prevalence and associated disability.

---

55. For early work on **wicked problems**, see H.W.J. Rittel and colleagues [1425]. See again Susan Neiman [1152]. A number of volumes have addressed the '**science of evil**', see Michael Shermer [1261], Michael Stone and Gary Brucato [1426], Roy Baumeister [1427], and Simon Baron-Cohen [1428].

56. See earlier footnote on Richard Wrangham and Dale Peterson's 'Demonic Apes' [1288] and related volumes. For more on the epidemiology of trauma, see C. Benjet and colleagues [1429] and S.L. Desmarais and colleagues [1430].

57. For more on subtle (but not necessarily harmless) forms of **perpetration**, see Jan-Willem van Prooijen and Paul A. M. van Lange [1431]. For more on reward brain circuitry and altruistic punishment, see B. Knutson [1432], and for more on cruelty as rewarding, see University of South Africa psychologist Victor Nell [1433]. Joshua Greene argues that common-sense punitive judgements reflect emotional outrage, rather than rigorous reasoning [1408].

However, this condition has received relatively scant clinical or research attention; clearly we need to be more adept at diagnosing and treating those with very high levels of aggression. As for the rest of us, Aristotle's advice makes good sense: "Anybody can become angry – that is easy, but to be angry with the right person and to the right degree and at the right time and for the right purpose, and in the right way – that is not within everybody's power, and is not easy".[58]

Hierarchy and status

Hierarchy and status are a component of social structures across the animal kingdom. Social structures can act in a protective way: in the hierarchy of a democratic society the state ensures that individuals have access to resources, in the hierarchy of families parents take responsibility for looking after their children. At the same time, hierarchy and status come with a range of complications: fighting for status may be stressful, and loss of status even more so. Furthermore, those occupying the lower rungs may be harmed in a range of different ways. Countries like South Africa have gross social inequality, which in turn causes major health disparities. Furthermore, in day-to-day interactions between people of different status, there may be social slights and 'micro-aggressions'.[59]

The psychobiology of social hierarchy has been studied in a range of animal models. Social defeat in rodents, for example, is accompanied by massive changes in brain circuitry and chemistry. When animals are subjected to ongoing social defeat, they also demonstrate a range of behavioural changes, including increased avoidance of threat (consistent with anxiety) and diminished exploratory activity (consistent with altered mood). In humans, there is parallel evidence that social discrimination is a chronic stressor that may lead to a range of alterations in the brain–mind. Such changes may in turn be associated with poorer physical and mental health.[60]

Is knowledge of the psychobiology of hierarchy at all useful in informing our approach to the big questions? For a social species such as our own, competition with conspecifics is clearly an inevitable component of life. On the one hand, this can certainly result in negative experiences and emotions, with loss of status being particularly unpleasant. On the other hand, healthy competition may be useful and productive in a whole range of ways, and an argument can be made

58. For more on intermittent explosive disorder, see K.M. Scott and colleagues [1434]. For more on the **philosophy of anger and aggression**, see Seneca [1435], Mary Midgley [1154], Richard Bernstein [1436], Martha Nussbaum [559], Myisha Cherry and Owen Flanagan [1437], and Gavin Rae and Emma Ingala [1438].

59. For a thoughtful discussion of **hierarchy**, see Daniel Bell and Wang Pei's [1439]. The term 'micro-aggressions' was coined by Harvard University psychiatrist Chester Pierce [1440].

60. From more on animal models of **social defeat**, see C. Hammels and colleagues [1441]. For more on the neurobiology and health impact of social discrimination, see M. Berger and Z. Sarnyai [1442], and Y. Paradies and colleagues [1443]. See also Robert Sapolsky [1444].

that we therefore need to welcome it into our lives. The harms associated with gross social inequality, however, clearly necessitate a change in social policies; the Nobelist Amartya Sen and Martha Nussbaum have put forward a useful conceptual framework that emphasizes the importance of measuring well-being and of developing policies to enhance this (the capability approach).[61]

Cronyism and other 'ism's

Primates, including humans, form groups; these provide members with several advantages. At the same time grouping may lead to erroneous or unfair ingroup biases. At the broadest level, nationalism and tribalism may erroneously regard other groups as essentially inferior. At a narrower level, there are phenomena such as nepotism, where family and friends are unfairly favoured. Within families, it turns out that step-mothers provide their step-children with less food than their biological children. People may be sexist, racist, age-ist, anti-Semitic, or Islamophobic. All these forms of cronyism can be accompanied by serious forms of aggression and perpetration that lead to great harm, or by more minor but still harmful insults and slights.[62]

Neuroscience tells us that there is a particular neurocircuitry and neurochemistry involved in ingroup bias. At a hormonal level, for example, oxytocin mediates increased attachment to the ingroup, but at the same time mediates increases aggression to the outgroup. At a psychological level, various trust paradigms have shed light on mechanisms involved in ingroup bias. At the same time, psychobiology provides evidence of how similar we all are: race turns out to be largely a social construct. We've been sleeping with each other, with other hominids, and even exchanging DNA with other species, for a good while now: we're all pretty much a mixture. It turns out, therefore, that humans living on different continents have greater genetic overlap than do chimpanzees living in different areas of Africa. There is some evidence that male and female humans have different brain–minds, but variation within each gender also ensures significant overlap.[63]

---

61. For more on the philosophy of inequality, see Robert Nozick [659] and Thomas Nagel [1445]. For more on the **capability approach**, which emphasizes both capacities and opportunities, see Martha Nussbaum and Amartya Sen [1446]. Alain de Botton's volume 'Status Anxiety' addresses issues of status and well-being [1447]; the epidemiologist Michael Marmot provides an interesting critique, emphasizing that from a public health perspective the capability approach is particularly important [1448]. See also previous footnote on the intertwining of individual and community eudaimonia. For more on an evolutionary approach to hierarchy, see Christopher Boehm [1449].
62. See Joshua Greene [1450]. For more on **pragmatism and racism**, see Philip Kitcher [176] and Terrance MacMullen [1451].
63. For more on the neurobiology of **ingroup versus outgroup bias**, see J.H. Egito and colleagues [1452] amd M. Kavaliers and colleagues [1453]. For more on interspecies transfer of DNA, see David Quammen [1454]. For more on chimpanzee versus human genetic variation, see R. Bowden and colleagues [1455]. For more on the **neurosexism** controversy, see Debra Soh [1456], Gina Rippon [1457], Angela Saini [1458], Rebecca Jordan-Young [1459], Roy Baumeister [1460], Simon Baron-Cohen [1461], and Deborah Tannen [1462].

What are the implications of cronyism for the big questions of life? Do we emphasize the importance of affirmative action, for example, to help ensure social justice? In the South African setting this seems an absolute necessity, and yet ironically it may also perpetuate apartheid categories. Do we try to increase unity within our societies, perhaps promoting nationalism? China uses this kind of unity to great advantage, and yet it's difficult to expunge the memories of how nationalism ran amuck in the 20th century. In the section on disgust, we suggested that a balance between conservativism and progressivism may at times be useful. These two stances may well entail different metaphors of caretaking, and at times one may be more appropriate than the other. The perils may lie in the particulars.[64]

Forgiveness and reconciliation

The pass system in South Africa, which forced black people to carry identification at all times, and which enforced a range of inequalities, was surely evil. When apartheid came to an end, Archbishop Desmond Tutu was appointed to lead the South African Truth and Reconciliation Commission, and he used his position to advocate strongly for the importance of forgiveness. As noted earlier, one of his metaphors was of a surgical intervention: only by piercing the abscess could the truth emerge, and forgiveness ensue. This approach raises a number of questions, though. Should we forgive evil? Is it possible to genuinely do so?[65]

Neuroscience has gradually started to shed some light on brain–mind mechanisms underlying forgiveness and reconciliation. From a mechanistic point of view, it seems that particular neurocircuitry is engaged by the brain during forgiveness. From an evolutionary perspective, it can be argued that primates need to forge alliances, and that this requires letting bygones be bygones. There is a range of data pointing to associations between forgiveness and increased health, both physical and mental. At the same time, the psychotherapy literature has emphasized that forgiveness is a complex process: it's premature to simply turn the other cheek, forgiveness makes more sense in the context of apologies and reparations, and forgiveness should not necessarily entail forgetfulness.[66]

When it comes to the big questions and hard problems of life, appropriate responses seem inevitably to involve a degree of balance. Decision-making

64. See Jonathan Haidt [1256], who makes a **plea for moderation**. Even revolutionaries may need to be bear moderation in mind; I'm reminded of the feud between Camus and Sartre, about the endorsement of violence by socialist movements. Camus emphasized human fallibilism and moral humility, argued that freedom is important but relative, and advocated embracing limits and moderation [1463].

65. See earlier footnote on the discussion of forgiveness in the context of South Africa's Truth and Reconciliation Commission. For more on forgiveness in the context of Nazi Germany, see Hannah Arendt's 'The Human Condition' [1300].

66. For more on the **neuroscience of forgiveness**, see Donald Pfaff [1464], D.J. Stein and D. Kaminer [1465], D. Kaminer and colleagues [1466], and Frans de Waal [1467]. For a lighthearted perspective on the importance of not forgiving, see Sophie Hannah's "How to Hold a Grudge" [1468].

around forgiveness and reconciliation, like other important decisions, requires judicious weighing up of relevant facts and values. And as with other decision-making, given that our thoughts and feelings about these matters are embodied within our own brain–minds and embedded within particular social structures, there are no abstract rules that provide easy answers; instead we need to muddle our way through the world. To draw on Aristotle yet again, forgiveness should be with the right person and to the right degree and at the right time and for the right purpose, and in the right way. Forgiveness can, however, be a crucial process that contributes to growth and well-being. However, for those who have been victims of injustice, forgiveness ought not to entail forgetfulness: memory of injustice can be a powerful spur to address ongoing prejudice, and to prevent future iniquity.

Self-deception and anosognosia

It's rare that people think of themselves as evil. On the contrary, humans often signal their virtue to others. It's possible, then, that when people commit evil, or when they virtue-signal, there is some degree of self-deception. We automatically are able to provide justifications to explain the value of our behaviour, but closer examination may reveal that we are really causing significant harm. The technical term for lack of insight is anosognosia; this is seen in a diverse spectrum of human behaviour, ranging from rationalization of the harm we cause others, to denial of ill health in the presence of a major disease.[67]

The term anosognosia was coined by the Hungarian neurologist, Joseph Babinski, who noticed that people with strokes were sometimes not aware that parts of their body were paralysed. Neuroimaging of such individuals may find evidence of right parietal lobe damage, but a range of other lesions have also been documented. There has been less work on neuroimaging of day-to-day self-deception. One possibility is that attention to particular matters simply means that others are overlooked. Another is that self-deception involves specific psychobiological mechanisms. From an evolutionary perspective it's been argued that deceiving others is often useful, and this is best done if the person doing the deceiving is genuinely not aware of the deception.[68]

67. For more on **virtue signalling and moral grandstanding**, including their positive impacts, see Geoffrey Miller [1469] and Justin Tosi and Brandon Warmke [1470]. Jonathan Sacks articulately distinguishes between the righteous and the self-righteous; "The righteous are humble, the self-righteous are proud. The righteous understand doubt, the self-righteous only certainty. The righteous see the good in people, the self-righteous only the bad. The righteous leave you feeling enlarged, the self-righteous make you feel small" [1471]. See also earlier footnote on narcissism.

68. For more on **anosognosia** and its neuroanatomy, see P. Landi and colleagues [1472], D.C. Mograbi and R.G. Morris [1473], and L. Pia and colleagues [1474]. For more on **self-deception** and its neurobiology, see Daniel Goleman [1475], Carol Tavris and Elliot Aronson [1476], Michael Myslobodsky [1477], Robert Trivers [1478], Christopher Booker [1479], Tasha Eurich [1480], and David McRaney [1481]. Discussion of self and self-deception entails the use of metaphors (see later for more on metaphors of self), but self-deception may be a very different sort of phenomenon from intentional deception of others [1482, 1483].

What are the implications of the psychobiology of self-deception for answering the big questions? Knowing oneself has been emphasized by the ancient Greeks and by modern psychotherapy. This certainly seems like valuable advice; in particular being mindfully aware of our habits (which are ordinarily so automated) may be an important element of changing them. At the same time we need to be cautious here. First, it seems that a certain amount of self-deception may be useful for people. Second, it seems that good self-insight is not always possible, our brain–minds seem to entail blindspots. Third, to paraphrase Marx, sometimes what is most important is not to understand oneself, but to change. It is noteworthy that many useful psychotherapies are not based on insight.[69]

The big question of morality

In the natural world, life can be 'nasty, brutish, and short'. There are, for example, the hunters, and there are the hunted. Hunters have to work extremely hard to catch their prey, and the hunted work hard at not being caught. A kill is neither good or bad, it just is what it is. Infectious agents are always actively reproducing and mutating; we have billions of bugs living in us at any given time, and while we get on well with most, others can cause tremendous harm. There is a constant war going on, between our immune systems and bugs trying to penetrate our defences. During the inevitable pandemics that inflict us, we are extremely aware of our vulnerability, but when the pandemics recede so does our mindfulness of this aspect of life.

In the case of primates, we see the emergence of particularly complex constellations of competition and collaboration, which provide the basis for a proto-morality. Here, hurting goes beyond hunting of other species. Hurting can be done to maintain hierarchy, to bring deviants back into line, and perhaps even for pure pleasure. In the case of primates and *Homo sapiens*, inflicting harm on others may involve a good deal of planning. 'Macachiavellian' is a word coined by the primatologist Dario Maestripieri; in his book on the topic he emphasizes how the evolutionary success of rhesus macaques and *Homo sapiens* is dependent on our ability to compete and to collaborate in extraordinarily complex ways.[70]

In thinking about morality, we necessarily use moral metaphors. Evil itself is a word that conjures up the devil, fighting against the good. It's a construct from the supernatural world that we adopt in the secular world when we want to emphasize how truly bad something is. Evil is not simply a matter of pain

---

69. For more on the importance of **knowing oneself** in both philosophy and psychoanalysis, see R.G.T. Gipps and M. Lacewing [245]. See also later footnote on philosophy of self-knowledge.
70. See Dario Maestripieri on '**Macachiavellian Intelligence**' [1285]. A genre of self-help books that focus on how to compete and collaborate better has emerged; I earlier mentioned Dale Carnegie's "How to Win Friends and Influence People" [1484], today we have the work of authors such as Robert Cialdini [1485].

being inflicted; our metaphors of evil emphasize that the pain is wholly inexplicable. When the devil gets the better of God, we experience a sense of personal despair, a loss of trust in the world, and the lack of a future (the so-called depressive triad). The term evil encapsulates this shattering of ourselves and our world. Unfortunately, such experiences are not rare in nature: the evolutionary theorist George Williams has argued that while perhaps the physical world is morally indifferent, the natural world is grossly immoral.[71]

It may be useful, however, to sharpen our metaphors of evil. A first kind of evil seems to me to be done when people have routine behaviours that facilitate hurt when there are readily available ways to minimize this. I would argue that when a lion kills an impala, there are no ways to minimize pain to the impala, and a judgement of immorality is inappropriate. However when humans eat animals, the routine behaviour obscures the hurt that is done when we, say, buy cheaper chicken (chicken that is cheaper by virtue of having been mass-produced in inhumane factories). The decision to buy and eat such chicken may be done with deliberation, and so may reflect one or other error in judgement; after all, we have moral fallibility. Alternatively, we may have deliberated and judiciously decided not to buy and eat such chicken, but we give in to our urges to do so; we'll discuss such weakness of will in more detail later on.[72]

A second kind of evil is done when people act precisely in order to take pleasure from the pain of others. In movies this is a popular trope; the sadistic murderer gleefully laughing at the suffering of others. And it does seem that the phenomenon of 'sweet revenge' exists. While one argument in the literature has been that in order to inflict harm we must first dehumanize others, it's also been pointed out that empathy (e.g. for a perceived victim) may increase the harm inflicted (e.g. on a perceived perpetrator). Fortunately, regular torture of individuals in order to enjoy their pain seems to be a comparatively unusual human phenomenon. It's notable that even though we may be a relatively aggressive species, many of those who perpetrate aggression experience subsequent distress.[73]

While cruelty is a real phenomenon, Roy Baumeister argues that the trope of the sadistic murderer harming an innocent victim with evident joy comprises the widespread *myth of pure evil*. In this myth the perpetrator of evil is someone who is essentially bad, and who is very different from us. In reality, however, in the intricate web of our social lives, decision-making is complex and may

---

71. For a view of evil as encapsulating the shattering of ourselves and our world, see again Susan Neiman [1152]. **George Williams** collaborated with Randy Nesse to develop the field of evolutionary medicine, for his view see his paper on ethics and sociobiology [1486]. To sustain his argument he points to a range of phenomena in the natural world (e.g. rape, infanticide), noting that we are not the only species to kill wantonly, and entirely destroying the idea of the 'noble savage'. Amongst his points is that higher male mortality reflects the malign influence of sexual rivalry.

72. See D.J. Stein [1487].

73. For the argument regarding dehumanization, see David Livingstone Smith [1488]; for the argument that empathy may lead to harm, see Paul Bloom [1489].

contribute to preventing or causing harm to others. Put differently, many of our social decisions fall somewhere between our polar exemplars of banal and sadistic evil, and may involve some part virtue and some part vice.[74]

Another metaphor that we use when guiding our moral decision-making is that of the family and of parenting. Take the example of weaning. From one perspective, it is in both the mother and the infant's interest for the infant to become a strong and independent individual; while there may be some distress during weaning, this is transient and can be ignored. From a different perspective, the best way to ensure that the infant does become strong and independent is to ensure that distress during weaning is minimized; ensuring that this occurs gradually and caringly is therefore appropriate. And from a third perspective, each of these approaches is an overly individualistic one which ignores the valuable role of a range of other caregivers. Deciding which metaphor, which combination of metaphors, or which specific aspects of these particular metaphors to use, in any one case, is a matter of practical judgement and balance.

In summary, primate life in general and human life in particular involves a range of complex phenomena, including aggression, hierarchy, cronyism, forgiveness, and self-deception. Cognitive-affective science has contributed to understanding how these complex phenomena are embodied in our brain–minds and embedded in our social interactions. Moral decision-making is also embodied and embedded, and involves using different metaphors for thinking and feeling our way through these complex phenomena. We are, however, able to carefully weigh up our different metaphors, the principles that they encompass and the particular circumstances at hand, in a deliberative way, and so to determine better paths to take. Over time, within both individuals and societies, there may be moral progress.[75]

6.5 Conclusion

Cognitive-affective science suggests that our embodied sense of morality has its roots in long-evolved cognitive-affective processes that involve nurturance and cooperation. Neuroscience is also consistent with the hypothesis that different kinds of evil exist; severe impulsive-aggression (e.g. in intimate partner violence) differs from everyday slights (e.g. in the reinforcement of social hierarchy). In many cases of evil, however, there is a banality, with perpetrators apparently having little insight into the damage they are inflicting.

Psychiatric research indicates that psychopaths behave in a particularly unusual way, and that their brain–minds have specific abnormalities. On the other hand, we are an extraordinarily Machiavellian species, with expertise in a range of deceptive and manipulative behaviour, and a penchant for exacting sweet

74. For more on these sorts of **complexities of evil**, see again Mary Midgley [1412], Roy Baumeister [1427], Zygmunt Bauman and Leonidas Donskis [1490], and Jonathan Glover [1491].

75. Finding 'balance' and 'better paths' again relies on the use of metaphors. We've briefly discussed balance and will return to these metaphors later.

revenge. As always in psychiatry, differentiating the pathological from the normal is a fuzzy business, but there is an argument that much evil is indicative of bad rather than mad behaviour. Notably, our knowledge of the psychobiology of psychopathology rarely leads to exculpation of psychopathic behaviour.[76]

While it's possible that some evil is explained by psychopathy, a medical metaphor is only appropriate for some phenomena. The moral metaphor of evil is useful in describing that which absolutely shatters ourselves and our world. In philosophy, we return then to the question of being moral, as a particular way of thinking-feeling, and a particular set of habits and practices. Morality is embodied in our brain–minds; while moral decision-making is typically automatic, humans are able to engage in deliberate moral decision-making; to carefully weigh up the use of different metaphors and models; to consider relevant principles, particulars, facts, and values; and to reasonably decide on the right way forwards. Value pluralism emphasizes that there is more than one good way to act. But at the same time, moral naturalism holds that there can be better ways of acting, and moral progress.

---

76. It is perhaps useful to think of a spectrum between those have **no moral culpability** for crime (e.g. someone whose crime is entirely due to a frontal lobe tumour) and those who bear **full moral responsibility** (say, a person who makes inappropriate deductions on their tax forms). It seems reasonable to view psychopaths as closer to those who bear full responsibility, given that their brain-mind alterations are relatively subtle, and given that current interventions focus on limit-setting. However, in particular cases of crime, there may be good arguments for the use of both a MEDICAL and a MORAL metaphor, for both exculpation and responsibility (and see earlier related discussion of addiction). For a discussion of Aristotle's remarkably prescient discussion of these and related issues, see G. Pearson [1096] as well as Tom Angier, Chad Meister, and Charles Taliaferro [1492]. For more on the growing field of 'neurolaw', see J.A. Chandler [1493].

Chapter 7

How can we know what is true, then?

Having dealt with the big questions of pleasure, of suffering, and of morality, we can now move on to discuss the big question of what is true and false. After all, how do we know that the arguments in the previous chapters have any validity? And if we are going to addressing the truly big question of the meaning of life, we surely need to have some sense of how truth works, and what a true answer might look like. We've turned repeatedly to evolutionary theory for answers, but evolutionary theory accounts for the very way our brain–minds work: this paradox might appear to undermine confidence in our conclusions.

In this volume we have not referred much to God. As an adolescent thinking through the big questions, the Jewish God seemed to be a particularly abstract conceptualization, and this was deeply attractive to me. I'm reminded of Albert Einstein's comment that "My God is the God of Spinoza". Furthermore, despite all the many and obvious ills of organized religion, religions provide people with community, which is profoundly important. Still, in the age of neurophilosophy and neuropsychiatry, we cannot but go with a naturalistic approach, and in this chapter we are concerned with empirical truth.[1]

The question of truth has of course been a major focus within philosophy, particularly epistemology. The literature on the philosophy of natural and social science, while vast, has focused on the narrower question of whether science can reveal the truth about reality. A long tradition, dating back to at least Aristotle, and given a great deal of impetus by the Enlightenment, answers in the affirmative. But there is also a long and strong tradition arguing that science is simply another story, albeit written in a rather distinctive genre. In exploring these debates, we'll focus on the philosophy of biology, neuroscience, and psychiatric science, in particular.

1. There is debate about the similarities and differences between **Maimonides and Spinoza's** view of nature and of God. For some, Maimonides' 'negative theology', in which the essence of God cannot be articulated, is a protomodern forerunner of Spinoza's views, as well as of current views that metaphors of God are necessarily anthropomorphic (see later discussion of spirituality). For others, their views are quite different. See Joshua Parens [1494] and K. Seeskin [1495].

Problems of Living. https://doi.org/10.1016/B978-0-323-90239-7.00010-9

7.1 Philosophy and the truth

We've talked a good deal about the evidence from neuroscience in this volume. Can we rely on those findings? There is a growing literature on the philosophy of biology in general, as well as the philosophy of neuroscience in particular, which addresses this issue. However, these fields often refer back to earlier and perennial debates in philosophy of science and language. Let us therefore return to the topics covered in Table 1, but now adding a range of considerations about the philosophy of biology and neuroscience, as well as about the relationship between different disciplines, including that between the sciences and the humanities, and that between science and philosophy (Table 11).

Clearly there has been real progress in the biological sciences. Charles Darwin and Arthur Wallace provided a comprehensive explanation of how distal or evolutionary mechanisms work, emphasizing that mechanisms that promote successful survival and reproduction are replicated. Francis Crick, James Watson, and Rosalind Franklin provided a key foundation for explaining the molecular basis of this replication, emphasizing that deoxyribonucleic acid (DNA) is structured as a double helix. It seems unlikely that any future work will be as seminal, but it's difficult to make predictions, especially, as the proverb goes, about the future. More certainly, as biologists answer one question, so additional ones emerge, so the end of biological science is certainly not yet in sight.[2]

Describing scientific progress in terms of the discovery of more and more encompassing laws may be relevant to physics, but does not seem entirely accurate when it comes to the biological sciences. Certainly important principles have been discovered (e.g. parental DNA is recombined during replication), and interesting regular relationships have been described (e.g. mammals with a smaller body size have shorter lifespans). But biology is in many ways also focused on the study of particular irregularities. The biology of marsupials, for example, provides important information on divergent evolutionary pathways within mammalian biology, and understanding the biology of disease has been enormously helpful in understanding how normal physiology works.[3]

The idea that biology is a social activity is certainly correct. Darwin himself was very much a product of his times, as some of his comments about Africans demonstrate. And the 'lessons' that social Darwinists drew from evolutionary theory matched their social zeitgeist. There is an argument that neurobiology takes an individualistic stance, that it is, for example, focused on individual

---

2. Several have contributed to the aphorism that it's difficult to make predictions, especially about the future, see quoteinvestigator.com. George Bernard Shaw is said to have toasted Albert Einstein with the words "Science is always wrong. It never solves a problem without creating 10 more". For a contrary view, taking a mysterian approach, and emphasizing the end of science, see John Horgan [1507].

3. For more on **philosophy of biology**, see Elliott Sober [1508], Kim Sterelny and Paul Griffiths [1509], David Hull and Michael Ruse [1510], and Peter Godfrey-Smith [1511].

TABLE 11 Classical, critical, and integrative philosophical positions revisited.

	Classical position	Critical position	Integrative position
Historical sources	Plato, Descartes, early Wittgenstein, logical positivism, analytic philosophy	Vico, Herder, later Wittgenstein, Kuhn, continental philosophy	Aristotle, Dewey, Bhaskar, pragmatic realism, critical realism, embodied realism
Philosophy of science	Science describes the world objectively, and deduces covering laws	Science depicts the world subjectively, constructing a narrative	Science is a social activity, and it discovers real structures, processes, and mechanisms
Philosophy of language	Meaning requires verification	Meaning is based on validation	Language is embodied and embedded, but does reference real phenomena
Philosophy of biology	Biological laws can explain all biological phenomena	Biological constructs are socially constructed	Biological science is a social activity, and it discovers real structures, processes, and mechanisms
Philosophy of neuroscience	Neuroscience laws can explain all mental phenomena	Neuroscience constructs are socially constructed	Structures, processes, and mechanisms found by neuroscience have crucial, but also crucially limited, explanatory power[a]
Philosophy of psychiatry	Psychiatric laws can explain all psychiatric phenomena	Psychiatric constructs are socially constructed	Biological and psychological structures, processes, and mechanisms help explain psychiatric phenomena[b]
Relationship between various disciplines	Multidisciplinarity: There are different disciplines with physics at the apex	Interdisciplinarity: Boundaries reflect professionalization, even this term reflects the power of universities	Transdisciplinarity: Key to understand interlevel structures, processes, and mechanisms[c]

Continued

TABLE 11 Classical, critical, and integrative philosophical positions revisited—Cont'd

	Classical position	Critical position	Integrative position
Relationship between sciences and humanities	Sciences are about real knowledge	There are many different forms of knowledge, scientific and humanistic	Activities in both the sciences and the humanities are both theory-bound and value-laden
Relationship between science and philosophy	Reason (or philosophy) is what underlies all disciplines, which discover the laws of the world	Boundaries between philosophy and other fields are socially constructed	Philosophy is a rigorous social activity, which helps explain how rigorous science can progress
Philosophy of philosophy	Philosophy makes progress as it converges on an authoritative account of its subject	There is no progress or convergence in philosophy, only changes in philosophical fads	Philosophy progresses by providing better models and more sophisticated disagreements
Human nature	Human nature is essentially bestial	Human nature is culturally produced	Humans are part of nature, our nature is shaped by genes and culture
Answering the big questions	Propositions about the big questions typically do not make sense	There are multiple answers to the big questions, each is acceptable	Careful weighing up of principles and particulars, and of facts and values, leads to better answers

[a] *For more on* ***philosophy of neuroscience****, see John Bickle [1497], Max Bennett, Daniel Dennett, Peter Hacker, and John Searle [1496], and Suparna Choudhury and Jan Slaby [1498]. See also later footnote on use of 'mechanisms' in neuroscience.*

[b] *As noted in a previous footnote, although this volume is primarily concerned with philosophy of psychiatry, there are important overlaps between work on philosophy of psychiatry, philosophy of psychology, and philosophy of cognitive science, as exemplified in the earlier discussion of the body–mind problem.*

[c] *Aristotle described a* ***hierarchy of the disciplines****, but also saw philosophy as a universal field that brought these together. See Joe Moran's volume 'Interdisciplinarity' [1499] for more on this point, and on the view that 'discipline' is more of a tough term, while 'inter' is more of a tender term. The term transdisciplinarity was introduced by Piaget; for more background, see Jay Bernstein [1500]. Another key contributor, Basarab Nicolescu, argued that transdisciplinarity goes beyond interdisciplinarity and multidisciplinarity, by studying that which is between, across, and beyond disciplines [1501]. See also B.C. Choi and A.W. Pak [1502]. Note also recent interest in* ***'convergence science'*** *[1503–1505], including convergence science and psychiatry [1506].*

psychopathology rather than on social determinants of disease. These are crucially important points, and help deflate inappropriate hubris about the insights of scientists and about the march of scientific progress. At the same time, seeing neurobiology merely as a social activity seems to ignore the real discoveries that have been made, with incremental advances in our understanding of biological structures, processes, and mechanisms.[4]

The approach here again attempts to draw on the strengths of both the classical position (in praise of biology) and the critical position (emphasizing that biology is a social activity), while avoiding their weaknesses. We cannot appeal to 'the view from nowhere' about how biology works, we can only think through how biological scientists have engaged with the world and whether this is reasonable, much as we deliberate about whether our common-sense knowledge of the world is well founded.[5] Our experience of biologists as social actors, as well as of the explanatory success of biology, are only explicable if biology is a social activity that advances our knowledge of the real structures, processes, and mechanisms of the world. This is the so-called 'No Miracles Argument', a phrase that derives from the contributions of philosopher Hilary Putnam to pragmatic realism, and his statement that realism "is the only philosophy that doesn't make the success of science a miracle".[6]

Table 11 contrasts the classical, critical, and integrative positions on a range of other issues, including the philosophy of neuroscience, the relationship between different scientific disciplines, and the relationship between science and the humanities. Throughout, an integrative position emphasizes that while the sciences are social activities, they discover and explore real structures, processes, and mechanisms. The fact that reality is complex, with different phenomena emerging at different levels of organization, has important consequences. First, the sciences do not necessarily reduce to one another in general, and mental

4. That said, Darwin was an abolitionist [1512].

5. Many have emphasized the **continuity between everyday and scientific knowledge,** including Dewey [120]. Both involve, for example, not only 'knowledge that' but also 'knowledge how' [380] or 'tacit knowledge' [1513], as well as imagination. Like Dewey, Thomas Kuhn noted that scientific knowledge was embodied [61]. For more on 'knowledge how', see Adam Carter and Ted Poston [1514], John Benson and Marc Moffett [1515], and Jason Stanley [1516]. Continuity between everyday and scientific knowledge is consistent with a **naturalistic and empirical approach to thinking about science**; see, for example, Paul Churchland [237], Peter Carruthers, Stephen Stich, and Michael Siegal [1517], Ángel Nepomuceno Fernández, Lorenzo Magnani, Francisco Salguero-Lamillar, and Cristina Barés-Gómez [1518], Ronald Giere [1519], Alison Gopnik [1520], Arnon Levy and Peter Godfrey-Smith [1521], Lorenzo Magnani, Nancy Nersessian, and Paul Thagard [1522], Mary Morgan and Margaret Morrison [1523], and Paul Thagard [1524].

6. Rather than using the phrase '**view from nowhere**', Daniel Dennett, a student of Quine and Ryle, insists that we stick with cranes and avoid skyhooks [1525]. Alternatively, some (including Nietzsche) have emphasized that we must live without scientific foundations [119] or moral fundamentalism [1526]. As Hannah Arendt wrote, "even if there is no truth, man can be truthful, and even if there is no reliable certainty, man can be reliable" [1300]. For a contrary view in moral realism, arguing that morality is built into the structure of the universe, see George Ellis [1526]. For the No Miracles Argument, see Putnam [1527].

phenomena are not wholly explained by neuroscience in particular. Second, the sciences themselves cannot necessarily be neatly unified; rather a plurality of constructs, methods, and explanations is needed. As we've seen, this position has Aristotelian roots, has been extended by a range of philosophical work on realism (e.g. pragmatic realism, critical realism, embodied realism), and was pioneered in psychiatry by Jaspers.[7]

From this perspective, the two cultures of the sciences and the humanities each provide important contributions to our knowledge of the world. We should be cautious about seeing them as entirely reconcilable; they entail different spheres of knowledge, so offering complementary perspectives. But we can also note many overlaps; for example, both the sciences and the humanities are social activities that are theory-bound and value laden.[8] In the realm of psychiatry, the humanities may play a particularly important role in increasing our capacity for understanding (or verstehen), and so contributing to our knowledge of people and psychopathology. At the same time, scientific explanations of mental disorders may well help us to better understand the experience of suffering from them; the pioneering psychiatrist-philosopher Kenneth Kendler refers to this as 'explanation-aided understanding'. Thus, while all of medicine entails both science and art, this is particularly true of psychiatry.[9]

These points in turn raise the question of the relationship between philosophy and science. If science in general, and neuroscience in particular, discover and explore real structures, processes, and mechanisms, what does philosophy do? One view is that doing philosophy is an entirely different thing from doing science; the two bear no relationship at all. Within this view, positions range

---

7. In analytic philosophy, the Stanford School, which includes Patrick Suppes, Ian Hacking, John Dupré, and Nancy Cartwright, has argued for the **disunity of science**, and for both ontological and methodological pluralism. In continental philosophy, in opposition to post-Kantian idealism, a different form of realism that resonates with pluralism has recently emerged; **speculative realism** argues that there is a mind-independent reality, but that this is far weirder and less knowable than earlier realists have suggested [1528]. For more detail on points in this paragraph, see my 'Philosophy of Psychopharmacology' [53], as well as earlier footnotes on conceptual and explanatory pluralism, on Jaspers' emphasis on such pluralism, and on Kendler's more recent contributions. Previous footnotes on reductionism and scientism, on essentialism and emergence, and on multiple causality including top-down causation, are also of relevance here.

8. The **two cultures** of science and humanities have often been contrasted [1529]. In a subsequent essay, C.P. Snow suggested that biology might provide the best model for interdisciplinary work, partly because it asks the big questions [1530]. Edward Wilson echoes this view, arguing that biology is a discipline that can bring consilience [1531]. This view has on occasion received support from some in the humanities; see Jonathan Gottschall [1532]. However, Wilson's aim of unification seems overreaching; see, for example, Stephen Gould [1533] and Jerome Kagan [1534]. Along these lines, Moran contrasts 'hard' Darwinians (e.g. Wilson, Dawkins) who believe that everything, including culture, can be explained by evolution, and 'soft' Darwinians (e.g. Stephen Gould) who see life as a complex process and avoid such reductionism [1535].

9. See K.S. Kendler [345]. See earlier footnotes on Ken's mentorship of me, on the erklären–verstehen distinction, on the disease-illness distinction, and on medicine as a science–humanity (as emphasized by Hippocrates in ancient times and Osler more recently).

from those who see philosophy as deeper than science (with science leading us only to disenchantment),[10] through to those who see science as ultimately subsuming philosophy (just as metaphysics has increasingly given way to physics, so will epistemology increasingly yield to the cognitive sciences).[11]

For Wittgenstein, who falls more towards a view of science as entailing disenchantment, science and philosophy are two distinct forms of understanding; scientific understanding is obtained by testing hypotheses, while philosophical understanding 'consists in seeing connections'.[12] For Jaspers, who takes a more celebratory position regarding science, science and philosophy are wholly different, with the sciences a tool to be employed by philosophy. Perhaps in the middle of this spectrum on the respective value of philosophy and science, Bernard Williams argues that philosophy should not try to be an extension of the natural sciences, rather it is part of "a wider humanistic enterprise of making sense of ourselves".[13]

An alternative approach to the relationship between philosophy and the sciences is to regard them as equally profound and irreplaceable, with both overlapping

---

10. A view of **philosophical understanding as deeper than scientific knowledge** dates back at least to Vico [1499]. Bertrand Russell argues, "the greatest men who have been philosophers have felt the need both of science and of mysticism: the attempt to harmonize the two was what made their life, and what always must, for all its arduous uncertainty, make philosophy, to some minds, a greater thing than either science or religion" [1536]. Taking a perhaps more nuanced view, William James opines, "Philosophy is at once the most sublime and the most trivial of human pursuits. It works in the minutest crannies and it opens out the widest vistas" [1537]. For Mary Midgley, in her last volume, "[T]he reason why some philosophers are remembered is not that they have revealed new facts, but that they have suggested new ways of thinking which call for different ways of living—paths which were really needed" [1538].

11. For more on **physics and metaphysics**, see Don Ross, James Ladyman, and Harold Kincaid [1539] and Tim Maudlin [1540]. For more on **epistemology and cognitive science**, see Piaget [1541], Konrad Lorenz [1542], and Robert McCauley [1543]. David Livingstone Smith has put forward '**biophilosophy**' as an analogue of 'neurophilosophy'; that is, using biology "to constrain, guide, and inspire philosophical theorizing" [1544]. He points to the work of Ruth Millikan [1545] as paradigmatic in using biological models for philosophical ends.

12. Max Weber famously charged science with '**disenchanting**' the world, and many have agreed. Keats, for example, suggested that Newton had destroyed the poetry of the rainbow by explaining it as a product of the refraction of white light [1546], and advocated for '**negative capability**', "that is when man is capable of being in uncertainties, Mysteries, doubts, without any irritable reaching after fact & reason" [1547]. Wittgenstein, similarly, noted that "To experience wonder, man – and perhaps populations – must be awakened. Science provides a means to fall asleep again" [1548]. See also earlier footnote on John McDowell's criticism of a 'naturalism of disenchanted nature'. On the other hand, the notion of science leading to disenchantment may itself be a myth [1549]. Certainly, some views of science have emphasized its close connection to mystery, wonder, and awe. See later footnotes on the awe of science and on the grandeur of science, as well as on 'essentially contested constructs'.

13. See Søren Overgaard, Paul Gilbert, and Stephen Burwood's 'Introduction to **Metaphilosophy**' [1550] for a useful introduction to these issues. For a broad range of definitions of philosophy, see David Edmonds and Nigel Warburton's 'Philosophy Bites' [1551]. For these citations, see Wittgenstein [39], Jaspers [47], and Williams [1552]. Wittgenstein did, however, see philosophy as an activity saying "Philosophy is not a body of doctrine but an activity", and in that sense there is an overlap with science [38].

and distinctive features. The golden mean often hearkens back to Aristotle; indeed Aristotle viewed metaphysics (or 'first philosophy') and physics (a 'second philosophy' devoted to the study of nature) as overlapping, but also emphasized the importance of employing methods appropriate to the question under study. Those who regard philosophy and science as lying on a spectrum have, however, conceptualized this spectrum in different ways. For some philosophy and science use similar methodologies,[14] for others they have similar ends (e.g. philosophy of science underlabours for science; it clarifies conceptual issues so that they can be empirically investigated, and it helps explain how science in fact works).[15]

I take the position that philosophy is well grounded in science (the humanities are not always), but also tackles issues that science cannot (just as the humanities do). For the pioneering developmental psychologist, Jean Piaget, "Philosophy is synonymous with science or reflection upon science, and philosophy uninformed by science cannot find truth". As Putnam puts it, philosophy has a double face: the scientific (looking towards the natural and human sciences), and the moral face (which "interrogates our lives and our cultures as they have been up to now, and that challenges us to reform both").[16] Or as

---

14. Strict naturalists may focus on **continuity of methods**. Quine recommended the "abandonment of the goal of a first philosophy prior to natural science", together with a "readiness to see philosophy as natural science trained upon itself and permitted free use of scientific findings" [1553]. For Sellars, philosophy is the endeavour "to see how things in the broadest possible sense of the term hang together in the broadest possible sense of the term" [193], while "In the dimension of describing and explaining the world, science is the measure of all things, of what is that it is, and of what is not that it is not" [420]. Notably, right-wing Sellarsians defend this 'scientia mensura', while left-wing Sellarsians focus more on ideas such as the space of reasons; for example, Ruth Millikan, who studied under Sellars before he moved to Pittsburgh, has proposed a 'biosemantics', which explains intentionality using an evolutionary framework, and is considered to fall in the former category [1554]. For arguments regarding methodological similarities between philosophy and science, see D. Papineau [1555] and Susan Haack [165].

15. Soft naturalists may focus more on **continuity of ends**. For Wundt, philosophy was "the general science whose task it is to unify the general pieces of knowledge yielded by the particular sciences into a system free of contradiction" [1556]. For the less ambitious view that philosophy underlabours for science, see Bhaskar [29]; his phrasing derives from John Locke's "Essay Concerning Human Understanding" [685], where Locke modestly compared himself to 'the incomparable Mr. Newton' and the other great scientists of his day, and said "it is ambition enough to be employed as an under-labourer in clearing the ground a little, and removing some of the rubbish that lies in the way to knowledge". This language brings to mind Socrates's description of himself as a midwife who assists with the birth of discoveries (a description that continues to influence more recent views of the relationship between philosophy and science [176, 1557]).

16. The **continuity of philosophy and science** has been suggested by many others after Aristotle, including some of our favourite philosophers. Dewey, for example, writes "The problem of restoring integration and cooperation between man's beliefs about the world in which he lives and the values and purposes that should direct his conduct is the deepest problem of modern life. It is the problem of any philosophy that is not isolated from that life" [1558]. Piaget's phrase is from his 'Insights and Illusions of Philosophy' [1559], while Putnam's phrase is from his 'Naturalism, Realism and Normativity' [1560]. Piaget has similarly said that "Philosophy occupies itself with the study of knowledge as a whole, while science occupies itself with specific questions. Or, philosophy tends towards the study of being qua being and science towards the study of specific beings" [1561]. For more on the interaction between science and philosophy, see Freidel Weinert [1562].

Moran states, philosophy not only forms part of the traditional search for a wide-ranging knowledge, but it also involves a more radical questioning of the nature of knowledge itself and our attempts to organize and communicate it.[17] Furthermore, both disciplines emphasize rigorous practices, acknowledge their own fallibility, and continuously strive for improvement.[18]

These considerations raise the question of whether philosophy makes progress? There are certainly reasons to be dubious about the claim that philosophy advances as straightforwardly as some sciences. Some might go so far as to agree with Russell's view that "Philosophy, from the earliest times, has made greater claims, and achieved fewer results, than any other branch of learning". On the other hand, it seems overly harsh to claim that philosophy makes no progress at all. Peter Unger, for example, has argued that philosophy doesn't establish or refute any nontrivial ideas, rather only fashions in philosophy change.[19] The ongoing engagement of philosophy with science and the development of an integrative realism, which successfully provides a way of seeing science as not only a social activity, but also as discovering the structures, processes, and mechanisms of the world, seems to me to indicate progress in epistemology. Similarly, philosophical work that successfully negotiates a naturalistic but nonreductionistic path avoiding both moral absolutism and moral nihilism seems to me to reflect progress in ethics. Given how dappled, complex, and fuzzy the natural and human world are, progress in philosophy as well as in science does not mean that all philosophical and scientific disagreement disappears. To at least some extent,

---

17. See Joe Moran [1499]. Many have written elegantly on the **relationship between science and philosophy**. Rickert noted that "Philosophy has to be characterized as the scientific activity, which, in the stream of restlessly forward-moving development, endeavors to make a stop and find a resting point. It stands still in order to bring to light the significance for the meaning of life of what has been hitherto achieved" [1563]. Feigl noted, "[T]here is no sharp line of demarcation between (good) science and (clearheaded) philosophy. Every major scientific advance involves revisions of our conceptual frameworks; and doing philosophy in our days and age without regard to the problems and results of the sciences is – to put it mildly – intellectually unprofitable, if not irresponsible" [1564]. For Rebecca Goldstein, "perhaps we can ... divide our intellectual terrain up into facts about being – and here it's science that most reliably informs us – and facts about mattering, about which philosophy most reliably informs us" [1216].
18. A related interesting issue is the nature of and role for **experimental philosophy**. See, for example, Joshua Knobe, Tania Lombrozo, and Shaun Nichols [1565], Eugen Fischer and John Collins [1566], and Justin Sytsma and Walter Buckwalter [1567].
19. Russell's **criticism of philosophy** is from his 'The Problems of Philosophy' [1568]. Peter Unger's quote is from his "Empty Ideas: A Critique of Analytic Philosophy" [1569]. Wittgenstein was also not a particularly pro-progress writer; his best defence was, "Philosophy hasn't made any progress? If somebody scratches the spot where he has an itch, do we have to see some progress? Isn't genuine scratching otherwise, or genuine itching itching? And can't this reaction to an irritation continue in the same way for a long time before a cure for the itching is discovered?" [1548]. For more on Wittgenstein's view of progress in general, see R. Read [1570]. Perhaps more humorously, the difference between the scientist and the philosopher has been put as: "Some people regard the former as one who knows a great deal about a very little, and who keeps on knowing more and more about less and less until he knows everything about nothing ... Then there are the latter specimen, who knows a little about very much, and he continues to know less and less about more and more until he knows nothing about everything" [1571].

disagreement can be seen as reflecting healthy reasoning-imagining in the scientific and philosophical community about reality and about 'essentially contested constructs'.[20]

What would an integrative approach to claims about progress in philosophy look like? Such a position would accept that philosophy itself is a social activity, and that the 'resources' it produces do need to be understood against the context of both conceptual debates and socio-political conflicts. At the same time, such an approach would view philosophy as making progress by providing exemplars of human problems, outlining key conceptual issues, and offering potential models and solutions that remain useful over generations. Some may counter that given continual changes in human contexts, such exemplars and solutions do not necessarily entail progress.[21] Still, as we understand more about the natural world, so some philosophical ideas lose value, and other issues become more sharply defined. Other advances may be linked to advances in our appreciation of human nature: as we debate human practices and values we can at the least develop sharper definitions of where and why we disagree. As in the case of moral progress, the aim of philosophical progress is not necessarily convergence on a single authoritative account, but rather may involve increasingly sophisticated engagement with perennial and persistent questions.[22]

7.2 Psychiatry and truth

We've talked a lot about psychiatric science and its evidence base in this volume. Can we rely on these findings? There is a growing literature on the philosophy of psychiatry which raises this question amongst others. Once again, the

20. For a discussion of progress in philosophy as entailing **greater sophistication in philosophical disagreement**, see D.J. Chalmers [1572]. See also H. Cappelen on this topic [1573]. The phrase '**essentially contested construct**' is from the work of Walter Gallie [1574]. Earlier on, Bertrand Russell concluded his 'Problems of Philosophy' by writing, "Thus, to sum up our discussion of the value of philosophy; Philosophy is to be studied, not for the sake of any definite answers to its questions, since no definite answers can, as a rule, be known to be true, but rather for the sake of the questions themselves. Because these questions enlarge our conception of what is possible, enrich our intellectual imagination and diminish the dogmatic assurance which closes the mind against speculation; but above all that because, through the greatness of the universe which philosophy contemplates, the mind also is rendered great, and becomes capable of that union with the universe which constitutes its highest good" [1568].

21. For a view of **philosophy as involving model-building**, with better models over time, see T. Williamson [1575]. For a pluralist approach that attempts to avoid both absolutism and relativism regarding philosophy, but which stresses the particularity and relatively short life expectancy of philosophical solutions, see John Kekes [1576]. How far one leans towards seeing philosophy as progressive may be partly a matter of temperament: Philip Kitcher has pointed out that whereas he, following Dewey, sees the possibility of a renewal of philosophy, neopragmatists such as Richard Rorty can only envisage a burial [176].

22. For more on **progress in philosophy**, see Daniel Stoljar [1577], Russell Blackford and Damien Broderick [1578], Massimo Pigliucci [1579], and Timothy Williamson [1580]. On a personal note, The Stanford Encyclopedia of Philosophy has been a marvellous resource while working on this volume, and has left me with a sense of forward movement in philosophy.

TABLE 12 Beyond scientism and skepticism in psychiatry.

	Scientism	Skepticism	Integrative position
Historical roots	Kraepelin	Foucault	Jaspers
Contemporary psychiatry	Psychiatry is a science focused on brain disorders (analogous to physics)	Psychiatry is one of several social practices that structure our subjectivity	Both mechanism and meaning are key; science is theory-bound and value-laden
Future of psychiatry	Psychiatric essential natural kinds will be better delineated, so providing precise treatment targets	Pessimistically psychiatry will gain more power, optimistically it will transform into a humanity	Psychiatry is in some key ways both a science and a humanity; psychiatry can improve over time

debates in this area often reflect long-standing debates in philosophy of science and language. Let us therefore return to some of the topics initially raised in Table 2, reframing them in terms of scientism and skepticism about psychiatry (Table 12). I will contrast these positions in order to consider two related issues, in particular. First, has psychiatry made progress towards truth? Second, what sort of thing is psychiatry; is it really a science or what?[23]

It seems to me clear that over the last century, psychiatric science has indeed made progress. If you had obsessional symptoms and saw Freud, you would have received an insight-oriented psychotherapy, and it is unlikely that there would have been symptom resolution. If you were treated more recently, you might have received specific pharmacotherapy and psychotherapy which have been shown in randomized controlled trials to be efficacious for OCD. At the same time, you and your clinician would have explored your own experience of the illness, and there would have been shared decision-making. If you had bipolar disorder and saw Kraepelin, you might have been confined to an asylum for many years. If you were treated more recently, you might again have received effective evidence-based interventions. And again this treatment would have been individualized, so that the illness and its treatment made sense for you.[24]

At the same time, there are many gaps in our knowledge and in our interventions. We don't know what causes OCD or bipolar disorder, and many patients

23. There has been substantial growth in the literature at the **intersection of philosophy and psychiatry** in recent decades; for useful compendia, see John Sadler, Werdie van Staden, and KWM Fulford [1581] and KWM Fulford, Martin Davies, Richard Gipps, George Graham, John Sadler, Giovanni Stanghellini, and Tim Thornton [1582].

24. This is easier said than done, see Kay Jamison [1583].

with these conditions don't respond well to current treatments. Diagnoses comprise very heterogenous entities; OCD and bipolar disorder have dramatically different presentations and courses in different individuals. Treatment is a hit and miss affair; we aren't able to predict who will respond to which intervention, and so we try first one treatment then another. The majority of our medications have been found serendipitously; the hopes of translational science, that basic laboratory-based science will lead to new clinical pharmacotherapy and psychotherapy, have been fulfilled only occasionally.[25]

These points suggest that a scientistic position is overly reductionistic and Panglossian, while a sceptical position avoids reality and is overly pessimistic. The scientistic idea that psychiatric disorders are essential natural kinds that can be defined in terms of their necessary and sufficient characteristics has not worked out and is unlikely to do so. At the same, in contrast to skepticism, we have made partial progress in understanding the structures, processes, and mechanisms that underlie mental disorders. Indeed, it seems to me that a naturalistic approach that is able to integrate mechanism and meaning is the only position that doesn't make the kinds of therapeutic success that have occurred in psychiatry, miraculous. One could perhaps even make the claim that progress in psychiatry contributes to advancing philosophy of science and psychology, by forcing these fields to move beyond overly simplistic accounts of science and of the brain–mind to more complex ones that fully address the dappled nature of reality.[26]

Clearly much more needs to be done to further the progress of psychiatry. Later in this chapter we'll discuss issues around the future of psychiatric classification, biological psychiatry, and psychotherapy in more detail. Here we can perhaps make the more general point that both the classical and the critical traditions have provided important resources for the field. The development of the classically influenced DSM and ICD systems has been crucial for doing psychiatric epidemiology, and undertaking clinical trials, despite the problems of thresholds, reification, and causality discussed earlier. Similarly, despite its shortfalls, a critical tradition has been key in highlighting issues such as the negative impact of labelling and stigma, and the positive value of shared decision-making. Those who have focused on erklären and those who focused on verstehen, those who have viewed psychiatry as a science and those who have focused on psychiatry as a humanity, have all provided partial but important contributions.[27]

---

25. The World Mental Health Surveys consortium has provided more reassuring epidemiological data, suggesting that just staying in treatment is important; people who do so often improve eventually [1584].

26. See Nancy Cartwright's volume 'The Dappled World: A Study of the Boundaries of Science' [175] and K.S. Kendler [342]. For more on the value of psychiatric phenomena for philosophy, see my 'Philosophy of Psychopharmacology' [53].

27. The current focus on the importance of **decolonization of knowledge** in general, and the work of the psychiatrist Frantz Fanon, who worked in Africa, in particular, may similarly provide some useful insights [1585]. See also A.A. van Niekerk on this topic [1586].

At the same time, a second general point is that psychiatry has not always been served well by schisms in the field, where perspectives clash with rather than complement one another. We earlier discussed one major psychiatric schism of the 20th century: the collision of psychoanalytic and biological psychiatry. The advent of DSM-III was potentially useful in providing a framework for collaboration, but for psychoanalytic psychiatrists this was a reductionistic manual that did not provide sufficient attention to key mental phenomena, while biological psychiatry has increasingly criticized the manual for not being based on neuroscience. A biopsychosocial approach also seems potentially useful for integration, but its failure to bring this about perhaps suggests this has at times been more of a slogan than a rigorous conceptual framework.[28]

We have emphasized the lesson from contemporary cognitive-affective science that human thoughts and feelings are structured by metaphors. Some schisms in psychiatry may reflect conflicting explanations; for example, clinicians may hold that children develop fear because of fear conditioning (as Watson thought) or that childhood fears reflect unconscious Oedipal conflicts (as Freud thought). However, the schism between those who focus primarily on erklären versus those who focus on verstehen may to some extent reflect a reliance on different metaphors. When we refer to erklären, we use a range of metaphors of causality (often involving 'seeing' the mechanism), while when we refer to verstehen we use a different range of metaphors (often involving 'hearing' the meaning). Jaspers' advocacy for explanatory pluralism in psychiatry is consistent with a view that different explanatory metaphors have different emphases and uses, complementing rather than contradicting one another. The fact that we use different metaphors does not, however, entail that 'anything goes'; some accounts are more powerful than others.[29]

Two important metaphors or conceptual frameworks are currently guiding the future of psychiatry: that of clinical neuroscience and that of global mental health. Clinical neuroscience is the current incarnation of biological psychiatry; the term is used as part of the National Institute of Mental Health's Research Domain Criteria (RDoC) framework, which has evolved over the past decade to guide 'discovery research' (i.e. more basic investigations). Global mental health is a newly emergent field that has drawn on earlier work in cross-cultural psychiatry, and which has had a particular focus on 'implementation research' (how to take what we know into the field). While there are clearly multiple differences between clinical neuroscience and global mental health, my view is that they can be synergistic, and that psychiatry should draw on the strengths of both (Table 13).

---

28. See earlier footnote on the biopsychosocial model.

29. See earlier footnotes on behavioural versus psychoanalytic explanations of childhood fears, on explanatory pluralism, on Jaspers' views, and on metaphors of causation. See also footnote in next section for more on metaphors of causation.

TABLE 13 Clinical neuroscience and global mental health.

	Clinical neuroscience	Global mental health	Integrative position
Historical roots	Basic neuroscience, biological psychiatry	Cross-cultural psychiatry	Need for multidisciplinary approaches
Current focus	Focus on laboratory knowledge, focus on translational of dimensional research from bench to bedside	Focus on understanding social determinants, emphasis on human rights	No health without mental health, a range of mechanisms are relevant to this slogan
Future focus	Discovery research will find innovative targets for personalized psychiatry for individuals	Implementation research will scale up feasible population-relevant interventions	Integration of discovery and implementation research, personalized public mental health[a]

[a] *D.J. Stein and G. Wegener [1587] and S. Desmond-Hellmann [1588].*

A brief consideration of diagnosis, aetiology, and treatment from the perspective of clinical neuroscience and global mental health may help exemplify their potential synergy. For clinical neuroscience, a major gap in psychiatry is that our diagnostic systems are not aetiologically based and that our treatments are not sufficiently personalized; psychiatry will advance by understanding how brain mechanisms lead to symptoms, by developing biomarkers that are useful for diagnosis and treatment stratification, and by developing treatments that address those mechanisms that are involved in a particular individual's symptoms. For global mental health, on the other hand, a major gap in psychiatry is underdiagnosis and undertreatment; psychiatry will advance by understanding the social determinants of mental disorders, by developing interventions that are feasible and acceptable across the world, by scaling these up for delivery by nonspecialized health workers. An integrative perspective argues that psychiatry has a range of gaps; that advances in psychiatry require both discovery and implementation research, that clinical neuroscience and global mental health can join forces to drive such research forwards, aiming for a personalized public health that addresses more precisely a range of individual and social determinants of mental illness.[30]

30. See D.J. Stein and colleagues [1589]. There is, however, some skepticism about notions of personalized public health; see, for example, D. Taylor-Robinson and F. Kee [1590].

In summary, I would like to go beyond scientism (i.e. it is key to recognize important gaps in psychiatry) and skepticism (i.e. it is key to recognize those advances that psychiatry has made). These sorts of consideration apply to a range of areas in the health sciences: much has been done, and much remains to be done. It is unlikely that a specific paradigm shift will advance the field of psychiatry anytime soon; rather we can expect and should support iterative progress. We do understand much more about the brain–mind and its alterations than we did a century ago, and we can intervene much more effectively for brain–mind disorders. At the same time, our knowledge of relevant psychobiological mechanisms is thin, our application of what we do know is often suboptimal, and we need to take every possible opportunity to further advance the field.[31]

7.3 Neurophilosophy and neuropsychiatry

How does modern neurophilosophy and neuropsychiatry inform our understanding of the psychology and neuroscience of science in general, and the scientific validity of psychiatric science in particular? A few points can be emphasized in response to this question.

First, the brain–mind is embodied somatically and embedded socially even when it is doing science. Importantly, therefore, biological science and psychiatric science are social activities. When it comes to knowledge of other humans (erklären), the process is necessarily one that involves understanding (verstehen). There is no view of scientific truth 'from nowhere', we have to work hard to develop new technologies that extend our appreciation of the world, to develop rigorous experimental designs that control for confounders, and to combine quantitative and qualitative research methods to maximize knowledge and understanding.[32]

Given this point, it is unsurprising that science is a slow process. Our senses allow us to appreciate only a small part of the physical world. Our human biases mean that we perceive only a limited part of the psychological world. Physical science is still grappling with basic debates about the nature of the world, and psychiatry is forever in danger of its schisms splitting the field. Science faces multiple problems including the problem of reproducibility, and psychiatric science must deal with multiple issues including undue influence from and conflicts of interest with the pharmaceutical industry and from governments.[33]

31. M. Maj [1591] and K.S. Kendler and M.B. First [1592].

32. Given the complexity of psychiatric phenomena, **triangulation of data** from different sources and methods may be particularly useful; see H. Ohlsson and K.S. Kendler [1593], K.S. Kendler (who considers a coherence theory of truth in relation to these conditions) [1594], D.A. Lawlor and colleagues [1595], and Owen Flanagan [1596].

33. For more on ongoing debates in physics about the **ultimate nature of reality**, see Jimena Canales [1599], as well as Adam Becker [1597] and the review of his volume by Christopher Fuchs [1598]. For more on schisms in psychiatry, see again Tanya Luhrmann [259]. For more on the **replication crisis**, see M.R. Munafò and colleagues [1600] and Chris Chambers [1601]. For more on conflicts of interest, see M. Maj [1602].

Second, the brain–mind has the capacity to engage successfully with important aspects of the word. Technology allows us to go beyond the limitations of our own senses to appreciate microscopic and cosmological aspects of the physical world, and human interaction may increase our insights into ourselves and others. We have survived as a species and as individuals because we are so successful in navigating both the physical and the social world. As science advances we have greater and greater explanatory power, and a whole range of new interventions has opened up; we are able to sequence whole genomes, and to develop smart phone apps that can assess a range of different conditions. So while science is hard work, the future continues to be exciting.[34]

Can we say that the brain–mind is wired to do science in general and psychiatric science in particular? I would answer 'absolutely'! The psychologist Alison Gopnik has emphasized the extent to which babies can be considered budding scientists, working to understand both the physical world and the social world. Professional scientists have developed ways of extending our observations, for example, introducing new technologies that allow us to see brain anatomy and function. They have also developed ways of ensuring rigour, and minimizing bias, for example, emphasizing the fallibility of our theories, being transparent about our methods, and introducing ways of working that focus specifically on rigour (e.g. careful study design, appropriate statistical analyses, sharing of study data).[35]

Third, the brain–mind can go horribly wrong at explaining the world, and our attempts to improve it can fail miserably. There is general agreement that some areas of study, such as astrology and phrenology, are pseudoscience rather than science. However, informed laypersons as well as expert scientists may disagree about the questions of whether other areas of study, such as psychoanalysis, are valid sciences. Pseudoscience can emerge in any area of science, but medicine seems particularly prone, with all sorts of quackery having thrived in the past, and many sorts of quackery continuing to do so. This makes the field of evidence-based medicine, which focuses on bringing rigour to our evaluations of data, particularly important. We'll discuss the demarcation problem (differentiating science from pseudoscience), and some of the misguided paths that psychology and psychiatry in particular have taken, in the next section.[36]

Work on the cognitive-affective science of science may be helpful in understanding how we go wrong. One reason for our mistakes, perhaps, is that we are good at inventing causal accounts, but less so at rigorously evaluating

34. I was fortunate to be a member of a Lancet Commission that focuses on mapping a way forwards for mental health, see V. Patel and colleagues [1603].

35. See previous footnote on work of authors such as Giere and Gopnik on the cognitive-affective science of science.

36. For more on **quackery**, see Hugo Mercier [1604], Sara Gorman and Jack Gorman [1605], Lydia Kang and Nate Pedersen [1606], Paul Offit [1607], Ben Goldeacre [1608], Simon Singh and Edzard Ernst [1609], R. Barker Bausell [1610], Michael Shermer [1611], and Timothy Caulfield [1612].

them. An extraordinary number of different accounts of mental symptoms make sense to people (e.g. they are a 'build-up of toxins' that requires expurgation; they are a 'defence' against Oedipal drives; they are a 'curse' that requires exorcizing), but far fewer are supported by rigorous evidence. In contemporary neuroscience, the metaphor of 'mechanism' is often used, and we often use the metaphor of 'causation as manipulability' in our work on interventions to determine causality. When done rigorously, this is persuasive. That said, this metaphor seems far from ideal; 'mechanisms' seem better suited to explanations of phenomena in molecular biology than to explanations of some phenomena in psychobiology.[37]

Fourth, the human brain–mind has capacity for self-reflection and is able to make corrections about its own constructs. This is quite a remarkable fact and helps account for the success of science and ongoing scientific progress. Copernicus's idea that the earth does not lie at the centre of the Universe was a remarkable one, that worked much better than the older view. Darwin's idea that species emerged via a process of natural selection was another extraordinary contribution, with enormous explanatory power. Willis's idea that the brain had analogous components across different species and that specific disturbances in the brain accounted for specific neuropsychiatric symptoms was of huge importance for the development of neurology and psychiatry.[38]

The cognitive neuroscience of science has shed some light on the phenomenon of shifts in scientific theory, but these remain poorly understood. Thomas Kuhn made an important contribution to the philosophy of science by pointing out how particular paradigms both encourage research, but also constrain it. From time to time, a new paradigm emerges that provides much more powerful explanations and that works better. Timothy Wilson has pointed out that when it comes to our understanding of the social world, we are able to develop new narratives, which allow us to see things in an entirely different way. Again, from time to time, we seem to develop more powerful ways of understanding

37. We earlier noted that Aristotle was a pluralist regarding causation, emphasizing multiple kinds of causation. More recently, Elizabeth Anscombe emphasized that in ordinary language there are multiple ways of thinking about causation, and Lakoff and Johnson have more recently described multiple **metaphors of causation**; see Lakoff and Johnson [159] and Anscombe [474]. There is ongoing debate about the **metaphor of mechanism** in biology (see Ingo Brigandt and Alan Love [273], Brian Henning and Adam Scarfe [1613], and D.J. Nicolson [1614]), in neuroscience (see Carl Craver [1615] and William Bechtel [1616], in psychology (see R. Rorty [74]), and in medicine (see Olaf Damman [1617]). The **causation as manipulability metaphor** is found in the work of Georg von Wright [1618] and has been expanded in more recent work by Judea Pearl [1619]. For more on causality as manipulability, see again James Woodward [363]. See earlier footnote on 'difference-makers', a looser term than 'mechanisms'. For a potentially useful attempt to characterize natural kinds without resort to the notion of mechanism, see M.H. Slater [1620].

38. See earlier footnote on Copernicus, Darwin, and Freud; Willis may be a more appropriate fit than Freud for a triad of truly great scientists.

ourselves and others that work better. Humans, although not always a likeable species, are not entirely incorrigible.[39]

7.4 Sharpening our thinking about scientific progress

With these concepts in mind, let us try to sharpen our thinking about the nature of scientific progress. As noted previously, the devil is often in the detail. Here we discuss in turn the nature of human nature, progress in psychiatric classification, progress in biological psychiatry, progress in psychotherapy, and the pseudoscience versus science demarcation.

The nature of human nature

Debates about human nature, like other perennial conceptual questions, can be approached from the classical, critical, and the integrative positions. A classical position regards human nature as an essential natural kind, using evidence from studies of humans across time and place, to make hypotheses about necessary and sufficient criteria for operationalizing this natural kind. Physical anthropologists and evolutionary psychologists in particular have emphasized universal features of human behaviour across different cultures and societies. Views have ranged from the idea that humans are essentially bestial to views that humans are essentially a symbol using species.[40]

A critical position has objected that human nature is not an essential natural kind, and that human culture plays a key role in shaping who we are. This view may reject a naturalistic approach, and may emphasize that human nature is fluid rather than fixed. Human nature is not something that resides within us, but rather is something we construct. Indeed, a full understanding of humans requires engaging with humans. Such engagements will differ from time to time, and from place to place, underlining the point that human nature is flexible and malleable, and that historical and geographic context is key for an understanding of particular individual human beings.[41]

---

39. For more on scientific paradigms, see Kuhn [61]. For more on **developing new narratives**, see Timothy Wilson [1361]; note that his techniques also involve acting differently, with consequent changes in perspective. The efficacy of redirection presumably reflects key ways in which human memory works; memory retrieval seems to involve memory alteration (see Frederic Bartlett's pioneering work on schemas and on memory [1621]). See also earlier footnote on narrative therapy.
40. For more on **universal perspectives on human nature**, see work in philosophy (e.g. Mortimer Adler [1622]), evolutionary psychology (e.g. Edward Wilson [1623, 1624] and David Barash [1625]), and anthropology (e.g. Donald Brown [1626]).
41. For more on the **critical perspective on human nature**, see work in philosophy (e.g. D. Hull [1627]), genetics (e.g. Richard Lewontin, Steven Rose, and Leon Kamin [1628]), and history (e.g. Joanna Bourke [1629]). While the critical position is often associated with the left (e.g. see Richard Lewontin [1630] and Paul Gross and Norman Levitt [1631]), it may also be taken by conservatives (e.g. see Roger Scruton [1632]). Hull himself suggested that the species category (including the species of *Homo sapiens*) may be a natural kind that entails variability, but others have emphasized that the concept of a universal human nature is simply not useful (e.g. John Dupré who focuses on human nature as a process rather than as a substance [1633]). For reviews of this issue, see Elizabeth Hannon and Tim Lewens [1634], Maria Kronfeldner [1635], Agustín Fuentes and Aku Visala [1636], Joel Kupperman [1637], John Kekes [1638], David Buller [1639], and Janet Richards [1640].

Mary Midgley is the Oxford moral naturalist who has been most irritated by scientism. In her volume on 'Beast and Man: The Roots of Human Nature', she contrasts those who say there is no such thing as human nature (existentialists and social scientists) with those who say there is such a thing as human nature and it's nasty and brutish (e.g. Thomas Hobbes and some reductionist evolutionary theorists). She argues for a more nuanced perspective; that we do have a nature and that it's more benign than that put forward by Hobbes, and that other animals are much more sophisticated than some have thought and so we are not unlike them. I too would argue for an integrative view of human nature, emphasizing that human nature is a fuzzy natural kind, that human nature is the sort of thing that is partly moulded by genes and partly moulded by our context, and so is continuously evolving. Furthermore, human nature may be an 'essentially contested concept'; our deliberations may not reach immediate resolution, but such deliberations are nevertheless useful.[42]

A notion of using culture to overcome nature is often used; in this metaphor, culture is civilizing. An alternative metaphor is that we are noble savages, and that culture acts to spoil us. However, given that culture is part of human nature, both these accounts seem flawed to me. We make progress not by being more or less 'cultured' (as against natured), but rather by adopting different cultures. In the previous chapter, for example, we mentioned the notion of weakness of will. Weakness of will is neither simply genetic, nor simply a reflection of social determinants. Rather weakness of will emerges from our genetic endowment and is shaped further by human culture; it's nature–nurture intertwined. We can combat weakness of will both by changing our own brain–minds (e.g. using techniques of precommitment that ensure we do not have to face temptation), or by changing aspects of social life (e.g. using 'sin taxes' to help us to restrain our unhealthy choices).[43]

---

42. See Midgley [1641]. Arthur Schopenhauer agreed with Hobbes, noting that "Man is at bottom a savage, horrible beast" [932]. Later on Midgley specifically opposed the views of Richard Dawkins [580]. Somewhat similarly, Peter Singer's response to sociobiology is that "Human nature is not free-flowing, but its course is not eternally fixed. It cannot be made to flow uphill, but its direction can be altered if we make use of its inherent features instead of fighting against them" [1089]. See earlier footnote for debates about sociobiology. **Integrative perspectives on human nature** that take a nuanced approach, eschewing an essentialist and reductionist view of human nature as well as avoiding a view that human nature is merely a social construction, arguably date back to Aristotle and include Spinoza, Hume, and Dewey (see previous footnote on nature–nurture). The idea of human nature as an 'essentially contested concept' is that of Kronfeldner (see previous footnote), who in turn draws on the work of Gallie (see earlier footnote).

43. Within the context of evolutionary theory, the first to put forward the idea that the ethical man could transcend nature was Thomas Huxley, 'Darwin's bulldog', in his 'Evolution and Ethics' [1642]. Huxley was appropriately arguing against social Darwinism, but this dichotomy seems difficult to support. Indeed, in an essay on Huxley's work, Dewey argued that ethics is part of our nature [1643]. Later developmental psychologists, including Piaget and Lev Vygotsky, emphasized the interactions of nature and nurture (and see earlier footnote on nature–nurture). Many authors have emphasized the **co-evolution of genes and cultures**, or the point that human nature has evolved precisely to respond to nurture; see Matt Ridley [586], Roy Baumeister [1644], Luigi Cavalli-Sforza and Marcus Feldman [1645], William Durham [1646], Dale Goldhaber [1647], Tim Lewens [1648], and Jesse Prinz [1649].

It's notable that in discussions of human nature there is an intertwining of explanatory pluralism (there are different perspectives on how best to approach human nature) and value pluralism (there are different views of what human nature should best approach). The comparison of moral and scientific progress was made early on by Aristotle, who ended a discussion of eudaimonia by saying "So much for our outline sketch for the good. For it looks as if we have to draw an outline first, and fill it in later. It would seem to be open to anyone to take things further and to articulate the good parts of the sketch. And time is a good discoverer or ally in such things. That's how sciences have progressed: it is open to anyone to supply what is lacking". For Aristotle, the search for human nature is on ongoing journey.[44]

Progress in psychiatric classification

Earlier on we suggested that recent editions of the DSM have constituted a clear advance in psychiatric classification, but that from the important perspectives of both clinical neuroscience and global mental health, they are deeply flawed. We noted that key problems included the threshold problem, the reification problem, and the causality problem. DSM does not provide a clear-cut distinction between psychopathology and normality, rather it relies on the so-called clinical criterion; the clinician must make a judgement call regarding how much distress and impairment exists. DSM categories provide one particular way of dividing up the world of psychopathology, but other equally valid approaches exist. Finally, DSM categories do not seem to map on to any theory about the aetiology of mental disorders; while we infer underlying dysfunction, we often have little understanding of it.

While these criticisms make a good deal of sense, I worry about throwing the baby out with the bathwater. The difficulty facing nosologists is that there are currently few better ways of defining thresholds, choosing criteria, and reflecting causality: if there were, then psychiatric nosologists would very likely have used them. Revisions of the DSM have been rigorous processes in which the evidence base has been continuously reviewed, searching for better ways of defining thresholds, choosing criteria, and incorporating causality. At the end of the day, however, the evidence base is what it is. While biological psychiatrists have searched for biomarkers for decades, few exist. We can only console ourselves that psychiatry is in the same position as many other areas of medicine

44. This reading of Aristotle's view of human nature is based on Nussbaum [300]. An earlier footnote indicated that for Aristotle human function involves a life of practical judgement; but practical judgement too can be considered an 'essentially contested construct'. The ideas that human nature is continuously being moulded, and that we are involved in an **ongoing search regarding the nature of human nature**, are relevant to current debates about both biomedical treatments and, especially, enhancements: potential interventions should be neither overly idealized, nor summarily dismissed, instead they require rigorous evaluation using practical judgement [53, 1650].

(e.g. the classification of headache used in neurology follows the same template as the classification system used in psychiatry).

That said, there is certainly scope for improving our evidence base, and for better work in the science of nosology. One of the advantages of having two different approaches to psychiatric nosology, the DSM and the ICD frameworks, is that different diagnostic criteria sets can be rigorously investigated to compare their diagnostic validity and clinical utility. This helps address the problem of reification; that once diagnostic criteria for a condition are provided, they take on a life of their own, and other approaches to defining that entity are paid less attention. During debates over DSM-5, there was initially a call for entirely new diagnostic paradigms, but unfortunately, we are simply not at the point where this is possible. Still, iterative progress is an important and potentially attainable goal.[45]

Progress in biological psychiatry

Failure to find biomarkers has led biological psychiatrists to suggest that an entirely new way of working is necessary. As discussed earlier, the Research Domain Criteria (RDoC) framework was put forward by Tom Insel, then Director of the National Institute of Mental Health in the US. Insel is a consummate clinician–scientist: his work on OCD was a major contribution, and his work on the neurobiology of pair-bonding in voles was inspirational. The RDoC framework emphasizes that behaviours fall along different neurobiological dimensions that can be studied in the laboratory and in the clinic. As we noted earlier, reward circuitry and fear circuitry, for example, underlie a range of approach and avoidance behaviours seen in the clinic, and can be studied in detail using animal models. The hope is that such studies will yield new treatment targets, and that instead of clinical studies using very heterogenous patient populations, they will allow a more focused approach that targets specific biological pathways underlying particular symptoms.[46]

Another important development in biological psychiatry has been a focus on increasing the sample sizes of studies. The argument is that small sample studies yield results that are not particularly reliable, with other researchers unable to replicate many initial findings. In neurogenetics, as sample sizes increase, one can begin to tease out the role of the hundreds of genes that contribute to any particular trait. Similarly, brain imaging researchers across the world have increasingly begun to share data, allowing pooled analysis of thousands of scans. While treatment trials of new medications in academic centres have long had large enough sample sizes, the individuals that enter these trials may not be that representative, thus there has also been interest in 'real-life' studies of treatments in the community using 'typical' patients.[47]

45. See again work of S.E. Hyman [308] and K.S Kendler [309], and K.S. Kendler and M.B. First [1592].
46. See B.N. Cuthbert and T.R. Insel [1651].
47. See M. Zwarenstein [1652], P.M. Thompson and colleagues [1653], and P.F. Sullivan and colleagues [1654].

The RDoC model undoubtedly has a number of advantages, and RDoC has already informed thousands of studies. That said, there are a number of concerns that can be raised. First, ideally clinical assessment systems are 'clinician friendly', easy to use and helpful in the clinical setting. The technical term for this is 'clinical utility', and it's far from clear that the RDoC system fits this bill. Second, RDoC may be criticized as being 'old wine in new bottles'; biological psychiatry has long been aware that different symptom dimensions involve different biological pathways. Furthermore, while RDoC may represent the most up-to-date constructs in biological psychiatry, RDoC behavioural dimensions are complex, and to date this sort of approach has not yielded specific and sensitive biomarkers for mental disorders. Third, RDoC may neglect important factors that are relevant to clinical decision-making; these include the social context within which symptoms manifest, and the social determinants of illness. While it clearly important to increase sample sizes, it is notable that several advances in psychiatry have come from careful clinical observations of just a few individuals: once again, we should be careful not to throw the baby out with the bathwater.[48]

Going forwards it seems to me that we need to retain both the DSM/ICD approach and the RDoC approach, with both a focus on careful observation of individual cases as well as a focus on big data approaches. Although DSM/ICD categories are heterogenous, they have proved to be very helpful for studying the epidemiology of mental disorders, for clinical communication, and for undertaking treatment trials. Similarly, while large studies are needed to demonstrate small effects, there is simply no need to do a randomized trial, using a large group of controls, to show that a parachute is effective in saving the life of someone falling from an aeroplane. As noted earlier, it's difficult to make predictions, especially about the future. At this stage of the game, it's not entirely clear what research strategies will yield the most bang for the buck, and to what extent advances in translational neuroscience will fully explain the structures, processes, and mechanisms underlying psychiatric disorders. My approach would therefore be to back a range of different horses in the race to advance the psychobiology of these conditions, bearing in mind that these are not simply brain disorders, but rather brain–mind disorders.[49]

Progress in psychotherapy

Psychotherapy can point with pride to the introduction over the past several decades of many interventions that have a solid evidence base demonstrating their efficacy. We earlier noted how cognitive-behavioural therapy (CBT), the

48. See D.J. Stein [1655].

49. For more on **trials of parachutes**, see R.W. Teh [1656]. The point that it's difficult to make forecasts, particularly about the future, has been falsely attributed to Niels Bohr and Yogi Berra, see earlier footnote. For more on the complexity of biological research, see again K.S. Kendler [1057] and H. Walter [1657].

psychotherapy with the largest and most persuasive evidence base, has strong roots in animal behavioural work done in South Africa in the 1950s. CBT has developed specific treatment methods and manuals for a broad range of conditions, including mood and anxiety disorders, as well as substance use disorders. Furthermore, CBT-based therapies have been adapted for use around the world; such adaptations have been shown to be feasible and acceptable, and many can be delivered by nonspecialized health workers; this has been a major advance for the field.[50]

At the same time the argument for evidence-based psychotherapy is not universally accepted. A number of valid criticisms are worth noting. First, patients who enter clinical trials may be very different from those found in everyday practice; clinical trials, for example, often focus on individuals who suffer from only one condition—such individuals are not representative of the broader patient population, where comorbidity is the norm. Second, evidence-based guidelines are useful for making decisions about first-line psychotherapy in treatment-naïve patients with a single disorder, precisely because clinical trials focus on these sorts of patients. The more complex a particular patient—who may not have responded to multiple treatment trials and who may have a range of comorbidities—the less helpful are the guidelines.

These points help explain the 'clinical-research gap', with researchers providing evidence for one intervention, and clinicians using a different approach in practice. This gap may be particularly large in psychotherapy, where clinicians often maintain that each patient requires a highly individualized approach (making treatment manuals redundant), and where there is a tradition of drawing evidence from single patients rather than from controlled trials (a tradition beginning with Freud, and maintained by many of his followers). A key overlooked aspect of nonevidence based psychotherapy is the fact that any intervention may have adverse events. In the case of pharmacotherapy, each medication undergoes rigorous scrutiny for such adverse events, both during initial research, and after marketing. In the case of psychotherapy, such potential adverse events are simply not an issue for many therapists, despite clear evidence that psychotherapy can, on occasion, cause great harm.[51]

What about the fact that much psychotherapy is aimed not at specific mental disorders, but rather at 'problems in living'? Can we really make progress in providing help for this sort of issue? It would seem to me that an Aristotle providing guidance to a student in ancient Greece about their romantic life, and a contemporary therapist providing guidance to a young patient with similar questions, likely have a great deal in common. There are a number of 'common factors' that successful psychotherapy shares; Aristotle and our contemporary therapist may develop an alliance with the student, they may create an expectation that the problem can be resolved, and they may suggest useful health

50. See D.J. Stein and colleagues [298].
51. See S. Dimidjian and S.D. Hollon [1658].

promoting actions. An important early contributor to the literature on 'common factors' in psychotherapy, Jerome Frank, saw problems of living as sometimes leading to 'demoralization', and argued that a range of different kinds of intervention were able to restore morale.[52]

The effectiveness of both Aristotle and the contemporary therapist may suggest that the Dodo Bird Verdict, that "Everybody has won and all must have prizes", also applies to psychotherapy. However, this is not necessarily the case. First, the assessment of the modern therapist is hopefully more comprehensive and accurate: if the student is struggling in their romantic life because they have symptoms of, say, obsessive–compulsive disorder (OCD), a different approach to intervention is required. Second, we now have evidence that particular techniques are more efficacious for particular disorders (e.g. exposure therapy is more effective for OCD than relaxation-based therapies). That said, in line with the evidence that key 'common factors', found across different times and different places, are useful in resolving problems and restoring morale, I can't help but wonder if Aristotle would not also have recognized OCD and the need for a different approach in such a case, and if he would not also have had a large number of fit-for-purpose health-enhancing techniques at his disposal.[53]

Science vs pseudoscience

While the scientific question of how to speed up progress in any particular scientific field is difficult, so too is the more conceptual question of how to differentiate scientific explanations and approaches from pseudoscience; the so-called demarcation problem. The term 'pseudoscience' was used by the historian James Andrew in order to refer to alchemy, and was subsequent employed by critics of psychoanalysis. The issue of whether psychoanalysis is a science or pseudoscience is central to the 'Freud Wars' discussed earlier, and so an ongoing question of interest for contemporary psychiatry.[54]

Today, there are many schools of psychology, as well as new approaches to self-help, coaching, and management. Sometime in my 40s, I switched career paths to become a Chair of a Department. So I veered a little from reading the literature on self-help books, and immersed myself in books on management and leadership. I was amazed how little these volumes relied on empirical evidence. Business schools do some research, but the quantity and quality is far

---

52. Key contributors to an early literature on **common factors in psychotherapy** include Saul Rosenzweig [1659] and Jerome Frank [1660, 1661]. For a more recent review, see R. Mulder and colleagues [1662]. See also earlier footnote on psychotherapy integration, and next footnote on efficacy of psychotherapies.

53. The **Dodo Bird Verdict** that "Everybody has won and all must have prizes" (see Lewis Carroll's 'Alice's Adventures in Wonderland' [301]), was used by Saul Rosenzweig to suggest that all psychotherapies are equally effective (see reference in previous footnote). For views that this verdict remains substantively true, see R. Budd and I. Hughes [1664] and B.E. Wampold [1663].

54. See James Pettit Andrews [1665].

below that of faculties of health sciences. How do we separate the wheat from the chaff, how do we know which fads to follow and which to discard?

It turns out that posing the demarcation question is a whole lot easier than answering it. Differentiating science from pseudoscience is made difficult by several issues that emerge in areas such as psychoanalytic and leadership theory. First, people who believe in pseudoscience can be really smart, and their theories can be really sophisticated. Second, differences between science and pseudoscience can be subtle; both may involve positing complex theories, and obtaining evidence for these theories. Third, there may well be good evidence for some components of theories, and there may be significant social support for their views.[55]

Notably, both science and pseudoscience are social activities, often entailing complex hierarchies with different levels of authority, and with different types of mentorship. That said, there seems to be a greater emphasis in science on the values of transparency and fallibility; there is open encouragement to achieve better explanations than those provided in the past, and those who succeed in this aim are given prestige. In contrast, in pseudoscience, there seems to be a greater focus on opacity (it is not clear why some people reach positions of authority, and some practices of the field are kept hidden) and infallibility (the founder of the field is often charismatic, and viewed as a genius). At the risk of offending their supporters, fields with the later characteristics include both psychoanalysis and scientology.[56]

The best-known solution to the demarcation problem is Karl Popper's notion of falsification. For Popper, science encourages refutation of its theories, and should a particular experiment demonstrate that a particular theory fails, a process of developing a different theory ensues. He argued that psychoanalysis and Marxism, because they do not allow for the possibility of falsification, are pseudosciences. However, this seems incorrect; when it comes to psychoanalytic

55. For more on **pseudoscience**, see Martin Gardner [1666], Richard Rosen [1041], L. Laudan, who includes a discussion of Aristotle's views of the demarcation problem [1667], Michael Ruse [1668], Michael Shermer [1611], Steven Weinberg [1669], Massimo Pigliucci [1670], Michael Gordin [1671], Philip Kitcher [1672], Massimo Pigliucci and Maarten Boudry [1673], Bryan Farha [1674], Scott Lilienfeld, Steven Lynn, Jeffrey Lohr [1675], Will Storr [1676], Allison and James Kaufman [1677], and Lee McIntyre [1678]. Remarkably, despite their criticism of outdated metaphysics, some members of the Vienna Circle were fascinated by both psychoanalysis and the paranormal [1549]. For a more recent, but equally fascinating exemplar of pseudoscience, see M. Browne on vaccine skepticism [1679].

56. See earlier footnote on transparency and fallibilism. See also earlier footnotes on the Freud Wars, and the efficacy of psychoanalysis, for references regarding these criticisms. Work in the sociology of knowledge and science has long pointed out the importance of rigorous critique (see **Karl Mannheim** [1680]) and of 'organized skepticism' (see **Robert Merton** [1681]). Mannheim was another early German philosopher interested in bridging scientific and hermeneutic approaches (see, e.g., earlier footnotes on Cassirer and Jaspers), and Robert Merton (born Meyer Schkolnick), was a Columbia University thinker, influenced by Durkheim and Parsons, who played a pioneering role in sociology.

theory, for example, the issue is not that the theory cannot be tested, but rather that there is ongoing contestation about the extent to which empirical data have supported theory. Conversely, the fact that astrological theory has been subjected to empirical testing doesn't seem to make it any less of a pseudoscience. It's noteworthy that Popper changed his mind about evolutionary theory, early on arguing this "is not a testable scientific theory", and then later conceding that it is just "difficult to test". It seems that his falsifiability rule of thumb may be more useful in some areas of work (say, where there is a focus on developing more comprehensive and more predictive laws) than others (say, where the focus is on delineating structures, processes, and mechanisms to be able to provide more powerful causal explanations).[57]

Popper's ideas gave impetus to a range of subsequent work; amongst the most influential was that of Thomas Kuhn. Kuhn argued that Popper's work focused on the rare revolutions in science when there was a shift in scientific paradigms, and ignored the fact that the vast majority of 'normal science' addressed small puzzles rather than fundamental theories. For Kuhn astronomy solves puzzles while astrology does not; for Popper this proposal is deeply problematic because it replaces 'a rational criterion of science' with a 'sociological one'. Indeed, one difficulty with Kuhn's account is that it runs the risk of relativism; holding that for those engaged in puzzle-solving activity, any one account of the world is as good as any other. There is ongoing debate about the extent to which Kuhn was a relativist, but much of his writing seems to lean in this direction. Realists are committed to the idea that some scientific theories have more explanatory power, or work better, than others; again, per the 'No Miracles Argument' this seems to provide the only reasonable explanation of, say, progress in biology, where the discovery of DNA-related structures, processes, and mechanisms has led to more robust accounts of the physiology of organisms.[58]

In demarcating science from pseudoscience it may be useful, then, to consider both the explanatory progress that has been made by the field (there should not simply be puzzle-solving, rather such puzzle-solving should provide a progressively better account of the facts), as well as the values of the field (are practitioners transparent about their methods, open about the fallibility of their theories, and enthusiastic about obtaining progressively better accounts). It might also be helpful here to return to Lakoff and Johnson's focus on graded categories. There are some exemplars that everyone agrees are science (e.g. astronomy), and some exemplars which pretty much everyone agrees are

---

57. See **Karl Popper** [1682]. The point that it's difficult to overthrow a well-established theory was early on made by Max Planck, who stated that, "A new scientific truth does not triumph by convincing its opponents and making them see the light, but rather because its opponents eventually die and a new generation grows up that is familiar with it" [1683].

58. As earlier noted, **Thomas Kuhn** is well known for his 'Structure of Scientific Revolutions' [61], which introduced the idea of scientific paradigms, but there is some debate about the extent to which he is a relativist; see, for example, Errol Morris and criticism thereof [62]. Susan Haack has pointed out how Kuhn's own writing waivers on this score [120].

pseudoscience (e.g. astrology). But there are more fuzzy exemplars, and it may be difficult to find necessary and sufficient criteria that resolve the demarcation in such cases. Analogously, within a particular field, there may be agreement about prototypic explanations, and disagreement about more atypical exemplars. Careful consideration of conceptual thinking, empirical data, and scientific values is needed to determine whether a particular discipline is scientific, as well as to determine when to shift paradigms within a particular discipline.[59]

Some fields may have been developed scientifically, but conceptual and empirical work has advanced, and so they must now fall by the wayside (e.g. phrenology). This seems to me the case for psychoanalysis: it might well have been scientific for its time, but it is difficult to support its key theses from a contemporary perspective. Charitably, then, psychoanalysis may not be so much a pseudoscience, as a somewhat dated set of hypotheses; these have been very productive, they continue to have some support, but it is now time to move beyond them.[60]

Neuroscience teaches us that humans are wired to tell stories, wired to create meaning, wired to develop cultures. This may account in part for why it is so difficult for us to distinguish science and pseudoscience. Each tells a story about the world, that seems coherent, and that is believed by its practitioners. If one does find an error in a story that does not necessarily invalidate it, after all science is full of discarded hypotheses. Nevertheless, failure to discard outdated science can be extremely dangerous. My dad was a wonderful man ahead of his time in many ways. Long before vegetarianism became faddish, he grew his own vegetables. Long before the concept of antibiotic stewardship arrived, he was cautious about their use. However, he believed that vaccines were dangerous; this took things too far, and when I got to medical school I made sure that I received mine.[61]

---

59. This point may mean that philosophy of sciences moves from addressing more general questions to more local ones; see D. Ross [1684].

60. For particularly comprehensive work on the history of psychoanalysis that while not uncharitable nevertheless demonstrates how psychoanalysis has so often viewed critique as heresy, see Paul Roazen [1685]. Physics is sometimes referred to as the queen of the sciences, and Frederick Crews has termed psychoanalysis the queen of the pseudosciences, as it is the only one to indicate that criticism of its theory is indicative of psychological 'resistance'. Several authors have put forward multi-criteria solutions to the **demarcation problem** (see, e.g. A.A. Derksen's paper on the seven sins of pseudoscience [1686]). The proposal here is not dissimilar from that of Paul Thagard's early two criteria approach (focusing on scientific progression, and on concerns and behaviours of the community of practitioners, although he later expanded this into a multicriteria proposal [1687]). The idea that science is a graded category is not dissimilar to Dupré's characterization of science as a Wittgensteinian family resemblance concept [174], and other work on the disunity of science (see previous footnote on the Stanford school). It's also relevant to mention Putnam's early critique of falsificationism [1688], and his later view that neither epistemic virtues nor ethical values can be reduced to a single principle, for example, a principle such as 'try to falsify your theories' [225].

61. See earlier footnotes for references on meaning-making and on vaccine skepticism.

The big question of truth and psychiatry

The world of psychiatry and psychology has had its share of disastrous pseudo-scientific fads. One is the belief that all mental symptoms are due to childhood abuse. There is a large literature showing that early childhood abuse is associated with neurobiological sequelae and increased risk for a range of mental disorders. However, there is also a large literature showing that if a psychotherapist keeps asking about childhood trauma, then it is quite possible for false memories to be created, with the potential for the psychotherapy to then be extremely harmful. This seems to be a lesson that we need to repeatedly reinforce: while Freud moved away from his original theory that trauma was at the heart of neurosis, his subsequent equally erroneous view that repressed wishes underlie all psychopathology and need to be uncovered persists.[62]

Earlier we talked briefly about the pros and cons of medicalization. I worry that a second mistake in psychiatry is overtreatment. It is instructive that work on psychosurgery led to a Nobel prize. While this was thought of as a good biological psychiatry at the time, frontal lobotomy turned out to be ineffective, and to have serious adverse events. Today we see ongoing medicalization of psychological trauma: as noted earlier there are important advantages in ensuring recognition and treatment of posttraumatic stress disorder (PTSD). At the same time, there is also the real possibility that by framing the world as one in which trauma is highly prevalent and PTSD commonly occurs, we help facilitate the emergence of PTSD. More broadly, framing all human problems in terms of a brain disorder may well foster unnecessary prescription of pharmacotherapy.[63]

An equally important mistake in psychiatry is, however, underdiagnosis. A range of structural and attitudinal barriers prevent people with real mental disorders from being recognized, and from receiving appropriate intervention, contributing to the 'treatment gap' noted earlier. Thus, as I've emphasized, where trauma has led to PTSD, it makes good sense to diagnose and treat this condition, as robustly as we can. There is now growing acceptance of the construct of neurodiversity, which may be useful in fostering an inclusive approach

---

62. For more on the harm caused by the **psychotherapy of repressed memory**, see Paul McHugh [1689], Mark Pendergrast [1690], and Frederick Crews [1691].

63. A range of authors have accused psychiatry both of disease-mongering (e.g. Ray Moynihan [1692]) and overprescribing pharmacotherapy (e.g. Elliot Valenstein [1693]). See also earlier footnote on neuroreductionism and neuroscientism. There is an interesting debate between Robert Whitaker and Ronald Pies on the notion of '**chemical imbalance**' in psychiatry. Pies points out that no psychopharmacologist would reduce mental illness to a chemical imbalance, but Robert Whitaker argues that this is precisely what organizations such as the American Psychiatric Association have done [1694]. Pies is one of my favourite authors, and I'm hesitant to disagree with him. My own take on this issue is that the notion of a chemical imbalance was put forward by psychopharmacologists as a heuristic framework, that because the metaphor of imbalance is so commonly used it is easily reified, and that this reification has both advantages (patients easily understand and accept it) and disadvantages (it's a vast oversimplification). See previous footnote on the metaphor of imbalance.

to difference. At the same time we need to be careful to help people, rather than focusing all our efforts on the idea that the only source of distress is lack of social acceptance.[64]

To reiterate, in psychiatry we always need a balance to ensure that we avoid both underdiagnosis and overdiagnosis. In science, we need to aim for practices that are closer to transparent and fallible science and further from pseudoscience. The more the field uses authority as gospel, and the more it insults criticism (e.g. by calling it evidence of repression), the more its debates entail conflicts of interest, the more it has cult-like features, then the more it has in common with pseudoscience. Admittedly all the professions have some cult-like features such as lauding our founders, putting neophytes through rigorous training, and rituals that maintain the tradition. So it's a matter, as scientists and professionals, of striving to ensure that scientific values and epistemic virtues, such as transparency and fallibility, are upheld.[65]

Part of the answer to pseudoscientific fads is to be fully aware of progress when it does occur. While the clinical-research gap is understandable, ongoing efforts to minimize it are important. With regards to psychoanalysis, as I've said, there is evidence that the field has aimed to be a science. Certainly its theories were productive and useful, and for a time the best and the brightest were psychoanalysts. While we ought to retain some of its ideas, it is also time to move on, and to advance an integrative psychotherapy that rests on stronger theory and better evidence. Psychodynamic psychiatry may be a useful compromise, insofar as it attempts to integrate lessons gleaned from psychoanalysis with a range of contemporary disciplines.[66]

7.5 Conclusion

Cognitive-affective science has taught us a great deal about more focused questions, such as colour perception, as well as about larger questions, such as the nature of human nature. That our brain–minds, so similar in many ways to those of nonhuman primates, have been able to develop science, including cognitive-affective science, is quite astonishing. That said, highly sophisticated cognitive-affective skills are required to function in the world and in primate society, and extending these to scientific understanding was a logical next step. There is growing understanding of the psychobiology of science itself, and it would be remarkable if such understanding eventually helped contribute to conversations about how to further advance neuroscience.

---

64. For more on **neurodiversity**, see Steve Silberman [1695] and Thomas Armstrong [1696]. For a discussion of some controversies in the area of gender dysphoria, for example, see J. Drescher and colleagues [1697].
65. See earlier footnotes on epistemic virtues and vices, as well as on phronesis in science and in medicine.
66. See previous footnotes on integrative psychotherapy and on psychodynamic psychotherapy.

When we think of a successful science that has dramatically increased our understanding of the world, and our place in it, we think of astronomy, cosmology, and evolutionary biology. We don't immediately think of psychiatry. However, a close examination of brain–mind dysfunction potentially sheds a great deal of light on humans. Furthermore, while psychiatric science may be slow, it has made a number of useful contributions, including the introduction of remarkably effective pharmacotherapies and psychotherapies. Epistemic virtues such as transparency and fallibility are key to keeping such science moving forwards.

Scientism argues that physical sciences are the most fundamental and important source of knowledge, and that all other knowledge can be reduced to them. This view ignores the important role of psychology, the arts, and the humanities as a whole in informing us about who we are. Skepticism regards science as only one story amongst many, surviving only because of the particular power configurations of our time. This view is unable, however, to account for the way in which scientific accounts, of both the physical and the human world, have had increasing explanatory power, working better and better over time.

In this volume, we have emphasized that science is a practical activity that is theory-bound and value-laden. Acknowledging this point is perhaps particularly important in the human sciences, where our view of the nature of human nature may in turn influence our perspectives and our behaviours. On the other hand, we have emphasized that science is in fact progressing in understanding the structures, processes, and mechanisms of both the physical world and the human world. This is an achievement full of awe and wonder, and the ongoing journey of figuring out humans, promises to remain exciting and inspiring.[67]

67. Science has been described as a quest by the physicist Marcelo Gleiser, amongst others; he writes "The quest is what makes us matter: to search for more answers, knowing that the significant ones will often generate surprising new questions" [1698].

Chapter 8

The meaning of life

We've covered a number of big questions, including questions about goodness and about truth. We're well placed, then, to move to the 'really hard question' of the meaning of life. Given that 'meaning' may mean different things, this question is likely to involve a range of different and complex issues, and it seems unlikely that any single authoritative answer is possible. Given how many have contributed to this literature over the ages it also seems overambitious to try say anything entirely new. Still, living as we are in the age of neurophilosophy and neuropsychiatry, perhaps there are some useful things that can be said from that sort of perspective. We'll begin as usual with a discussion of general philosophical perspectives and general psychiatric perspectives, before moving to the neuropsychiatric.[1]

8.1 Philosophy and the meaning of life

Prominent exponents of analytic philosophy, which has dominated the field for many decades, have regarded the question of the meaning of life, as a meaningless one. However, this question may have been asked with greater insistence in recent times, and a growing body of philosophy has emerged in response. As noted the question might mean different things to different people: determining the meaning of life could, for example, refer to making sense of life (is it understandable?), finding purpose in one's life (what is its goal?), or finding aspects of one's life that have significance (is it valuable?).[2]

---

1. Earlier we referred to David Chalmers's notion of the **'hard problem' in philosophy of mind** [1601]. Owen Flanagan refers to the question of meaning as the **'really hard' question** [1699].
2. Ayer, for example, writes that "there is no sense in asking what is the ultimate purpose of our existence, or what is the real meaning of life" [1700]. For more on the growing importance of the question of the meaning of life, see Iddo Landau [1701]. For discussion of the different **meanings of meaning**, including objective versus subjective approaches, see Charles Ogden and Ivor Richards [1702], Terry Eagleton [1703], J. Seachris [1704], and F. Martela and M.F. Steger [1705]. For collections focused on the **meaning of life**, see Michael Hauskeller [1706], Stephen Leach and James Tartaglia [1707], Elmer Klemke and Steven Cahn [1708], David Benatar [1709], Paul Wong [1710], Joshua Seachris [1711], David Friend [1712], and Will Durant [1713]. The topic has also been addressed in continental philosophy [1714]. A range of popular writing also focuses on meaning or purpose rather than happiness; see, for example, Emily Estfahani Smith [1715] and Daniel Pink [1716].

Problems of Living. https://doi.org/10.1016/B978-0-323-90239-7.00011-0

One useful contrast is between approaches that focus on the understandability of life (or determining the *meaning of life*) versus approaches that focus on what makes a life worthwhile (or finding *meaning in life*). English speakers may correctly understand the Spanish phrase 'bona sera', but it's not clear that a similar logic is applicable to determining the meaning of life. Making a determination of whether a particular life is worthwhile seems a much more tractable question. As Albert Einstein put it, "To inquire after the meaning or object of one's own existence or that of all creatures has always seemed to me absurd from an objective point of view. And yet everybody has certain ideals which determine the direction of his endeavours and his judgments".[3]

The idea that human life can be made understandable using some sort of dictionary of cosmic meaning has long been criticized. Early on Ecclesiastes pointed out: "All is meaningless. What does man gain by all the toil at which he toils under the sun? A generation goes, and a generation comes ... There is no remembrance of former things". In the philosophical tradition, the point that from the perspective of the universe our lives have no meaning perhaps begins with Schopenhauer, who writes darkly that "The world is just a 'hell' and in it human beings are the tortured souls on the one hand, and devils on the other". Such views can be termed existential nihilism.[4]

Some who despair of translating the meaning of life have taken a more lighthearted approach. The novelist Douglas Adams, for example, irreverently declares that the meaning of life is '42'. The philosopher Paul Feyerabend claims to be a Dadaist; a Dadaist is "utterly unimpressed by any serious enterprise and he smells a rat whenever people stop smiling and assume that attitude and those facial expressions which indicate that something important is about to be said ... a worthwhile life will arise only when we start taking things lightly". Thomas Nagel revisits Sisyphus and suggests that Sisyphus's defiance and scorn in the face of his fate is overly dramatic; given the cosmic unimportance

---

3. For more on this contrast, see Iddo Landau [1717]. See **Albert Einstein's** 'The World as I See It' [1718]. Note that while some hold that a scientific account of the universe necessarily shows there is no meaning of life, others emphasize that such an account does not impact on humans finding meaning in life. The Nobelist Stephen Weinberg, for example, writes that "The more the universe seems comprehensible, the more it also seems pointless" [1719], while Kurt Baier writes that "acceptance of the scientific world picture provides no reason for saying that life is meaningless, but on the contrary every reason for saying that there are many lives that are meaningful and significant" [1720].

4. Ecclesiastes is authored by **Kohelet**, who advises a golden mean (e.g. there is "A time for silence and a time for speaking; A time for loving and a time for hating") and living with enjoyment (e.g. "There is nothing worthwhile for a man but to eat and drink and afford himself enjoyment with his means"). For more on Kohelet, see Steven Cahn and Christine Vitrano [1721]. See also the view that the term 'hevel' is mistranslated as 'meaningless' or 'vanity', that rather Kohelet is emphasizing the shortness of life, and arguing that our mortality gives meaning to life, and that happiness lies in living: in work, love, and joy [1722]. For this citation, see Schopenhauer's 'Parerga and Paralipomena: Short Philosophical Essays' [1723].

of our situation, approaching the absurdity of life with irony rather than despair or heroism is more seemly.[5]

Simon Critchley argues that like the number 42, all answers to the question of the meaning of life will on closer examination be found to be rather disappointing. He concludes that the question of the meaning of life is an important one, and the fact that humans have been asking it for so long emphasizes that we humans are rightly perplexed by our lives; so the point is to keep asking the question, without necessarily expecting an answer.[6] For Victor Frankl, "Challenging the meaning of life is the truest expression of the state of being human". Put differently, it's one of those 'essentially contested' issues, which is useful to keep discussing.[7]

While this point may be true for humankind in general, from the perspective of any particular individual, determining what makes his or her life worthwhile is a pressing question. It is perhaps no different from many other questions about what he or she values, and so can potentially be answered in a straightforward way, with no need to get mired in the muddy sands of existential nihilism. Indeed, much of the current literature on the philosophy of the meaning of

---

5. For more on **irreverent and ironic approaches** to the meaning of life, see Douglas Adams [1724] and again Paul Feyerbend [162]. For a volume that uses the number '42' in particular to reflect on life, see Mark Vernon's '42: Deep Thought on Life, the Universe, and Everything' [11]. Monty Python has an answer to the meaning of life that is slightly longer than '42': "Well, that's the end of the film. Now, here's the meaning of life. (An envelope is handed to her. She opens it in a businesslike way.) Thank you, Brigitte. (She reads.) … Well, it's nothing very special. Try and be nice to people, avoid eating fat, read a good book every now and then, get some walking in, and try and live together in peace and harmony with people of all creeds and nations" [1725]. For more zestful advice, see Nietzsche's view that it is not so much the meaning of life that we should focus on, but rather the degree of our affirmation of life [1726]. For additional irreverent and ironic approaches, see T. Nagel [1728] and John Gray's 'Feline Philosophy: Cats and the Meaning of Life' [1727].

6. See Simon Critchley [1729, 1730]. The term **'perplexed'** here seems to me spot on and is redolent of the view of ancient Greek philosophers, that 'philosophy begins in wonder'; in being perplexed about the virtues (e.g. Socrates) and about nature (e.g. Aristotle) [1731]. The word has been used by a by a range of authors speaking to the question of how to live life, including Maimonides and Auden. For Jaspers, "Man is reduced to a condition of perplexity by confusing the knowledge that he can prove with the convictions by which he lives" [47]. For Nagel "Certain forms of perplexity – for example, about freedom, knowledge, and the meaning of life – seem to me to embody more insight than any of the supposed solutions to those problems" [1732]. See later footnote for Anton van Niekerk's emphasis on 'mystery'.

7. See earlier footnote for references on 'essentially contested categories', as well as earlier footnote on Keats' notion of 'negative capability'. Many authors write articulately on the **importance of asking if not answering the question of the meaning of life**. Arendt writes "Men, if they were ever to lose the appetite for meaning we call thinking and cease to ask unanswerable questions, would lose not only the ability to produce those thought-things that we call works of art but also the capacity to ask all the answerable questions upon which every civilization is founded" [880]. Mark Vernon writes about the quest for meaning, noting that "Meaning is not found by dwelling in the regions that one believes one understands, and erecting walls around them, material or metaphysical, in order to pretend they are all that is. Paradoxically perhaps, the desire for meaning is satisfied by dwelling on the thresholds of ignorance" [1733].

TABLE 14 Philosophical approaches to the meaning of life.

	Objectivism	Subjectivism	Integrated approach
Philosophical roots	Marx	Sartre	Aristotle[a]
Conceptualizing a meaningful life	Meaningful life is about conditions x, y, and z (e.g. increasing happiness of others)	Meaningful life is a matter of our constructing our meanings	Meaningful life entails engaging with worthwhile projects, that individuals find purposeful
Assessing a meaningful life	Meaning can be operationalized (e.g. happiness achieved)	Meaning varies from individual to individual and is imponderable[b]	Metaphors concerning truth, beauty, goodness can be reasonably debated and progress achieved

[a] *A view of Marx as an objectivist and Sartre as a subjectivist with regards to the question of what make life meaningful is taken from Thaddeus Metz [1737].*
[b] *The term* ***'imponderable'*** *is from Wittgenstein [39], where he argues that while there are expert judgements about feelings, they are based on imponderable evidence, which cannot be formulated by science, but might be depicted by art.*

life focuses on this question of meaning in life, debating different approaches to deciding what makes a life worthwhile. Once again I want to contrast three positions that have been taken, a contrast that reflects perennial debates in philosophy of science and language, as well as in ethics (Table 14).[8]

An objectivist approach argues that meaning inheres in specific qualities of life: some lives are more meaningful than others. Objectivists may however differ on how they conceptualize such specific qualities. Some might turn to science for answers, suggesting that scientific knowledge can provide a basis for evaluating different qualities relevant to the meaningful life. Some might turn to classical positions in ethics; one idea, for example, is that promotion of maximal happiness is what is most meaningful. There are, however, a number of difficulties with this position. Perhaps the most important is that it seems to

8. Another position within **existential nihilism** is to avoid the negativity of authors such as Kohelet and Schopenhauer, as well as the irreverence/irony of authors such as Adams and Feyerabend, and to respond with neutrality, see James Tartaglia [1734]. Tartaglia is scathing about moving the conversation away from the meaning of life and towards a discussion of meaning in life (or what makes a life worthwhile). While existential nihilism seems difficult to counter from a naturalist perspective, see A.B. Trisel for an attempt to do so [1735]. Mary Midgley regards the view that the universe has no meaning as scientistic, arguing, "To find the universe meaningful is not to decode an extra, cryptic message hidden behind it, but simply to find some continuity between its patterns and those of our lives – enough continuity to confirm that our presence here makes sense. *The point is not that this world belongs to us, but that we belong to it*" [1736].

entail a 'view from nowhere'; only an absolutely independent view can resolve differences about what makes some lives more meaningful than others.[9]

A subjectivist approach to the question of what makes for a meaningful life holds that meaning inheres in what a person makes of life: people ultimately decided their own meaning. The subjectivist position may acknowledge aspects of existential nihilism; following Camus, the psychiatrist Irvin Yalom depicts the existential human dilemma as involving "a being who searches for meaning and certainty in a universe that has neither".[10] The subjectivist position may point out that science can tell us nothing about the meaning of our lives; as Wittgenstein said, "We feel that even if all possible scientific questions be answered, the problems of life have still not been touched at all". This position is consistent with a critical position in ethics, which argues that there are no foundations for determining values; it is up to each person to construct their own values. Once again, there are a number of difficulties with the subjectivist position. Perhaps the most important is that in this view there is no way of conceptualizing how a person may misjudge what is most meaningful. Even if there is no 'view from nowhere', there are surely more reasonable and less reasonable ways of judging the meaningfulness of a life.[11]

---

9. For a comprehensive discussion of objectivist and subjectivist positions, see again Thaddeus Metz [1737]. For an **objectivist position** based on neuroscience, see Paul Thagard [1738]. For objectivist approaches consistent with conservative positions, see bioconservative Jordan Peterson [1739] and Leonard Kass [1740]. Walter Veit has returned to the myth of Sisyphus to put forward an objectivist position: he argues that it is an empirical question as to whether revolt in the face of the absurd can be successful, see his paper on 'the only really serious philosophical question' [1741]. The phrase 'a view from nowhere' is from Thomas Nagel [1137].
10. For more on a **subjectivist position** that acknowledges existential nihilism, see Irvin Yalom [1742]. Earlier on, William James, who seems to have ignored the ancients when writing that "The greatest discovery of my generation is that human beings can alter their lives by altering their attitudes of mind" [1743] advised, "Believe that life is worth living, and your belief will help create the fact" [1744]. More recently, Richard Taylor has returned to the myth of Sisyphus to put forward a subjectivist position: he argues that if the gods implanted in Sisyphus a 'compulsive impulse' to roll stones, then Sisyphus's life would be filled with meaning [1745]. For views that moves from thought experiments to focus on real people's accounts of the meaning and purpose of their lives, see Paul Froese [1746] and Brian Little, Katariina Salmela-Aro, and Susan Phillips [1747]. For a self-help book based on existential philosophy, see Gordon Marino [1748].
11. While a subjectivist position will not look to science for meaning, it may regard science positively (e.g. Sean Carroll [178] and Michael Ruse [1749]) or not so much. It is worth citing in full **Wittgenstein's conclusion** of his 'Tractatus Logico-Philosophicus' [38]: "For an answer which cannot be expressed the question too cannot be expressed. The riddle does not exist. If a question can be put at all, then it can also be answered. Skepticism is not irrefutable, but palpably senseless, if it would doubt where a question cannot be asked. For doubt can only exist where there is a question; a question only where there is an answer, and this only where something can be said. We feel that even if all possible scientific questions be answered, the problems of life have still not been touched at all. Of course there is then no question left, and just this is the answer. The solution of the problem of life is seen in the vanishing of this problem. (Is not this the reason why men to whom after long doubting the sense of life became clear, could not then say wherein this sense consisted?) There is indeed the inexpressible. This shows itself; it is the mystical ... Whereof one cannot speak, thereof one must be silent". In discussing this passage, John Messerly notes that it is unclear whether Wittgenstein is saying that the question of the meaning of life is meaningless, or whether the answer is ineffable. However, "if the question is senseless, then we waste our time trying to answer it; and if the answer is ineffable, then we waste our time trying to verbalize it" [1750, 1751]. For more on Wittgenstein and the meaning of life, see Reza Hosseini [1752] and Isabel Cabrera [1753].

From an integrative perspective, the 'meaning of life' connotes a number of different issues, including the purpose of the universe and our own purpose within this universe, and both subjective and objective aspects of meaning in life need to be addressed. Susan Wolf, for example, persuasively argues that in a meaningful life, "subjective attraction meets objective attractiveness", that is, meaning derives from active engagement with projects of worth. Wolf emphasizes the insignificance of individual lives (using this to argue that meaning emerges from going beyond a focus on one's own fulfilment), and brings together intuitions that a meaningful life involves both 'finding one's passion' and being part of something 'larger than oneself'.[12] Such an integrative perspective brings together some of the strengths of the objectivist position (while there is no 'view from nowhere' about human meaning, there are human yardsticks), and the subjectivist position (there are different valid ways of adjudicating human meaning, and individual perspectives on this are key).[13]

Wolf indicates that her view is Aristotelian in spirit, and indeed there are overlaps between this sort of view of meaning in life, an emphasis on purpose in human flourishing, and a virtue ethics concerned with the nature of human nature. Indeed, from an integrative perspective, it may be useful to emphasize that when we talk about a meaningful life we are, as always when addressing complex constructs, continuously employing metaphors, and relying on family resemblances. Thus Wolf's concept of meaning resonates with the work of the

---

12. See Susan Wolf [1754]. Roy Baumeister has pointed out a number of differences between seeking meaning and happiness, for him a life is meaningful to the extent that it finds responses to questions of purpose, value, efficacy, and self-worth [1108]. Thaddeus Metz, too, notes a number of similarities and dissimilarities between meaning and happiness, and he sees 'purpose' as one of a number of constructs that bear family resemblance to meaning in life [1737]. Indeed, there seems to be some continuity between **notions of 'deep happiness', 'flourishing', and 'meaning'**. See, for example, Daniel Russell's eudaimonist volume, 'Happiness for Humans' [1755]. Bertrand Russell puts it elegantly, "To abandon the struggle for private happiness, to expel all eagerness of temporary desire, to burn with passion for eternal things – this is emancipation, and this is the free man's worship" [1756]. Peter Singer similarly defines objective value in terms of something 'larger than the self' [1351]. Daniel Dennett puts it pithily, "The secret of happiness: Find something more important than you are and dedicate your life to it" [1757].

13. See earlier footnote on the issue of 'moral saints', raised by Susan Wolf and others. A number of authors have **integrated aspects of the objectivist and subjectivist perspectives**, including Karl Britton who argues that a life is meaningful to the extent that the facts of the world are such that a person's life matters to themselves and others [1758]; Owen Flanagan who triangulates different perspectives including those of neuroscience, philosophy, and subjective experience [1699]; Wim de Muijnck who emphasizes that meaning entails a 'good fit' or fitting-in with an agent's embodied-embedded mode of being [1759]; and Todd May who argues that a meaningful live involves narrative values that are defined objectively but experienced subjectively [1760]. Such views have been described as hybrid by F. Svensson [1761]. Svensson's own account is a desire-based one, but he emphasizes that desires often concern aspects of the objective world; this view then also arguably integrates aspects of the objective and subjective perspectives. Some Confucian approaches to the meaning of life may also integrate aspects of the objectivist and subjectivist perspectives [1762].

Oxford moral naturalists. Philippa Foot, for example, speaks of 'deep happiness', which has as its objects "things that are basic in human life, such as home, and family, and work, and friendship". And Mary Midgley writes, "Our unity as individuals is not something given. It is a continuing, lifelong project, an effort constantly undertaken in the face of endless disintegrating forces". Terry Eagleton puts his Aristotelian conclusions about the meaning of life pithily: "The meaning of life is not a solution to a problem, but a matter of living in a certain way. It is not metaphysical, but ethical".[14]

Just as an integrative approach entails explanatory pluralism in science and value pluralism in ethics, so an integrative approach to the meaning of life emphasizes that there are multiple meanings in life. Given how different humans and their contexts are, there will be many worthwhile projects, and many meaningful lives. At the same time, by our own human standards, it is reasonable to say that some projects are more worthwhile than others, and that not 'anything goes'; this point is particularly important for each of us as individuals, facing the question of how most meaningfully to live our own life. Julian Baggini put things succinctly, writing "the only sense we can make of the idea that life has meaning is that there are some reasons to live rather than to die, and those reasons are to be found in the living of life itself".[15]

8.2 Psychiatry and meaning in life

Over the course of its history, psychiatry has taken various positions regarding human nature and the goals of life. As noted earlier, in his volume on 'Civilization and its Discontents', Freud paints a fairly bleak view of humankind, noting the extent to which our lives are full of suffering. In many ways this vision is a deeply tragic one; we are driven by unconscious drives, and even reason is not trustworthy. Nevertheless, Freud did hold that humans could gain self-insight through psychoanalysis, so that "where id was, there shall ego be", with a change from "neurotic misery to everyday suffering". For Freud meaning in life was, however, found in love and work.[16]

Cognitive-behavioural therapy (CBT) was developed in order to address anxiety and mood disorders; compared to psychoanalytic writing, the manuals of CBT are focused on practical rather than theoretical issues. That said, as noted earlier, CBT has much in common with the philosophical position of Stoicism; Stoicism and CBT emphasize the goal of obtaining control over

14. See Philippa Foot [1224], Mary Midgley [1736], and Terry Eagleton [1703].
15. See Julian Baggini [1763]. For Tolstoy, "A scholar is someone who acquires great knowledge from the books he has read. An educated person is someone who has mastered the most widespread current knowledge and scientific methods. An enlightened person who is someone who understands the meaning of his life" (see his 'A Calendar of Wisdom' [1764]).
16. See Freud [1033]. See also earlier footnote on how for Freud, "Love and work are the cornerstones of our humanness".

one's own thoughts and emotions. More recently, a 'third wave' within CBT has emphasized the importance of mindfulness. It's possible that for Stoicism and CBT the continuous striving for self-improvement gives life meaning, but it seems to me that they focus more on achieving tranquillity (ataraxia) than on finding meaning.[17]

Positive psychology and positive psychiatry have focused on human resilience rather than on human vulnerability to mental disorders. And for these schools of thought, a key component of resilience is having purpose in life. Positive psychology has drawn on empirical research on finding meaning in life, and has developed a range of practical and evidence-based ways of helping people to take a step back, to reflect on where they wish to go with their lives, and then to plan carefully on reaching their goals. As noted earlier, however, approaches that overemphasize positive thinking run the risk of being Panglossian. Furthermore, the evidence base for some recommendations of positive psychology is thin. Finally, a focus on subjective well-being may overlook key aspects of what is objectively meaningful.[18]

For Victor Frankl, while each person must find their own unique meaning, and while this may change over time, meaning is commonly found in work, in love, and in our response to suffering. A range of subsequent authors have further contributed to the idea that finding meaning in life is a key psychological task for humans. Abraham Maslow, for example, emphasized that once the basic needs were taken care of, humans could address higher-level issues such as self-actualization and 'peak experiences'. Similarly the psychoanalyst Eric Erikson discussed different challenges at different life stages, and argued that concerns about generativity become increasingly pressing with age.[19]

In the area of psychotherapy, a range of approaches, including Frankl's logotherapy, existential psychotherapy, and narrative therapy, emphasize the centrality of finding purpose and meaning. In addition, a 'second wave'

---

17. Stanley Messer and Meir Winokur, drawing on the work of Roy Shafer, have suggested that psychoanalysis has a **tragic and ironic world view** (emphasizing persistent conflict), while behaviour therapy has a **comic world view** (problems can be articulated and resolved) [1765]. That said, their depiction of behaviour therapy was not found accurate by all [1766]. See previous footnote for references on mindfulness.

18. For more on **measures of meaning in life** and related research, see Paul Wong [1710] and M. Brandstätter and colleagues [1767].

19. See again Victor Frankl's 'Man's Search for Meaning' [397], Abraham Maslow [1769], and Erik and Joan Erikson [1768]. Notably, Frankl's approach to finding meaning is not merely cerebral; he writes, "We needed to stop asking about the meaning of life, and instead to think of ourselves as those who were being questioned by life – daily and hourly. Our answer must consist, not in talk and meditation, but in right action and in right conduct. Life ultimately means taking the responsibility to find the right answer to its problems and to fulfil the tasks which it constantly sets for each individual". Frankl's work has also given impetus to empirical work on the association between measures of meaning in life and psychotherapy outcomes [1770]. For a more recent **philosophical perspective on the life cycle**, including an update of Erik Erikson's ideas, see Michael Slote [1771].

of positive psychology has attempted to incorporate a greater focus on the dark side of life, and on finding meaning. Michael Mahoney has used the term 'constructivism' in his work on how psychotherapy entails a process of finding meaning. Thus, for example, he argues that psychotherapy should encourage individuals to view themselves as active participants in their own lives, that humans are constantly finding patterns and creating meaning, and that a range of psychotherapy techniques help people to make new meanings as they develop.[20]

The University of Pretoria philosopher, Thaddeus Metz, has pointed out that there are views that psychotherapy should help clients to live objectively good lives, as well as views that psychotherapy should help clients to achieve subjective well-being. He argues that psychotherapy should instead aim to facilitate a meaningful life. This view seems to me partly consistent with the argument put forward earlier that psychotherapy involves objective and subjective elements, that it is both theory-bound and value-laden, and that it aims to ensure the flourishing of patients. Metz would point out, though, that at times a meaningful life may entail sacrifice, perhaps resulting in less pleasure and poorer health, and that such sacrifice may nevertheless be appropriate for a therapist to support.[21]

8.3 Neurophilosophy and neuropsychiatry

How does modern cognitive-affective neuroscience inform the meaning of life? A number of findings seem to me to be key.

First, we are primates. There is no getting away from our evolutionary past; primates are very focused on social relationships, including collaboration and care-giving. As humans, it is not surprising therefore that we find meaning in our close relationships, in our friendships, and in providing care to our children. In this context, meaning refers to our sense of value, to our subjective sense of what is most important to us in life. Positive psychology advises that we spend more time with friends and family, and the evidence suggests a close link between having good quality relationships and overall health.[22]

Second, as we've emphasized, humans have particular skills in meaning-making; our wetware enables theories about the patterns we see, and we love to tell stories that make meaning of things. In this context, meaning refers to

---

20. See earlier footnote for references to work on existential psychotherapy. For more on **'second wave' positive psychology**, see Itai Ivtzan, Tim Lomas, Kate Hefferon, and Piers Worth [718], B.S. Held [1772], and P.T.P. Wong [1773]. Note also the focus of the 'third wave' of CBT therapies, such as acceptance and commitment therapy (ACT) on finding meaning [1774]. See M.J. Mahoney and D.K. Granvold [1775].

21. See T. Metz on **meaning in life as the aim of psychotherapy** [1776, 1777]. A **meaningful life and a flourishing life** are only partly coterminous, if we accept that flourishing involves both pleasure and purpose.

22. See earlier footnote on loneliness and solitude.

making sense of our lives, even in the face of our experiencing life as perplexing. All cultures have 'origin myths' which provide an account of human beginnings, so helping to make sense of where we are going. And we use metaphoric thinking to structure our life; for example, 'hero myths' allow us to conceptualize the difficulties life brings, and encourage us to face these with resilience. We seem wired to make meaning of our experiences.[23]

Third, as admitted earlier, we sometimes go horribly wrong in our meaning-making. We might develop a view, for example, that the Gods demand that we sacrifice our children, and we go ahead with this evil activity in order to appease them. We might find meaning in exterminating members of minority groups or in submitting to the irrational demands of a dictator. Once we have a particular world view, we use this to explain our experiences, and to plan the future; while this is useful in some ways, it also ensures multiple biases, which in turn lead to all sorts of errors.[24]

Fourth, humans have the capacity to learn and to grow. We can be persuaded that our world views, or our ways of meaning-making, are flawed. We have the ability to develop fresh and creative ways of approaching the world, of understanding ourselves and others, and of finding meaning in life. Our neuroplasticity allows our cognitive-affective schemas to change (so that we see things differently, and perhaps more accurately), and it allows us to develop new habits that encourage more appropriate interactions with the natural and human world.[25]

8.4 Sharpening our view of the meaning of life

When trying to answer the question of the meaning of life, once again the devil is in the detail. We'll consider a few key issues that are relevant to the question of meaning. We'll consider in turn, the question of free will and willpower; what it means to 'find oneself'; the issue of individual differences in meaning-making; the concepts of truth, beauty, and goodness; and the nature of spirituality and generativity (Table 15).

---

23. See earlier footnote on homo symbolicus, and see again Camus for humans as driven by meaning-making in 'The Myth of Sisyphus' [975]. For Iris Murdoch, "Man is a creature who makes pictures of himself and then comes to resemble the picture" [1778]. Rebecca Goldstein refers to the importance of not only 'meaning' but also 'mattering' [57, 1779]. For more on hero myths, see Joseph Campbell [1780]. For more recent **perspectives on meaning-making and story-telling** that incorporate neuroscience and evolutionary theory, see Alex Rosenberg [405], Jesse Bering [1781], Christopher Booker [1782], Brian Boyd [1783], Gregg Caruso and Owen Flanagan [1784], Jonathan Gottschall [1785], Jonathan Gottschall and David Sloan Wilson [1786], Michael Shermer [1787], and Will Storr [1788].

24. See citations in previous footnote (e.g. Alex Rosenberg's work critiquing narratives), as well as previous footnote on cognitive-affective biases. Roy Baumeister has suggested that our meaning-making has evolutionary value, but see previous footnote on what he terms the '**myth of higher meaning**', the false idea that everything in the world ultimately makes sense, having either justification or a reason [1108].

25. See earlier footnotes on habits, on schema therapy, and on changing our narratives.

TABLE 15 Sharpening our view of the meaning of life.

	Objective position	Subjective position	Integrated position
Free will	Life is causally determined	We have moral responsibility	Compatibilism is consistent with human nature, spectrum of freedom of will
Weakness of will	Occurs when there is a failure in reason	Occurs when there is a moral lapse	Weakness of will is also a feature of human nature, spectrum of weakness of will
Finding oneself	Very important to find one's true self, which is often hidden	Everyone must create their own authentic self	Selves are complex: Finding/creating oneself are partly useful metaphors
Individual differences	The formula for finding the true self differs across individuals	Everyone creates their own authentic self	There are universal mechanisms, but finding meaning is something we each do
Truth, beauty, and goodness	Have clear objective definitions	Are entirely relative constructs	Are found through human activity, in a more or less reasonable way
Spirituality and death	Find the formula for maximal spirituality	There is no formula, live in the moment	Work towards wisdom, become generative, by engaging with the world

Free will and willpower

In thinking about the meaning of life, and in particular the question of how we should respond to the human condition, an immediate question is the age-old one of whether we really have any choice in doing so. In the time of neurophilosophy and neuropsychiatry, we are aware how important a role heredity and early childhood experience have in influencing our behaviour. We now have detailed explanations of how distal evolutionary and proximal neurobiological mechanisms underpin our thoughts and feelings, and how large a part nonconscious processes play in producing these thoughts and feelings. Furthermore, we understand more about the neuroscience of impulsivity (when willpower fails), 'wanting' (which is central in addictions), and compulsivity (e.g. in obsessive–compulsive disorder). Not surprisingly, cognitive-affective scientists have therefore suggested that our sense of free will entails some self-deception.[26]

This sort of view has, however, been opposed on a number of grounds. Even if we accept that behaviour is fully determined, such a determination is so complex that for practical purposes it's not necessarily possible or relevant to try dissect it out fully. As Wittgenstein put it, "The freedom of the will consists in the fact that future actions cannot be known now". Furthermore, as the philosopher Wilfred Sellars notes, "The past is not something with respect to which we are passive"; we are active and agentic beings. In her volume, "Are You an Illusion", Mary Midgley draws on Aristotle's view of different causes, to emphasize that our sense of free will is entirely reasonable, and that our explanations of neuronal activity and of human activity are entirely different. For Lakoff and Johnson, the notion of 'free will' is based on a flawed view of mental faculties as unembodied.[27]

Nevertheless, given that our behaviour not only has a causal basis but is also highly constrained, for life to be meaningful, it does seem that we need to retain some faith in free will, or at least to use this notion as a convenient metaphor for our sense of volition. If everything is preordained, not on basis of God's plan, but on the basis of our brain–mind and environment, our sense of our ability to really respond to life, to genuinely create our own meaning, seems diminished. As Isaac Bashevis Singer cleverly summed the matter up, "We must believe in free will, we have no choice".

From a psychological perspective, there is evidence that holding to the metaphor of free will has pro-social advantages; when we think of ourselves in a wholly deterministic way, we seem more inclined to act selfishly.[28] From a philosophical perspective, a compatibilist position is consistent with an integrative

---

26. See. Daniel Wegner on 'The Illusion of Conscious Will' [1791], J.A. Bargh [1789], Sam Harris [1790], and Heidi Ravven [1213].

27. For these views, see Wittgenstein's 'Tractatus Logico-Philosophicus' [1792], Sellars [1793] (and see earlier footnote on space of reasons), Midgley's 'Are you an Illusion?' [1794], and Lakoff and Johnson [159]. Sellars's remark is redolent of Kierkegaard's earlier "Life can only be understood backwards; but it must be lived forwards". Even earlier on, Locke said that to ask "whether a man's will be free as to ask whether his sleep be swift or his virtue square" [685]. Wittgenstein's views on free will similarly emphasize **'category mistakes'** (see earlier footnote), and are consistent with his anti-scientism (see earlier footnote); he writes: "Life is like a path along a mountain ridge; to left and right are slippery slopes down which you slide without being able to stop yourself, in one direction or the other. I keep seeing people slip like this and I say 'How could a man help himself in such a situation!' And *that* is what 'denying free will' comes to. That is the attitude expressed in this 'belief'. But it is not a scientific belief and has nothing to do with scientific convictions" [1548]. These views influenced key contemporaneous colleagues; see Gilbert Ryle [380], Elizabeth Anscombe [474], and G. von Wright [1795]. The importance of free will is key for existentialist thinkers such as Karl Jaspers [344]. More recent writers continue to emphasize the **reality of free will**; see Hannah Arendt [880], Robert Nozick [1421], Alfred Mele [1796], John Searle [1797], Daniel Dennett [1798], and P. Unger [1799].

28. This particular psychological argument, focused on the **pro-social benefits of free will,** has been made by Roy Baumeister. Free will, he argues counter-intuitively, is helpful in encouraging people to follow the rules [1800]. As Sheldon Kopp tells us, "You are free to do whatever you like. You need only face the consequences", see his "If You Meet The Buddha On The Road, Kill Him!" [1801].

position that acknowledges different kinds of explanation and understanding; on the one hand, we see ourselves and others as causally determined, on the other hand, we see ourselves and others as having causal control over our choices and hence moral responsibility. Such a view avoids a Panglossian position of limitless possibility, or an overly resigned position that everything is preordained, and instead emphasizes that moral responsibility is deeply interwoven into our human lives and interpersonal relationships.[29]

Certainly, no matter how determined human behaviour is, we do experience ourselves and others as having free will. Conversely, weakness of will, or what the ancient Greeks termed 'akrasia', is also very much part of human nature, and so something that we have a great deal of experience with, in both ourselves and others.[30] Our subjective experience suggests a spectrum between some behaviours which seem more voluntary, and others which seem less so; this seems consistent with neuroscientific accounts of mechanisms underlying impulsive, compulsive, and addictive behaviours, and suggests that the more causal influence we have over our behaviours the freer we are. While the notions of 'strength of will' and 'weakness of will' may be metaphoric they are therefore likely grounded in our brain–minds; if we are fortunate we have experiences of using free will to overcome weakness of will, and in helping others to do the same. We also respond to others as if they have free will and akrasia; thus, for example, for the most part we don't exculpate criminal behaviour on the basis of determinism.[31]

---

29. This particular philosophical argument, that our practices of determining moral responsibility are interwoven with our human lives and our interpersonal relationships, has been made by Peter Strawson [1802]. The **literature on free will** includes some of our favourite philosophers. Aristotle, for example, wrote, "Therefore these things [i.e. nature, habit, reason] must harmonize with each other. For people do many things contrary to their habits and their nature" [1208]. Spinoza's emphasis on the importance of accepting that we have no free will is notable [1213], but some have said he is a compatibilist who emphasizes the importance of imagination in making practical judgements [675]. Hume put forward an influential compatibilist position that attempts to reconcile determinism and liberty, although there is ongoing debate about his views [1803], and those of Dewey [1804]. Just as Hume's general approach is consistent with *soft naturalism* (see earlier footnote), so his view of free will is consistent with what has been termed *soft determinism*.

30. Bromhall usefully contrasts three views: that **akrasia** involves a failure of reasons, that akrasia involves a moral failure, and a pragmatic view that akrasia is part of human nature (Table 13 and K. Bromhall [1805]). While Aristotle has been read as saying that akrasia involves a failure of reasons, his view is more complex [538]. That decision-making necessarily involves intertemporal choice, is consistent with a view of akrasia as part of human nature (see earlier discussion of reward, and see George Ainslie [753]). For a direct comparison of Aristotle and Ainslie, see R. Uszkai [1806].

31. For more on the **philosophy and cognitive-affective science of free will and akrasia**, see Neil Levy [1023], Nick Heather and Gabriel Segal [1024], G. Ainslie [1807], John Baer, James Kaufman, and Roy Baumeister [1808], Roy Baumeister [1809], Andy Clark, Julian Kiverstein, and Tillmann Vierkantagain [1810], Jon Elster [1811], Jonathan Glover [1812], Ran Hassin, Kevin Oschner, and Yaacov Trope [1813], Richard Holton [1814], Robert Kane [1815], Alfred Mele [1816], Kevin Timpe, Meghan Griffith, and Neil Levy [1817], and Nick Trakakis and Daniel Cohen [1818]. Useful constructs in this literature include 'want/should conflicts' (e.g. K.L. Milkman and colleagues [1819]) and 'self-defeating behaviour as trade-offs' (e.g. R.F. Baumeister and S. Scher [1820]). See earlier footnote on culpability in psychopathy, which notes that judgements of culpability engage with a spectrum of responsibility.

There is some controversy about the neuroscience of willpower. One idea is that our brain–minds have a finite resource of willpower. When we resist urges, we gradually deplete this resource. We can build it up again by consuming sugar or getting rest. A contrasting view is that willpower is more like an emotion; it waxes and wanes, and tells us something important about our circumstances. In this view, we should be more accepting of our failures in will; if we are constantly giving into one or other temptation, this may be a sign that it's time to adopt a different life strategy. As ever, Aristotle, Hume, and Dewey may be useful here; they teach us the importance of developing healthy habits: with these in place, healthy behaviours can be performed more automatically, with less of a need to resist unhealthy ones. This is a lifelong journey, though; developing and maintaining the right habits, and strengthening our practical judgement.[32]

Finding oneself

In developing a response to the human condition in general, and to our own lives in particular, it would seem that one should begin with some self-understanding. As already noted, Socrates advised us to know ourselves, and after Freud this idea became a central one in psychotherapy. In our current therapeutic culture, the idea of finding one's 'true self' has become a key notion. The search may involve removing outer layers (or in psychodynamic parlance, defences), perhaps developed in response to traumatic experiences, in order to get to the inner and true core, sometimes conceptualized as the vulnerable self, the pure self, or the natural self.[33]

The folk psychology concept of an essential self resonates with a Platonic concept of the soul, with Descartes notion of a 'Pure Ego', and perhaps with Kant's description of a primordial and unified ego. In later times, the essential

32. Roy Baumeister puts forward the first view of **willpower** [1821], while Michael Inzlicht [1822] and Carol Dweck [1823] put forward the later one. This difference in view will lead to different self-help approaches; a positive psychology perspective might focus on building up willpower, while a negative psychology perspective may emphasize the value of having both, say, 'have to' and 'want to' goals. The failure of experiments on willpower to replicate further complicates the debate [1824]. Alan Watts, in his 'Wisdom of Insecurity: A Message for an Age of Anxiety' [1825] is also sceptical, noting, "I can only think seriously of trying to live up to an ideal, to improve myself, if I am split in two pieces. There must be a good 'I' who is going to improve the bad 'me'. 'I', who has the best intentions, will go to work on wayward 'me', and the tussle between the two will very much stress the difference between them. Consequently 'I' will feel more separate than ever, and so merely increase the lonely and cut-off feelings which make 'me' behave so badly". Along somewhat similar lines, George Ainslie sees willpower in terms of a bargaining stance during intertemporal choice, and notes that rule-based choice may generate rigidity and undermine appetite; pointing then to "the heart of a central human paradox: that the better the will is at getting rewards, the less reward it will finally obtain" [753]. See also Ainslie G: Willpower with and without effort, Behavioral and Brain Sciences (in press) on effortless habits, and earlier footnote on habits [1826].
33. See, for example, Alice Miller's 'The Drama of the Gifted Child: The Search for the True Self' [1827], and a response thereto (Martin Miller's 'The True 'Drama of the Gifted Child': The Phantom Alice Miller – The Real Person' [1828]).

self came to be seen as psychological; the physician–philosopher John Locke played a key role in arguing that personal identity inhered in a person's continuity of consciousness. This view would support the therapeutic quest for the true self. Even more recently, neurobiological underpinnings of the self have been emphasized. Classical views of the self tend to emphasize its unitary, conscious, and stable characteristics.[34]

From a critical perspective, a person's identity entails a particular narrative, we are continuously creating the story of our own lives. Hume pointed out that our notion of identity is a fiction; we are merely 'bundles of perceptions'. This view is prescient; contemporary neuroscience has emphasized that the self is not located in any specific brain region, and that our sense of self is to some extent an illusion. Derek Parfit, who regards himself as a 'Lockean', has made particularly compelling philosophical contributions to these debates; he argues that selves are less like tables and chairs, and more like football clubs; over time the self changes in multiple ways. Crucially, therefore, we are further from our future selves, and closer to others' current selves, than we typically imagine. Critical views of the self tend to emphasize the self as fractured, unconscious, and malleable.[35]

What might an integrative approach, drawing on these resources in philosophy and cognitive-affective science, conclude? A wetware metaphor of the brain–mind as embodied and embedded leads to a view of the self as similarly emerging from the body's sensorimotor system, as having a subjective and affective quality, and as being enacted in its engagement with the world. Self-awareness likely begins as a primitive sense of self based on proprioceptive experience, and over time schemas of the self become more elaborate. This view goes beyond an essentialist view of the self, but at the same time does not see the self as mere construction. In addition, this view takes a psychobiological approach; the self cannot be reduced to a neurobiological entity, but it also involves more than a

34. For more on debates in the **philosophy of folk psychology** and common-sense philosophy, see again Daniel Hutto [478], Scott Christensen and Dale Turner [1829], and Rik Peels and René van Woudenberg [1830]. The importance of psychological continuity in personal identity has been emphasized by a range of philosophers, including David Lewis, Robert Nozick, and Derek Parfit. Many have also written about narrative and personal identity; see, for example, Anthony Gidden's 'Modernity and Self-identity: Self and Society in the Late Modern Age' [1831].

35. For more on the **self in philosophy**, see John Perry [1832], Derek Parfit [670], Jonathan Glover [1833], José Bermudez, Naomi Eilan, and Anthony Marcel [1834], Owen Flanagan [1835], Jennifer Radden [1836], Rom Harré [1837], and Richard Sorabji [1838] (which although wide ranging is useful for outlining Aristotle's views); Bernard Williams [1839], Alva Noë [1840], and Julian Baggini [1841] (which argues that "The word I is a verb dressed as a noun"); Peter Goldie [1842], Patricia Churchland [1843], Shaun Gallagher [1844], Mary Midgley [1794], Georg Northoff [1845], and Markus Gabriel [1846]. For more on the **self in the cognitive-affective sciences**, see Joseph LeDoux [587], Roy Baumeister [1847], Francis Crick [1848], J. Panksepp [1849], Evan Thompson [1850], Bruce Hood [1851], Antonio Damasio [1852], Thomas Metzinger [1853], Todd Keenan and Julian Keenan [1854], Tilo Kircher and Anthony David [1855], V.S. Ramachandran and Sandra Blakeslee [1856], and D.J. Stein [1857]. Individuals vary in the extent to which they feel close to their future selves: a point that is related to earlier discussion of individual variation in temporal discounting, and that has received empirical study (see H. Ersner-Hershfield and colleagues [1858]).

Lockean emphasis on self as merely a psychological construct, rather it is multidimensional and ever-changing.[36]

From this integrative perspective, there does seem to be an aspect of self-insight that is important and useful. In our engagement with the world, we learn not only about external reality, but we also develop complex thoughts and feelings about ourselves. The value of paying attention to one's thoughts and feelings about the world and the self was emphasized by the Stoics, and it turns out that paying such attention is a key technique in contemporary cognitive-behavioural and mindfulness therapy. The French renaissance essayist Michel de Montaigne pointed out that writing is one way of figuring out what one thinks and feels, and more recent evidence from the world of clinical research suggests that writing may well have therapeutic value in certain contexts.[37]

On the other hand, 'finding oneself' seems a rather inapt and inept metaphor for the journey of life: it seem erroneous in key ways (we never got lost in the first place, and selves are not hidden), and it runs the risk of encouraging excessive introspection. Such navel-gazing may be problematic in key ways. First, it may lead to unnecessary amplification of one's own bodily signals; this phenomenon may contribute to a 'health paradox', in which objective measures indicate that people's health is improving, but people's subjective experience of their health is worsening. Second, given key constraints on our self-knowledge, such as those noted in the discussion of self-deception, conclusions that we formulate about our own selves are quite possibly either trivial or wrong. As Iris Murdoch put it, "'Self-knowledge', in the sense of a minute understanding of one's own machinery, seems to me, except at a fairly simple level, usually a delusion … Self is as hard to see justly as other things, and when clear vision has been achieved, self is a correspondingly smaller and less interesting object".[38]

---

36. Several authors have emphasized how the **self is embodied**; for a review, see Michele Maiese's 'Embodied Selves and Divided Minds'. Maeise refers to the debate between these emphasizing an essential self and those critiquing this, using the terms 'realist' and 'irrealist' [1859].
37. See **de Montaigne's** 'Essays' [1861], and **Pennebaker** and Roger Smyth's 'Opening Up by Writing It Down' [1860]. For more on de Montaigne, including his possible Marrano background, see Sarah Bakewell [1862] and Ann Hartle [1863]. He was likely one of the first to refer to the 'human condition', writing "Each man carries the entire form of the human condition". See also earlier footnote on O'Connor and Sontag on writing.
38. LeDoux suggests that **self-awareness** comes about in two different ways: we can access those aspects of our self that are stored in explicit memory, or we can monitor the bodily and behavioural manifestations of implicit processes [587]. Kahneman distinguishes between the experiencing self and the remembering self, emphasizing how different they are; oddly, "I am my remembering self, and the experiencing self, who does my living, is like a stranger to me" (see earlier footnote on hedonia, and his 'Thinking Fast and Slow' [435]). For more on the **health paradox**, see A.J. Barsky [1864]. Murdoch's view is from her 'Sovereignty of Good' [1226]. Note also George Ainslie's view of the self as a population of interests conducting limited warfare during intertemporal choice [1865]. See also previous footnote on self-deception. For more on the **philosophy of self-knowledge**, see Alex Byrne [1866], Quassim Cassam [1867, 1873], Annalisa Ursulu Renz [1868], Annalisa Coliva [1869], Bert Gerler [1870], Peter Carruthers [1871], and Stephen Heatherington [1872].

Once again, its seems useful to draw on Aristotle's approach of finding the golden mean. Aristotle acknowledges the importance of understanding the self (ignorance is not necessarily bliss), but at the same time he focuses primarily on practical engagement with the world. A bit of meditation may be useful, but if we want a better understanding of ourselves, it may be more useful to look at the consequences of our engagement with the world; indeed, thinking more about others may be an important way of learning more about ourselves. We can complete a personality questionnaire about how impulsive we are, but if we've had many jobs and multiple partners this is likely a better guide for self-evaluation. And certainly if we want to change the world, we need to replace self-focused contemplation with outwardly focused action.[39]

Individual differences and meaning

One of the flaws of self-help books is that they are generally addressed to a large audience, rather than to a specific person. In contrast, in today's world of health care, there is more and more of a focus on personalized medicine. The rationale is that each of us has an entirely different genome and has been exposed to an entirely different environment (the envirome or exposome). Thus each one of us will have a unique pattern of causal mechanisms and treatment targets, and a specifically tailored intervention will therefore be more useful than a nonspecific one. This idea has paid some dividends in oncology, but to date has had unclear success in psychiatry.[40]

The neuropsychiatric literature has, however, certainly reinforced the idea that there are important individual differences between us. Personality research, for example, has suggested that we differ on a number of key dimensions, including introversion versus extroversion. Psychobiological research has delineated how different neural circuits and neurotransmitter systems underlie this divergence. From an evolutionary viewpoint, such diversity makes good sense: different cognitive-affective styles may be more useful in different contexts. As mentioned earlier, there's also a growing literature that emphasizes our neurodiversity.[41]

What are the implications of this work for answering the big questions? Work on human nature has emphasized the universality of many aspects of being human; we are alike in many important ways. This surely contributes to explaining the ongoing relevance of the writings of the Axial Age to contemporary life. At the same time, work on human nature has emphasized our diversity and

---

39. The phrase 'ignorance is bliss' is from the poetry of Thomas Gray. For this reading of Aristotle, see again Edith Hall [516]. For a fascinating history of one **psychometric assessment**, see Merve Emre [1874]. See again Brian Little's work on fixed vs flexible traits [760]. See also Nietzche, who advised, What have you truly loved thus far? What has ever uplifted your soul, what has dominated and delighted it at the same time? Assemble these revered objects in a row before you and perhaps they will reveal a law by their nature and their order: the fundamental law of your very self. For your true self does not lie buried deep within you, but rather rises immeasurably high above you". Herder's famous mentee, Johann Wolfgang van Goethe, more humorously said, "Know thyself? If I knew myself, I'd run away" (i.e. run away from the temple at Delphi, where the inscription reminds us to enter with humility).

40. See previous footnote about related aspirations for, but concerns about the possibility of, **personalized public health**.

41. See previous footnote on neurodiversity.

divergence; not all answers are suitable for all people, and we need to be keenly aware of the relevance of explanatory and value pluralism. Each of us has to undergo the process of finding our own answers to the big questions of life; albeit drawing where possible on the assistance and resources of wise mentors, who can potentially help us to do so in as reasonable and judicious way as possible.

Truth, beauty, goodness

Philosophers have put a great deal of emphasis on truth, beauty, and goodness as components of a meaningful life. Various proposals have been put forward. Nozick, for example, writes that the search for truth, beauty, and goodness involves transcending one's narrow limits, and making outside connections. Metz links the search for meaning to truth, beauty, and goodness by arguing that a meaningful life is one which engages with the key conditions of human existence, and that contributions to truth, beauty, and goodness do just this. Others have emphasized the 'imponderability' of such notions, and have argued that our definitions of such concepts do them no justice.[42]

Modern neuroscience may provide some informative perspectives on these issues. We've discussed the neuroscientific bases of scientific thinking, of judgements of beauty, and of moral decision-making, all of which potentially contribute to advancing our understanding of these constructs. We've argued that in scientific work and in moral reasoning we draw on metaphoric thinking: this has a deeply subjective aspect, but at the same time different metaphors can be judiciously weighed up, and objective progress in truth and morality may therefore be possible. Perhaps something similar can be said about aesthetic appreciation, although this seems to me an area in which pluralism may be particularly important.

Aesthetics is far beyond the scope of this volume. I'd note, though, that a range of philosophers, including Aristotle, possibly Spinoza, Hume, and Dewey, have suggested that beauty has both objective and subjective elements. Thus aesthetic judgements, like scientific and ethical judgements, necessarily entail social practices, but at the same time are not merely relativistic. It's notable that pragmatist aesthetics may emphasize the continuity of artistic and everyday activity, embrace both the natural and the social aspects of aesthetic experience, and encourage a wide range of aesthetic experiences that inspire new worlds and new ideals. Thus pragmatism's aesthetic meliorism parallels its ethical meliorism.[43]

---

42. See Robert Nozick [1421] and Thaddeus Metz [1737]. The notion of 'imponderability' is from Wittgenstein's 'Philosophical Investigations' [39], see earlier footnote. See also earlier footnote on Bertrand Russell, Peter Singer, and Daniel Dennett's views on being part of something 'larger than oneself'.

43. For more on the **aesthetics of Dewey, Hume, Spinoza, and Aristotle**, see, respectively, T. Leddy [1875], E. Galgut [1876], L.C. Rice [1877], and D. Dutton [1878]. Paralleling work in pragmatic realism and moral naturalism is work in aesthetic pragmatism, aesthetic realism, and aesthetic naturalism; see respectively R. Shusterman [1879] (who notes parallels between aesthetic meliorism and ethical meliorism); D. Fenner [1880] and I. Morais [1881]. The aesthetic philosophy of Nelson Goodman may have been influenced by Dewey; both authors emphasize the overlap between art, science, and other human activities, and both authors emphasize the synthesis of reason and emotion in art [1882]. See also earlier footnotes on Aristotle on tragedy, and on Dickens on tragedy, and on the two cultures of the sciences and humanities.

Human flourishing involves engagement with the world, and when such engagement produces truth, beauty, and goodness it seems particularly worthwhile, and so meaningful. The principle of pluralism might suggest that people find purpose in activities that fall outside of concerns with these three large areas. That said, more expansive concepts of the true, the beautiful, and the good may cover most activities which provide people with purpose, for example, spending time with family and friends. We can accept that some lives are more meaningful than others, and that we can do more to make our own lives more meaningful. At the same time, care should to be taken to avoid overly constrained definitions of what comprises a meaningful life, given that a broad range of diverse but meaningful lives are possible.

The topic of human progress has come up several times in this volume. We've taken the position that humankind is making progress as measured by its own yardsticks, including both scientific progress and moral progress. At the same time, we've emphasized that overly sanguine optimism is inappropriate, rather we merely hope that we can contribute to making this world somewhat better. It seems to me that part of living a meaningful life means putting aside the knowledge that both we and the planet will one day die, and seeing oneself as contributing in a small but meaningful way to humanity. Some might argue that this is human self-delusion at its height, but an alternative perspective is that this the epitome of human courage. Like Sisyphus we know that the task is an impossible one, but we put a smile on it.[44]

Spirituality and generativity

The fact of death throws stark relief on the question of the meaning of life. In the past, supernatural explanations of a life after death have provided many with comfort. But this comfort turns cold in the age of neurophilosophy and neuropsychiatry, when the soul gives way to the brain–mind, and when spirituality and faith are accounted for by particular brain–mind states. Furthermore, the 'new atheists', who lean heavily on science in general, and evolutionary theory in particular, have taken a stance that is particularly critical of traditional religions, arguing that these are flawed in important ways.[45]

From a critical perspective, the foregrounding of scientific explanations in discussions of human spirituality inevitably leads to reductionism (ignoring the point that the subject matters of science and of spirituality are entirely different) and scientism (ignoring the point that the methods of the physical sciences are not the only way of approaching the big questions). The Oxford

44. See earlier footnote on 'The Myth of Sisyphus'.
45. See earlier footnote on philosophy as learning how to die. For more on the 'new atheism', see Christopher Hitchens, Richard Dawkins, Sam Harris, and Daniel Dennett [1883] and Nick Spencer [1884]. There is also a growing literature on the **neuroscience and evolutionary theory of religious and spiritual experience**; see, for example, Jesse Bering [1781], Robert Bellah [1885], Pascal Boyer [1886], Daniel Dennett [1887], Jay Feierman [1888], Robert Hinde [1889], Loyal Rue [1890], E. Fuller Torrey [1891], George Vallaint [1892], and Robert Wright [1893].

moral naturalists, influenced in part by Wittgenstein, have been scathing critics of scientism. Mary Midgley, in particular, has argued trenchantly that the new atheism is itself a form of faith. Even if scientism is avoided, too much trust in science is open to criticism; the philosopher John Gray, for example, argues that "the Gnostic faith that knowledge can give humans a freedom no other creature can possess has become the predominant religion".[46]

Can cognitive-affective science and psychiatry be at all helpful in finding a middle path forwards? A number of authors in these naturalistic traditions have focused on issues that seem closely related to spirituality and faith, including work on generativity and wisdom. Erikson's work on the task of generativity may be particularly useful; as we get older, so mentoring others becomes increasingly important to us, and we have the capacity to pass our knowledge on and to focus on providing care for others. Counter-intuitively, the empirical literature indicates that once we get to our 50s, our quality of life improves rather than decreases, perhaps reflecting the satisfaction of being able to guide others. From an evolutionary perspective, it's notable that humans are the only primates to live past menopause; perhaps genes contributing to generativity are selected because this stage is so useful to humans.[47]

There is a growing literature on the psychobiology of wisdom, which might help underpin some of what happens as we age. It seems that over time we often become more skilled in our practical judgements, and it is easier for us to

---

46. See earlier footnote on reductionism and scientism. See Mary Midgley [1894, 1895] and Terry Eagleton [1896]. For this citation, see John Gray's 'The Soul of the Marionette' [1897]. In a more conciliatory tone, Bertrand Russell earlier wrote, "Even the cautious and patient investigation of truth by science, which seems the very antithesis of the mystic's swift certainty, may be fostered and nourished by that very spirit of reverence in which mysticism lives and moves" [1536]. And in the famous words of Einstein, "But science can only be created by those who are thoroughly imbued with the aspiration toward truth and understanding. This source of feeling, however, springs from the sphere of religion. To this there also belongs the faith in the possibility that the regulations valid for the world of existence are rational, that is, comprehensible to reason. I cannot conceive of a genuine scientist without that profound faith. The situation may be expressed by an image: science without religion is lame, religion without science is blind" [1898]. Jonathan Sacks contrasts science and philosophy as left-brain activities, with religion and the arts as right-brain activities, arguing pithily that "Science takes things apart to see how they work. Religion puts things together to see what they mean", and "Science is about explanation. Religion is about meaning. ... Science tells us what is. Religion tells us what ought to be. Science describes. Religion beckons" [1471]. Distinctions between left-brain and right-brain processes are often erroneous (see Indre Viskontas [1899] and Christian Jarrett [1900]), but the idea that **religion should be conceptualized as one of the humanities** may be useful in moving away from impossible attempts to unify reason and revelation, and rather towards develop partnerships between them, facilitating rigorous reasoning-imagining when it comes to addressing aspects of the big questions of life. There is a vast literature on the relationship between science and religion, which goes beyond our purview here; for a realist perspective, see Susan Haack [120].

47. For more on quality of life and ageing, as well as its underlying psychobiology, see Jonathan Rauch [1901] and D.V. Jeste and A.J. Oswald [1902]. For more on evolutionary theories of menopause, see M. Takahashi and colleagues [1903] and R.A. Johnstone and M.A. Cant [1904]. See earlier footnote on work updating Erikson.

achieve the golden mean of expressing ourselves in the right way at the right time. We become more able to accept the tasks and absurdities of life, including its death and dying components. Generativity and mentorship certainly don't provide us immortality, but they exemplify the crucial significance of an 'ethics of care' as well as an 'ethics of responsibility', and they perhaps provide us a small glimpse of making a contribution that is not entirely transient.[48]

If we are fortunate then memories etched into our minds may include small acts of random kindness shown by strangers, larger sustained mentoring and support given by key individuals, and the joys of enduring bonds with family and friends. The African concept of ubuntu, where 'muntu ngumuntu ngabantu', that is we are who we are through others, encapsulates how life is made worthwhile by our lovers, our families, and our communities. As Jaspers wrote, "The individual cannot become human by himself". It's easy enough to blame the evils of the world on religion; it may be harder, but ultimately more worthwhile, to build moral and spiritual communities that sustain us, and help give us hope.[49]

Lakoff and Johnson speak passionately about how their work on embodied cognition is helpful for thinking about enhancing spiritualty. They argue, for example, that an embodied spirituality facilitates a focus on self-nurturance, on the nurturance of others, and on the nurturance of the world itself; that it requires "an understanding that nature is not inanimate and less than human, but animated and more than human". Further, they point out that the metaphors we use in framing spirituality have enormous potential to expand our understanding of the world and to enhance our appreciation of life; thus "An ineffable God requires metaphor not only to be imagined but to be approached, exhorted, evaded, confronted, struggled with, and loved".[50]

Awe is perhaps overlooked as one of the primary emotions; it encapsulates our sense of being connected to and drawing inspiration from nature. As Einstein argued, "The fairest thing we can experience is the mysterious. It is the fundamental emotion which stands at the cradle of true art and true science. He who knows it not and can no longer wonder, no longer feel amazement, is as good as

48. For more on the cognitive-affective **neuroscience and philosophy of wisdom**, see Robert Sternberg and Judith Glück [1905], Agustín Fuentes and Celia Deane-Drummond [1906], Barry Schwartz and Kenneth Sharpe [1907], T.W. Meeks and D.V. Jeste [1908], and Leslie Thiele [1909]. As regards the hope that work will bring some sort of immortality, Woody Allen understandably said, "I don't want to achieve immortality through my work … I want to achieve it through not dying". See earlier footnote on 'ethics of responsibility'; this term has been used in both the secular literature [1237] and the spiritual literature [927].

49. Jaspers writes, "The thesis of my philosophising is: The individual cannot become human by himself. Self-being is only real in communication with another self-being. Alone, I sink into gloomy isolation-only in community with others can I be revealed! in the act of mutual discovery" [1910]. See previous footnote on hope. Notions such as **'naturalized spirituality'** [1911] or **'secular faith'** [1912] may be useful in developing the sort of middle path sketched here.

50. See Lakoff and Johnson [159]. For more on **pragmatism and spirituality**, see again Philip Kitcher [176]; for more on **pragmatism and religion**, see A.A. van Niekerk [1913]. Pascal Boyer argues that metaphors of God that are more personal are more likely to be adopted [1886].

dead, a snuffed-out candle .... A knowledge of the existence of something we cannot penetrate, of the manifestations of the profoundest reason and the most radiant beauty, which are only accessible to our reason in their most elementary forms – it is this knowledge and this emotion that constitute the truly religious attitude". And while awe is of course an emotion that belongs to all people and all places, I would like to stake the claim that the wildlife reserves of Africa offer truly unique opportunities for communing with the God of Spinoza.[51]

8.5 Conclusion

Neuroscience emphasizes that humans are masters at finding patterns in life (and so find meaning in this sense), and that they experience certain things (e.g. having close relationships) as particularly significant (and in that sense meaningful). Unfortunately, however, neuroscience itself provides no better answer to the ultimate meaning of life than Douglas Adams's '42'. Indeed, we need to avoid the sort of neuromythology that reduces meaning to brain–mind activity; the search for meaning is an ongoing quest for humankind, and it involves perplexity and bemusement, as well as grandeur and wonder.[52]

Psychiatry perhaps does better than neuroscience, insofar as it often directly addresses the question of a meaningful life with patients. It has also contributed a range of techniques for assisting people to make their lives more meaningful, and to find purpose. Such techniques include taking a step back, reflecting on where we wish to go with our lives, and then planning carefully on reaching their goals. At the same time, we need to be wary of self-focused navel-gazing; meaning and purpose are found in engaging with the world, and we learn most about ourselves when we do this.

From a philosophical perspective, we've addressed a number of different issues that come up when considering the meaning of life. Our general approach

---

51. See Einstein's 'The World as I See It' [1718]. For more recent **writing on mystery, awe, and science**, see David Cooper [55], Richard Dawkins [1914], D.R. Hofstadter [1915], Frank Wilczek [1916], and physicist–novelist Alan Lightman [1917]. For **more on awe**, see K. Schneider [1918], Shaun Gallagher, Lauren Reinerman, Bruce Janz, Patricia Bockelman, and Jörg Trempler [1919], Paul Pearsall [1920], and D.J. Keltner and J. Haidt [1921]. See also earlier footnotes on Maimonides' 'negative theology' and Spinoza, and on the point that 'philosophy begins in wonder'. To avoid the Panglossian, it is perhaps also relevant to refer here to Schopenhauer's contrary view that "the astonishment which leads us to philosophize clearly springs from the sight of the suffering and the wickedness in the world, which, even if they were in the most just proportion to each other, and also were far outweighed by good, are yet something which absolutely and in general ought not to be" [910].

52. See previous footnote for more on 'perplexity'. The term 'grandeur' was used by both **Alexander von Humboldt** and Darwin to describe the awe in nature that science helped unveil [1419], and see previous footnote on Darwin. For more on von Humboldt as an author that combined elements of the Enlightenment and of romantism, see M. Dettelbach [1922]. The question of **why something, rather than nothing**, exists, always fills me with bemusement; for books that address this question, but with little risk of disenchantment in my view, see Robert Nozick [1421], John Leslie and Robert Kuhn [1923], Lawrence Krauss [1924], Jim Holt [1925], and Peter Atkins [1926].

to matters of truth, beauty, and goodness has been to emphasize that there is no 'view from nowhere' that helps us to decide these matters definitely, but that at the same time it is not true that 'anything goes'; we can engage reasonably with the relevant principles and particulars, and facts and values, and make progress. Many of us find purpose in working towards this sort of progress; sorely aware of our limitations and baffled by the apparent absurdity of life, we nonetheless find the courage to keep going, and we flourish in doing so.

Chapter 9

Conclusion: Metaphors of life

In this volume we have covered a fair bit of ground, aiming to address the big questions and hard problems in the era of neurophilosophy and neuropsychiatry. No definitive answers have been given, which may disappoint some, but which is perhaps reassuring given the fallibilist position that anyone confident they have conclusive solutions is on the wrong track. Similarly, one of our conclusions has been the value of each of us going through our own process of asking the big questions and considering the hard problems, imaginatively developing our metaphors for living, and becoming skilled in our practical judgements. If we are fortunate, we develop ways of thinking and feeling that move us towards truth, beauty, and goodness, and a flourishing life.

In this final chapter, we'll consider a number of different metaphors of life, and their potential value for addressing the big questions and hard problems. In a way, then, this chapter is an attempt to distil and pour some wisdom about life. However, it is also is somewhat whimsical, perhaps reflecting my mood at the time of writing it. I'd therefore want to emphasize that there are multiple meanings in life, that many metaphors may be helpful, and that this chapter will only be useful to some. There is a growing literature on the use of metaphor in psychotherapy, and this can potentially be consulted for a range of other ideas that may assist in thinking through the problems of living.[1]

Eagleton in his book on the 'Meaning of Life' discusses a group of jazz musicians playing music together, as epitomizing a meaningful life. There are several positive aspects of this metaphor, including the value of community, and the joy that life can bring. If the metaphor is extended to include all the preparation that goes into a jazz session: the years of intensive learning, the practicing of this particular group, the relationship-building that takes place, then this metaphor also gets at the developmental aspects of life. In impromptu jazz, there is also the point that we don't know where we are going, that we build off each other, and that we continuously create something novel. Finally, this metaphor

1. A range of **schools of psychotherapy have focused on metaphor**; see, for example, Jill Stoddard and Niloofar Afari [1927], R. Stott and colleagues [1928], Richard Kopp [1929], and Ellen Siegelman [1930].

Problems of Living. https://doi.org/10.1016/B978-0-323-90239-7.00012-2

of life resonates with others from the arts, including Shakespeare's "All the world's a stage, And all the men and women merely players".[2]

A commonly used metaphor of life is of God as our father and of humans as his children. This metaphor certainly brings great comfort to many. Notably, one version of the 'golden rule' principle emphasizes its grounding in a notion that we are all members of the family of God. In another variation of this metaphor, as John Updike puts it, "Ancient religion and modern science agree: we are here to give praise". He writes that "Without us, the physicists who have espouse the anthropic principle tell us, the universe would be unwitnessed, and in a real sense not there at all. It exists, incredibly for us ... What we beyond doubt do have is our instinctive intellectual curiosity about the universe from the quasars down to the quarks, our wonder at existence itself, and an occasional surge of sheer blind gratitude for being here".[3]

I particularly like versions of this metaphor of God as our father in which rebellion by the kids is acceptable. In Judaism, for example, there is a venerable tradition of arguing with God, accusing him of being unmerciful and unjust, and asking with him to change his ways. This seems to me to allow the most reasonable answer to the problem encapsulated in the Book of Job, the question of why bad things happen to good people: God needs to change! God might well retort, though, with Mahatma Ghandi's exhortation to "be the change that you wish to see in the world".[4]

Another potential metaphor is of life as a game, perhaps of chess or of soccer. What I like about this metaphor is its incorporation of the competitive aspect of life. For team sports, it also covers the social aspects of humanity: we are members of teams, of communities, of tribes. Again, if the metaphor includes all the preceding preparation that is required for a team to play with skill and creativity, then it gets at the developmental aspects of life. While this metaphor seems to work reasonably well, games tend to end in either a win, draw, or loss; life isn't

---

2. Eagleton emphasizes the Aristotelian point that individual and societal flourishing go hand in hand [1703], and see earlier footnote on this point. For another metaphor of life that draws on the arts, see **Pedro Tabensky's** comparison of painting and living; just as painting is a skills-based practice with the goal of a good artwork—a well composed unity, so living is a skills-based practice with the goal of a good life—which forms a coherent whole [937]. Doing philosophy is also a skills-based practice with the goal of the good, and spending a lot of time on footnoting reflects an attempt to engage with the community of philosophers, amongst other things.

3. See earlier footnote on the **golden rule and reciprocity**. For a more light-hearted perspective, see Aye Jaye's "The Golden Rule of Schmoozing" [1931]. This citation of Updike is in David Friend's "The Meaning of Life" [1712]. Robert Solomon takes a not dissimilar stance, framing spirituality in terms of gratitude [1911].

4. For more on **God as protagonist**, see Erich Auerbach [1932], Jack Miles [1933], and Eli Wiesel [1934]. The literature on the Book of Job is vast; see Susan Neiman [1935], Harold Kushner [1936], Terry Eagleton [1158], and Chad Meister and Paul Moser [1160]. Camus' words about his courageous protagonist, Dr Rieux, in 'The Plague' [1937] come to mind, "Rieux believed himself to be on the right road – in fighting against creation as he found it". For more on distinctively Jewish approaches to the meaning of life, see T. Metz. He considers whether the metaphor of the relationship between God and man as one of partners or lovers (rather than as a father–child or master–slave dyad) is a distinctively Jewish one [1938].

quite like that. Similarly a view of life as a business has some strengths, but given that business is typically profit driven, it suggests that the aim of life is to get to our deathbeds with greater productivity and more assets than others, using methods such as 'marketing oneself'. This doesn't seem a particularly edifying approach, and indeed it's been accused of encouraging alienation and anomie.[5]

A metaphor of life that has particular resonance with a number of philosophers and psychiatrists is of life as a story or narrative. Within the critical tradition there is the view that history is not so much an account of scientific or moral progress, but rather a series of narratives with changes in humans' understanding of themselves and of what matters. Within psychiatry there is the view that we are able to develop new narratives about ourselves and our world, and this then leads to entirely new perspective on our lives. Cognitive-affective science supports this view insofar as it emphasizes that humans are experts in meaning-making, and that thoughts and feelings are structured by metaphors. At the same time, life is more than merely a narrative; it involves engagement with the world, and this has real consequences. If we do go with this metaphor, it's as well to remember that our narratives can be terribly misleading, and that the notion that our lives tell a meaningful story may be overly optimistic.[6]

Indeed, it's worth thinking about what metaphors of life don't work well. One particularly well-known metaphor sees humans as engaged in an ongoing master–slave struggle. This Hegelian-Marxist narrative seems overly reductionistic (in that there is surely more to life than this), as well as overly optimistic (in that it promises an eventual utopia). Indeed metaphors and schemas that focus exclusively on having been traumatized or victimized are typically unhelpful. While life is certainly traumatic for many, such a metaphor may not be helpful in encouraging trust in others, or the possibility of posttraumatic growth. For Nobelist Joseph Brodsky, "The moment that you place blame somewhere, you undermine your resolve to change anything". This is not, of course, to deny the importance of acknowledging pain and injustice, and of the need for metaphors that help us to do so.[7]

Life is partly about mentoring and guiding others. Alison Gopnik, who has emphasized how philosophy has sadly neglected babies and related matters, argues that carpentering is a particularly inappropriate metaphor for parenting; carpenters shape inanimate objects in specific ways, hoping to achieve a

---

5. See Desmond Morris [879] and Gary Kasparov [1939]. Admittedly, there are businesses that focus on sustainability rather than profit, but they seem in the minority. Durkheim early on discussed the link between **modernity and anomie** (see earlier footnote); for more recent work on this issue, see Liah Greenfeld [1940] and Peter Berger and Thomas Luckmann [1941].

6. For more on the **meaning of life and narrative**, see Garrett Thomson [1942]. In addition, see earlier footnotes on Alex Rosenberg's work, on cognitive-affective biases, on the neuroscience and evolutionary theory of meaning-making, and on Baumeister's 'myth of higher meaning'. However, see also the footnote on changing the narrative of our lives.

7. For critiques of this kind of metaphor, see Didier Fassin and Richard Rechtman [1943], Theodore Dalrymple [1944], Jonathan Haidt and Greg Lukianoff [1945], Bradley Campbell and Jason Manning [1946], Frank Furedi on the **culture of fear** [1947], and Joseph Brodsky [1948]. See also earlier footnote on the therapy culture.

particular product that fits their schemes. She encourages us to instead think about parenting as closer to gardening; gardeners work in entirely different ways with different plants, hoping that each flourishes in their own way. Gardening involves lots of sweat and labour, digging and muckraking, and with one's plans invariably going awry. The self-help literature on parenting is filled with specific techniques for optimizing parenting; the gardening metaphor, which emphasizes instead the importance of providing nurturance, of loving relationships, and of building character, seems a better way to go.[8]

Another metaphor that seems to me flawed is one that regards life in terms of progressive improvement, perhaps using the idea that we are constantly evolving to a higher sort of being. A focus on progressive improvement may be complemented by other metaphors of evolution, depicting evolution in terms of a strict parent metaphor (encouraging survival of the fittest), a nurturant parent metaphor (what survives is that which is best nurtured), or a communitarian metaphor (what survives is that which sustains the interrelatedness of the planet). However, some evidence seems to indicate that it's a bit of an accident that *Homo sapiens* is around; it's not clear what we'll evolve into, and it's very likely that some 'lower form of life' will out-survive us on planet earth. A metaphor of progressive improvement seems redolent of the Panglossian positive psychology that we've criticized throughout the volume.[9]

That said, thinking of a particular individual's life in terms of the metaphor of growth, development, and progression is useful in a number of ways. It emphasizes our early dependence on others, the challenges and opportunities that come with getting older, including the generativity and wisdom that comes with maturity. It reminds us to make the most of each stage of our life, and it underscores how much we have in common with animal and plant life. The model of life as a cycle, with each particular individual moving through different stages, reflects our scientific knowledge of human growth and development, as well as our intimate experience of moving from childhood through to old age. As Shakespeare's lines go on to say, "And one man in his time plays many parts, His acts being seven ages". Notably, there are similarities between this metaphor, and the often-used metaphor of life as a journey, as both involve moving from start to finish, with different stages and stations along the way.[10]

The metaphor of life as a journey seems helpful in emphasizing not only that life has a beginning and an end, with different stages and stations, but also that life has its ups and its downs, and that although we may be headed in a particular

8. See **Alison Gopnik** [1949]. This metaphor is reminiscent of Phillipa Foot's use of the exemplar of flourishing plants to develop a naturalistic virtue ethics, as well as of similar cultivation exemplars in Confucian thought [1224, 1950]. For an argument that both the gardener and the carpenter metaphors are, however, straw men, see M. Angel [1951].
9. See earlier footnotes on **progress in evolution** (including the work of Steven Jay Gould and criticism thereof), and on metaphors of evolution (and the work of Lakoff and Johnson).
10. Again, for a philosophical perspective on the life cycle, see Michael Slote [1771]. This line is from Shakespeare's 'As You Like It'.

direction we are not always sure of our next steps. It encourages us to think about where we come from, where we are going, and how we get there. And in terms of the big questions and hard problems, a key answer that the metaphor helps emphasize is that life is not about the destination (we know only too well what that is), but rather about the journey. In the words of the Chinese thinker, Lao Tzu, "A good traveller has no fixed plans, and is not intent upon arriving". That said our attitude to journeying should be that of a 'pilgrim' (aiming to find meaning) rather than a 'nomad' (unable to settle on any particular meaning).[11]

It's notable that Aristotle's philosophy has been termed 'peripatetic', from the Greek word for walking. Edith Hall suggests that the Greeks thought of intellectual enquiry as a journey, and that this term encapsulates Aristotle's close interest in the natural world, and in human pleasure as a guide to living a virtuous and flourishing life. In his 'On the Soul', Aristotle talks about the need to examine earlier thinkers in order "to move forward as we try find the necessary direct pathways through (poros) impasses". A poros can refer to a passageway or a bridge. Similarly in his 'Physics', he invites his readers to go on the road (hodos) with him; together we will set from things that are more familiar, and from there we will progress to more difficult matters. In the East, the term 'tao' or 'dao', meaning 'the Way' or the path, is used to refer to Confucian teaching. For a while I thought of titling this volume 'Wetware's Way' to reference the guidance of a range of Axial Age thinkers.[12]

The Jewish tradition has a number of interesting concepts that parallel Aristotelian and Confucian metaphors of the journey of life. 'Halacha' or 'the way to walk' is used to refer to the formal laws of the written and oral Torah. Another term, 'derech eretz', literally 'the way of the world', refers to more informal 'common decency'; for example, the importance of greeting rituals. It can also be used to refer to human nature (e.g. we are judged in our home towns by our reputation, but elsewhere by our clothing; this is just the way humans are) or the natural world (e.g. the right mood is needed before having sex; this is the way of all animals). The aphorism of 'Torah im derech eretz' refers to the importance of engaging with the world of work together with the spiritual world. However, it has also been argued that 'derech eretz precedes Torah';

---

11. For a reflection on the value and limitations of the **travel metaphor** for thought, see Georges Van Den Abbeele [1952]. Tzu's aphorism is cited in Julian Baggini's 'What's It All About? Philosophy and the Meaning of Life' [84]. Or as Lewis Carroll puts it in 'Alice's Adventures in Wonderland' [301]: "Cat: Where are you going? Alice: Which way should I go? Cat: That depends on where you are going. Alice: I don't know. Cat: Then it doesn't matter which way you go". The distinction between a pilgrim and nomad perspective is from the work of Zygmunt Bauman, and is cited in A.A. van Niekerk [639].

12. See Edith Hall [516]. Rebecca Solnit covers the thoughts of a range of philosophers about walking in her 'Wanderlust: History of Walking' [1953], and notes that the origins of the term 'peripatetic' are likely more accidental. For Nietzsche, for example, "A sedentary life is the real sin against the Holy Spirit. Only those thoughts that come by walking have any value" [912]. For more on the **philosophy of walking**, see Frédéric Gros [1954] and Geoff Nicholson [1955]. For more on the 'tao', see Michael Puett and Christine Gross-Loh [542] and Jean-François Revel and Matthieu Ricard [1956].

being a 'mensch' (an accolade redolent of African ubuntu, Aristotelian virtue, and Confucian 'ren') is grounded in treating others with humanity.[13]

In modern philosophy, MacIntyre, who like the Oxford moral naturalists, played a key role in reviving interest in Aristotelian views, emphasized that the medieval notion of a quest encompasses a journey in which the goals are only discovered during the course of the adventure. As the adventurer engages with temptations and dangers, sustained by his or virtues, so the unfolding events and episodes of the journey lead to knowledge and understanding of both the character of the adventurer and of the goal of the quest. Just as Midgley sees the unity of a life as a continuing lifelong project, so for MacIntyre "The unity of a human life is the unity of a narrative quest". Additionally, he concludes that "the good life for man is the life spent in seeking for the good life for man".[14]

The journey metaphor includes the notion that different paths are open to us. "Two paths diverged in a yellow wood ... and I took the one less travelled by, and that made all the difference", as the poet Robert Frost put it. This sort of perspective helps us to accept some key constraints; as we move in one direction, for example, we necessarily cannot move in a different one. Life is unpredictable, and the counterfactual is imponderable. We can only move forwards with our own bodies (journeys are embodied), within a particular context (journeys are embedded). Accepting that there are no perfect solutions, and that life inevitably entails trade-offs, puts us in a position to better appreciate our current journey, and to see it as 'good enough'. This useful phrase of the paediatrician and psychoanalyst Donald Winnicott reminds us of steering a course between mediocrity on the one hand, and unrealistic and unrelenting standards on the other.[15]

At the same time, acknowledging that there are wrong turns and worse journeys may be important insofar as it helps us aim to be on a track that is true,

---

13. See earlier discussion of sages versus saints. See P. Rodman [1957].
14. See earlier footnotes on virtue ethics and for this citation to Midgley, and see MacIntyre [1207]. For **more on quests**, see again Joseph Campbell [1780], as well as philosophical and popular volumes which have emphasized related themes e.g. Phil Cousineau [1958], John Bly [1959], and Robert Pirsig [1960]. See earlier footnote on science as a quest, as well as earlier footnote on how Aristotle views flourishing in relation to the course of a life; for more on the issue of part-life versus whole-life happiness and meaning, see the contrasting views of F. Feldman [1961] and Thaddeus Metz [1737]. Relatedly, Anton van Niekerk argues that "The meaning which the reality of death invites us to find in the effort to establish its significance for our lives, is ... essentially a meaning that has to do ... with life as a whole" [1962]. MacIntyre's notion that "the good life for man is the life spent in seeking for the good life for man" is redolent of the concept of essentially contested categories, see earlier footnote. For a light-hearted argument along similar lines, see Sophia Hannah's "Happiness, a Mystery" [1963].
15. Some have suggested that choosing **mediocrity** is the greatest evil; for more on mediocrity see J.C. Hermanowicz [1964] and Daniel Milo [1965]. Christine Swanton, in her work on virtue ethics, contrasts virtuous and vicious acts, and notes that in between are a range of **good enough** acts [1965]. Nussbaum describes the constraints of life elegantly, writing "I must constantly choose among competing and apparently incommensurable goods and that circumstances may force me to a position in which I cannot help being false to something or doing something wrong; that an event that simply happens to me may, without my consent, alter my life; that it is equally problematic to entrust one's good to friends, lovers, or country and to try to have a good life without them – all these I take to be not just the material of tragedy, but everyday facts of practical wisdom" [527].

beautiful, and good. Some routes are simply better, and we may as well try to find the best routes for the particular terrain ahead. Indeed, as we travel, it's helpful to know as much about our means of transport, the route of the journey, and the terrain that we are traversing, as possible. That said, we need to be crucially aware of the importance of our own perspective of the journey on judgements about whether or not a particular path is best. Indeed, given that there may be a range of possible transport means and a variety of routes, it's helpful to take as much responsibility for the journey we can; we surely have some role in determining a range of matters, including direction and tempo.[16]

Even as we attempt to find the best routes, it's sensible to immerse ourselves fully in the experience. There will be beautiful scenes, and there will be impediments, for sure: the ups and the downs are inevitable. As Wittgenstein argued, "Life is like a path along a mountain ridge; to left and right are slippery slopes down which you slide without being able to stop yourself, in one direction or the other". These constraints reflect something about us as travellers, we enjoy some scenes, and we are threatened by others. These sorts of responses are helpful in providing important clues about the terrain, and in constructing our maps accordingly.[17]

From time to time, as Wystan Auden depicts in 'The Labyrinth', we will get lost and we will wish we were other sorts of beings, particularly beings who had a perfect view of the terrain. But experiences of being lost and forlorn may be valuable. As Irvin Yalom puts it, "The search for meaning, much like the search for pleasure, must be conducted obliquely. Meaning ensues from meaningful activity: the more we deliberately pursue it, the less likely are we to find it; the rational questions one can pose about meaning will always outlast the answers. In therapy, as in life, meaningfulness is a by-product of engagement and commitment, and that is where therapists must direct their efforts—not that engagement provides the rational answer to questions of meaning, but it causes these questions not to matter".[18]

The ultimate constraint in the journey of our life is the fact that it ends. The University of Stellenbosch philosopher Anton van Niekerk argues that the destination of death should be viewed as a mystery. van Niekerk points out that unlike puzzles which have a solution, mysteries are the sources of ongoing

---

16. For a volume on the value of **different perspectives when walking**, see Alexandra Horowitz [1966].
17. See earlier footnote for this citation of Wittgenstein. Henry Thoreaux wrote on the value of walking in nature, and emphasized the importance of '**sauntering**', "I have met with but one or two persons in the course of my life who understood the art of Walking, that is, of taking walks – who had a genius, so to speak, for SAUNTERING, which word is beautifully derived from idle people who roved about the country, in the Middle Ages, and asked charity, under pretense of going a la Sainte Terre, to the Holy Land, till the children exclaimed, 'There goes a Sainte-Terrer,' a Saunterer, a Holy-Lander. They who never go to the Holy Land in their walks, as they pretend, are indeed mere idlers and vagabonds; but they who do go there are saunterers in the good sense, such as I mean. Some, however, would derive the word from sans terre, without land or a home, which, therefore, in the good sense, will mean, having no particular home, but equally at home everywhere. For this is the secret of successful sauntering. He who sits still in a house all the time may be the greatest vagrant of all; but the saunterer, in the good sense, is no more vagrant than the meandering river, which is all the while sedulously seeking the shortest course to the sea" [1967].
18. For more on the unforeseen and the **value of getting lost**, see Rebecca Solnit [1968]. See Yalom [1969].

inquiry; their wonder draws us in, and while our insight may deepen over time, we never resolve them. Mysteries such as death teach us that the striving for knowledge ought to be tempered by a search for wisdom—"wisdom, not as a reachable *destination*, but as *orientation*, a quest and capacity to live within the fundamental uncertainty with which life confronts us without perishing because of it or surrendering our efforts to make do".[19]

While medieval quests are typically undertaken by individuals, the journey metaphor may also be appropriate and useful for considering groups. This volume has been particularly concerned with the field of psychiatry: with how far it has come, and with how much further it has to go. The end of any field of medicine is arguably to put itself out of business, just as Jonas Salk helped put an end to the scourge of polio. But evolutionary medicine has emphasized that humans are far from perfect, that our design reflects any number of trade-offs, and that there may be mismatches between our individual genome and our particular environments: there will therefore always be suboptimal health, and the need for healers. As long as there are peripatetic humans, so too will science, medicine, and psychiatry be 'on the road'.[20]

This volume has also been concerned with the journey of humankind as a whole, with how the trip has been, and with where we are headed. We've asked whether it's more appropriate to respond to these questions with optimism or pessimism. The journey metaphor might emphasize that the road is difficult and that the destination is unreachable, or it might emphasize the grandeur and the wonder of the trip. Just as individuals do well to accept the constraints of the journey as being informative, so it seems that humankind as a whole must take an approach that not only accepts that the human journey is tragic and futile in many ways, but also celebrates that there is much to be grateful for and hopeful about.

The metaphor of life as a journey helps reinforce a few points that may be useful in sharpening our view of the nature of human nature, and hence about the way we should live our lives, and with which I'd like to close.

First, as we start our journey, Aristotle reminds us that it's helpful to be guided by key virtues. Courage seems a particularly important virtue when travelling; we need courage to begin the trip, and we need courage to keep going. As another useful Chinese aphorism emphasizes, "A journey of a thousand miles begins

---

19. Anton van Niekerk writes, "**Mysteries** are sources of our deepest despair and our greatest joy. The mystery of death does not release us from the challenge of fighting death and fighting (or protesting against) the suffering that so often precedes death. Yet the mystery of death allows us to develop a better understanding of the value of that which, even if only for the briefest time, gives meaning to life. It is because there is certain death for us all that it makes sense to work hard, to love and to strive to do good – however clumsy the effort. Because death is a mystery, we can continue to resist evil, without succumbing to the inadequacy of most of our efforts to do so. In this sense, our understanding of death as mystery does bring peace of mind, not because it lets us rest in a fatalism concerning the unavoidable, but because it comforts us with the promise of meaning" [1962]. For more on metaphors of death, see Herbert Fingarette [1970]. For more on the philosophy and psychology of death, see Ben Bradley, Fred Feldman, and Jens Johansson [1971], Simon Critchley [976, 1730], and David Kessler [1972]. The contrast between mysteries and puzzles has also been drawn by Douglas Hofstadter (see previous footnote on mystery, awe, and science).
20. This phrase here is from Jack Kerouac [1973].

with a single step". And as Albert Einstein adds, "Life is like riding a bicycle. To keep your balance you must keep moving". Perhaps precisely because of his thoroughgoing disgruntlement about the journey of life, Schopenhauer encourages heroism, writing "A 'happy life' is impossible; the best that man can attain is a 'heroic life', such as is lived by one who struggles against overwhelming odds in some way and some affair that will benefit the whole of mankind, and who in the end triumphs, although he obtains a poor reward or none at all". If we've been mentored by travel experts, if we've spent time practicing how best to journey, and we've developed good travel habits, we'll naturally be better prepared. Good genes are helpful too, but it's useful to think of ourselves as having the freedom to go beyond what nature and nurture have wrought.[21]

Second, as we set about our journey, it's helpful to have good companions. This volume began with a reference to Socrates, who exemplifies the notion that wisdom emerges from dialogue with others. When I wrote those initial words, I hadn't anticipated relying so much on Aristotle, Spinoza, Hume, Dewey, and Jaspers, but time and again, these authors seem to have made truly important contributions to addressing the big questions and hard problems. We are fortunate to be able to have a conversation about their views, and possibly at times to stand on their shoulders. Each would agree that much of the value of life emerges from having lovers and loved ones, family and friends, mentors and mentees. And each would resonate with the modern constructs of an 'ethics of care' or an 'ethics of responsibility', and with the ubuntu notion that much of the value of life emerges from being a person through other persons. Stoicism bracingly takes the negative pathway even on this issue; as Seneca writes, "Soon we shall breathe our last. Meanwhile, while we live, while we are among human beings, let us cultivate our humanity". Some go even a step further, pointing out that travelling in close quarters with anyone can be difficult; "Hell is other people", as Sartre wrote, or at any rate, "you can't live with them, and you can't live without them".[22]

Third, as we plan our journey, it's useful to have a range of maps and other tools that might help us to do so. In this volume, we've been sceptical of the scientistic idea that a handful of techniques will lead to a happy or meaningful life. At the same time, to flourish we may need to hone a range of skills, and to continue to polish our habits. As thinking-feeling travellers, we face many big questions

---

21. See Walter Isaacson [1974] and Schopenhauer [1723]. For more on Schopenhauer and the Stoics, see H. Head [1975]. See also earlier footnote on hero myths.

22. Merton, who originated phrases such as 'organized skepticism' (see previous footnote) as well as 'role model' and 'self-fulfilling prophecy', wrote a wonderful volume on the long history of this metaphor: 'On the Shoulders of Giants: A Shandian Postscript' [1976]. See earlier footnotes on ethics of responsibility and ethics of care, and on ubuntu. Sarte's aphorism is from his play 'No Exit' [1977]. In keeping with the spirit of questioning the obvious, I'm reminded of Aldous Huxley's wry observation, "It is a bit embarrassing to have been concerned with the human problem all one's life and find at the end that one has no more to offer by way of advice than 'Try to be a little kinder'" [1978]. Seneca's words are from his 'On Anger' [1435]. The less charitably phrased aphorism at the end of the paragraph was apparently used by Desiderius Erasmus (in reference to women). A not dissimilar point is made by Schopenhauer and Freud, in their discussion of the 'porcupine's problem' (porcupines may huddle together for warmth, but their quills also encourage distance) [1979].

about the value of the trip, about the best approach, and about its destination. This volume has suggested that philosophy and science provide us valuable resources for continuously rethinking and reimagining our answers to such hard problems. A broad array of sciences are relevant to the big questions and hard problems, including those focused on proximal psychobiological mechanisms, as well as those focused on distal evolutionary mechanisms. Medicine may be more of a practical craft than a science, but it may be useful for precisely that reason; it necessarily entails addressing questions about what is most valuable. Indeed, exemplars from psychiatry may be particularly useful in helping to think through the big questions and hard problems. The approach in this volume to the big questions and hard problems has aimed at being both comprehensive and integrative, encompassing and bringing together a range of constructs and disciplines; indeed a broad array of good resources may be helpful for the journey of life. Again, however, in the spirit of avoiding the Panglossian, it's as well to always remind ourselves that such resources have key limitations: they provide only sketchy outlines, they may offer competing and contradictory views, and they are inherently fallible.[23]

Fourth, it's perhaps worth a word or two about the appropriate state of mind for a journey. In this volume we've been sceptical of the bromides of positive thinking, but at the same time we'd like to steer clear of absolute despair. Ancient Stoicism, and similar modern cognitive-behavioural therapy approaches, may help us steer a course between the Scylla of unbridled optimism and the Charybdis of relentless pessimism. Seeing ourselves as captains of our ships, embarked on a heroic quest, may help bolster the bravery that we need to begin the trip, and to keep going. Viewing ourselves as part of a community of travellers, in this together with others, may help encourage us to build our relationships with others, and to follow Emmanuel Levinas in seeing philosophy as not merely the 'love of wisdom', but rather as the 'wisdom of love'. Socratic acknowledgement of our fallibility, and a good amount of epistemic and moral humility, seem appropriate. At the same time, we want to avoid 'unrelenting standards', and to be open to seeing our journey as genuinely worthwhile, and as a meaningful contribution. Avoiding perfectionism may be helpful in allowing us to have a playful approach to the journey, to be open to experience the wonder of the journey, to be willing to get lost and to find new routes.[24]

---

23. For a remarkable account of **philosophical paradoxes**, aiming to support genuinely undogmatic pluralism, see South African-born Jeremy Barris [1980]. Jeremy is an old friend, whose thinking I've long admired.

24. See also earlier footnote on Levinas and ethics of responsibility. Bertrand Russell earlier wrote about **love and knowledge**, saying, "The good life is one inspired by love and guided by knowledge" and "Although both love and knowledge are necessary, love is in a sense more fundamental, since it will lead intelligent people to seek knowledge, in order to find out how to benefit those whom they love. But if people are not intelligent, they will be content to believe what they have been told, and may do harm in spite of the most genuine benevolence" [936]. For more on the importance of avoiding perfectionism when assessing meaningfulness, see Landau [1717]. See earlier footnotes on play, on awe, and on getting lost.

Fifth, and finally, as we journey forwards, the guiding principle of balance may be useful. From the time of the Axial Age, thinkers have emphasized moderation and balance: Aristotle's golden mean of moderation in the West, and the middle path in the East.[25] Balance seems useful for avoiding both scientism and cynicism, for valuing both reason and passion, for reasoning with principles and particulars as well as facts and values, for remaining hopeful in the face of the tragic, for valuing self-insight but not too much, and for balancing conservatism with progressivism. When we think about the journey of life, it's tempting to put a great deal of store in our own perspectives and intuitions; this volume has emphasized the value of being open about how little we know, and of drawing on contrasting but reasonable views. Balance seems reminiscent of the value of homeostasis in physiology and of the importance of trade-offs in evolutionary theory. Balance seems particularly important for human flourishing; a metaphor of balance is useful for staying on course, for reasoning about health, and for progress towards the true, the beautiful, and the good. That said, in the spirit of questioning all principles, let's close with a reminder of the ancient admonition: "Everything in moderation, including moderation".[26]

25. A number of Scandinavian concepts have recently become popular in the self-help literature; these include **hygge** (see Gunnar Gíslason and Jody Eddy [1981] and Meik Wiking [1982]) and **lagom**, which refers to balance and moderation (see Niki Brantmark [1983] and Anna Brones [1984].

26. See earlier footnote on transparency and fallibilism. For more recent useful **discussions of balance**, see Mike Martin [673], C.D. Ryff and B.H. Singer [711], Jonathan Haidt [974], and P.T.P. Wong [1773]. Mark Johnson has pointed out that metaphors of imbalance and journey are often combined to explain a psychological struggle [1275]. This admonition has been attributed to many, see www.wonderfulquote.com/l/moderation-quotes. In response, we may wish to point to the possibility and value of being a '**passionate moderate**', as exemplified by the philosopher Susan Haack [79].

References

[1] Dewsbury D. Psychobiology. Am Psychol 1991;46:198–205.
[2] Gardner H. The mind's new science: a history of the cognitive revolution. Basic Books; 1985.
[3] Boden M. Mind as machine. Oxford University Press; 2006.
[4] Dawson M. Mind, body, world: foundations of cognitive science. Athabasca Press; 2013.
[5] Poeppel D, Mangun G, Gazzaniga M, editors. The cognitive neurosciences. 6th ed. MIT; 2020.
[6] Hasler F. Neuromythology: a treatise against the interpretational power of brain research. Verlag; 2015.
[7] Stone I. The trial of Socrates. Anchor; 1988.
[8] Wilson E. The death of Socrates: hero, villain, chatterbox, saint. Profile Books; 2007.
[9] Morrison D. Cambridge companion to Socrates. Cambridge University Press; 2011.
[10] Hui A. A theory of the aphorism: from Confucius to Twitter. Princeton University Press; 2019.
[11] Vernon M. 42: Deep Thought on life, the universe, and everything. Oneworld; 2008.
[12] Klein D. Every time I find the meaning of life, they change it: wisdom of the great philosophers on how to live. Penguin; 2015.
[13] Armstrong K. The great transformation: the beginning of our religious traditions. Anchor; 2006.
[14] Bellah R, Joas H. The axial age and its consequences. Belknap; 2012.
[15] Provan I. Convenient myths: the axial age, dark green religion, and the world that never was. Baylor University Press; 2013.
[16] Hoyer D, Reddish J. Seshat history of the axial age. Beresta Books; 2019.
[17] Zimmer C. Soul made flesh: Thomas Willis, the discovery of the brain and how it changed the world. Heinemann; 2004.
[18] Makari G. Soul machine: the invention of the modern mind. Norton; 2015.
[19] Damasio A. Looking for Spinoza: joy, sorrow, and the feeling brain. Houghton Mifflin Harcourt; 2003.
[20] Goldstein R. Betraying Spinoza: the renegade Jew who gave us modernity. Schocken; 2006.
[21] Yalom I. The Spinoza problem: a novel. Basic Books; 2012.
[22] Potkay A. The passion for happiness: Samuel Johnson and David Hume. Cornell University Press; 2000.
[23] Edmonds D, Eidinow J. Rousseau's dog: a tale of two great thinkers at war in the age of enlightenment. Ecco; 2006.
[24] Rasmussen D. The infidel and the professor: David Hume, Adam Smith, and the friendship that shaped modern thought. Princeton University Press; 2017.
[25] Roberts R. How Adam Smith can change your life: an unexpected guide to human nature and happiness. Portfolio; 2014.
[26] Kant I. Critique of pure reason. Bell & Daldy; 1781.
[27] Vanzo A. Empiricism and rationalism in nineteenth-century histories of philosophy. J Hist Ideas 2016;77:253–82.
[28] Champagne M. A critique consistent with the critique. Philos Invest 2018;41:436–45.

[29] Bhaskar R. The possibility of naturalism: a philosophical critique of the contemporary human sciences. Routledge; 1979.
[30] Hanna R. Science, Kant, and human nature. Oxford University Press; 2006.
[31] Janik A, Toulmin S. Wittgenstein's Vienna. Simon & Schuster; 1973.
[32] Sigmund K. Exact thinking in demented times: the Vienna Circle and the epic quest for the foundations of science. Basic Books; 2017.
[33] Quine W. Two dogmas of empiricism. Philos Rev 1951;60:20–43.
[34] Rey G. The analytic/synthetic distinction. In: Zalta E, editor. The Stanford encyclopedia of philosophy. Stanford University; 2018.
[35] McGuiness B. Wittgenstein: a life. University of California Press; 1988.
[36] McDonald J. Russell, Wittgenstein, and the problem of the rhinoceros. South J Philos 1993;31:409–24.
[37] Edmonds D, Eidinow J. Wittgenstein's poker: the story of a ten-minute argument between two great philosophers. Harper Collins; 2001.
[38] Wittgenstein L. Tractatus logico-philosophicus. Kegan Paul; 1922.
[39] Wittgenstein L. Philosophical investigations. Macmillan; 1953.
[40] Sahlin N-E. The philosophy of F. P. Ramsey. Cambridge University Press; 1990.
[41] Chaterjee R. Wittgenstein and Judaism: a triumph of concealment. Peter Lang; 2005.
[42] Kripke S. Wittgenstein on rules and private language: an elementary exposition. Blackwell; 1982.
[43] Hofstadter D. Gödel, Escher, Bach: an eternal braid. Basic Books; 1979.
[44] Goldstein R. Incompleteness: the proof and paradox of Kurt Gödel. Norton; 2006.
[45] Urqhuhart A. Russell and Gödel. Bull Symb Log 2016;22:504–20.
[46] Levi A. Philosophy as social expression. University of Chicago Press; 1974.
[47] Jaspers K. On my philosophy, 1941. In: Kaufman W, editor. Existentialism from Dostoyevsky to Sartre. Meridian; 1956.
[48] Kierkegaard S. Concluding unscientific postscript to the philosophical fragments. Princeton University Press; 1846/1992.
[49] Sartre J-P. Existentialism is a humanism. Yale University Press; 1946/2007.
[50] Schacht R. Kierkegaard on 'Truth Is Subjectivity' and 'The Leap of Faith'. Can J Philos 1973;2:297–313.
[51] Crary A, Read R. The New Wittgenstein. Routledge; 2000.
[52] Calkins M. The persistent problems of philosophy. Macmilllan; 1907.
[53] Stein D. Philosophy of psychopharmacology: smart pills, happy pills, pep pills. Cambridge University; 2008.
[54] Soper K. Humanism and anti-humanism. Open Court; 1986.
[55] Cooper D. Senses of mystery: engaging with nature and the meaning of life. Routledge; 2017.
[56] Frankfurt H. Demons, dreamers, and madmen: the defense of reason in Descartes's Meditations. Princeton University Press; 1970.
[57] Goldstein R. Plato at the Googleplex: why philosophy won't go away. Pantheon; 2014.
[58] Berlin I. Three critics of the enlightenment: Vico, Hamman. Herder Princeton University Press; 2000.
[59] Cahill K, Raleigh T. Wittgenstein and naturalism. Routledge; 2018.
[60] Beale J, Kidd I. Wittgenstein and scientism. Routledge; 2017.
[61] Kuhn T. Structure of scientific revolutions. University of Chicago Press; 1962.

[62] Morris E. The ashtray (or the man who denied reality). University of Chicago Press; 2018.
[63] Chambliss J. The influence of Plato and Aristotle on John Dewey's philosophy. E. Mellen Press; 1990.
[64] Anton J. John Dewey and ancient philosophies. Philos Phenomenol Res 1965;25:477–99.
[65] Beebee H. Hume on causation. Routledge; 2006.
[66] Goodman N. Fact, fiction, and forecast. Athlone Press; 1954.
[67] Bhaskar R. A realist theory of science. Harvester Press; 1973.
[68] Cartwright N. How the laws of physics lie. Oxford University Press; 1983.
[69] Kitcher P, Salmon W. Scientific explanations. University of Minnesota Press; 1989.
[70] Beebee H, Hitchcock C, Menzies P. The Oxford handbook of causation. Oxford University Press; 2010.
[71] Illari P, Russo F, Williamson J. Causality in the sciences. Oxford University Press; 2011.
[72] Misak C. Verificationism: its history and prospects. Routledge; 1995.
[73] Ricoeur P. Hermeneutics and the human sciences. Cambridge University Press; 1981.
[74] Rorty R. The brain as hardware, culture as software. Inquiry 2004;47:219–35.
[75] Varela F, Thompson E, Rosch E. The embodied mind: cognitive science and human experience. MIT; 1991.
[76] Baptista T, Aldana E. Arthur Schopenhauer and the embodied mind. Ludus Vitalis 2018;49:153–81.
[77] Bloom A. The closing of the American mind. Simon & Schuster; 1987.
[78] Nehamas A. The art of living: Socratic reflections from Plato to Foucault. University of California Press; 1988.
[79] Haack S. Manifesto of a passionate moderate: unfashionable essays. University of Chicago Press; 1998.
[80] Solomon R. The joy of philosophy: thinking thin versus the passionate life. Oxford University Press; 1999.
[81] Kronman A. Education's end: why our colleges and universities have given up on the meaning of life. Yale University Press; 2007.
[82] Sokal A, Bricmont J. Fashionable nonsense: postmodern intellectuals' abuse of science. Picador; 1998.
[83] De Botton A. The consolations of philosophy. Vintage; 2001.
[84] Baggini J. What's it all about?: philosophy and the meaning of life. Oxford University Press; 2004.
[85] Evans J. Philosophy for life and other dangerous situations. Ebury Digital; 2012.
[86] Pigliucci M. Answers for Aristotle: how science and philosophy can lead us to a more meaningful life. Basic Books; 2012.
[87] Rorty R. The linguistic turn: recent essays in philosophical method. University of Chicago Press; 1992.
[88] Rorty R. Philosophy and the mirror of nature. Princeton University Press; 1980.
[89] Egginton W, Sandbothe M. The pragmatic turn in philosophy: contemporary engagements between analytic and continental thought. State University of New York Press; 2012.
[90] Bernstein R. The pragmatic turn. Polity; 2010.
[91] Rockwell W. Neither brain nor ghost: a nondualist alternative to the mind-brain identity theory. MIT Press; 2005.
[92] Popp J. Evolution's first philosopher: John Dewey and the continuity of nature. State University of New York; 2007.

[93] Brinkmann S. John Dewey: science for a changing world. Routledge; 2017.
[94] Wiseman R. Rip it up: the radically new approach to changing your life. MacMillan; 2012.
[95] Danziger K. The positivist repudiation of Wundt. J Hist Behav Sci 1979;15:205–30.
[96] Losch A. On the origins of critical realism. Theol Sci 2009;7:85–106.
[97] Menand L. The Metaphysical Club: a story of ideas in America. Farrar, Straus and Giroux; 2001.
[98] Nussbaum M. The therapy of desire: theory and practice in Hellenistic ethics. Princeton University Press; 1994.
[99] Sorabji R. Emotion and peace of mind: from Stoic agitation to Christian temptation. Oxford University Press; 2000.
[100] Carlisle C, Ganeri J. Philosophy as therapeia. Cambridge University Press; 2010.
[101] Wiley J. Theory and practice in the philosophy of Hume. Palgrave Macmillan; 2012.
[102] Walker M. Aristotle on the uses of contemplation. Cambridge University Press; 2018.
[103] Malcolm N. Ludwig Wittgenstein: a memoir. Oxford University Press; 1959.
[104] Egginton W, Sandbothe M. The pragmatic turn in philosophy: contemporary engagements between analytic and continental thought. State University of New York Press; 2004.
[105] Misak C. Cambridge pragmatism: from Peirce and James to Ramsey and Wittgenstein. Oxford University Press; 2016.
[106] Price H. Wilfred Sellars meets Cambridge pragmatism. In: Pereplyotchik D, Barnbaum D, editors. Sellars and contemporary philosophy. Routledge; 2016.
[107] Misak C, Price H. The practical turn: pragmatism in Britain in the long twentieth century. Oxford University Press; 2018.
[108] Price H, Blackburn S, Brandom R, Horwich P, Williams M. Expressivism, pragmatism and representationalism. Cambridge University Press; 2013.
[109] Dewey J. The reflex arc concept in psychology. Psychol Rev 1896;3:357–70.
[110] Piaget J. Origins of intelligence. International Universities Press; 1952.
[111] Seltzer E. A comparison between John Dewey's theory of inquiry and Jean Piaget's genetic analysis of intelligence. J Genet Psychol 1977;130:323–35.
[112] Gallese V, Lakoff G. The brain's concepts: the role of the sensory-motor system in conceptual knowledge. Cogn Neuropsychol 2005;22:455–79.
[113] Tambornino J. The corporeal turn: passion, necessity, politics. Rowman & Littlefield; 2002.
[114] Sheets-Johnstone M. The corporeal turn: an interdisciplinary reader. Academic; 2009.
[115] Marsh L. Dewey: the first ghost-buster? Trends Cogn Sci 2006;10:242–3.
[116] Dewey J. Experience and nature. Georg Allen & Unwin; 1896/1929.
[117] Dewey J. Soul and body. Bibliotheca Sacra 1886;43:239–63.
[118] Johnson M. Cognitive science and Dewey's theory of mind, thought, and language. In: Cochran M, editor. The Cambridge companion to Dewey. Cambridge University Press; 2010.
[119] Putnam H. The many faces of realism. Open Court; 1987.
[120] Haack S. Defending science – within reason: between scientism and cynicism. Prometheus; 2003.
[121] DeVries W. Wilfrid Sellars. McGill-Queen's University Press; 2005.
[122] Westphal K. Realism, science, and pragmatism. Taylor & Francis; 2014.
[123] Chang H. Pragmatic realism. Revista de Humanidades 2016;8:107–22.
[124] Brinkmann S. Psychology as a moral science: aspects of John Dewey's psychology. Hist Hum Sci 2004;17:1–28.
[125] Aristotle. Nicomachean ethics. 350 BC.

[126] Altham J, Harrison R. World, mind, and ethics: essays on the ethical philosophy of Bernard Williams. Cambridge University Press; 1995.
[127] Zagzebski L. Virtues of the mind: an inquiry into the nature of virtue and the ethical foundations of knowledge. Cambridge University Press; 1996.
[128] Sosa E. A virtue epistemology: Apt belief and reflective knowledge. vol. I. Oxford University Press; 2007.
[129] Battaly H. The Routledge handbook of virtue epistemology. Taylor & Francis; 2018.
[130] Cassam Q. Vices of the mind: from the intellectual to the political. Oxford University Press; 2019.
[131] Oreskes N, Edenhofer O, Macedo S. Why trust science? Princeton University Press; 2019.
[132] McCain K, Poston T. Best explanations: new essays on inference to the best explanation. Oxford University Press; 2018.
[133] Hume D. An enquiry concerning human understanding. A dissertation on the passions. An enquiry concerning the principles of morals. The natural history of religion. T. Cadell, A. Donaldson and W. Creech; 1777.
[134] Nietzsche F. Human, all too human: a book for free spirits. Cambridge University Press; 1986.
[135] Yeats W. The second coming. 1919.
[136] Russell B. The triumph of stupidity. 1933.
[137] Popkin R. The history of scepticism from Erasmus to Spinoza. University of California Press; 1979.
[138] Cohen L. I don't know: in praise of admitting ignorance and doubt (except when you shouldn't). Riverhead Books; 2013.
[139] Margolis J. Pragmatism ascendent: a yard of narrative, a touch of prophecy. Stanford University Press; 2012.
[140] Kotzee B. Dewey as virtue epistemologist: open-mindedness and the training of thought in democracy and education. J Philos Educ 2018;52:359–73.
[141] Medawar P. Advice to a young scientist. Harper & Row; 1978.
[142] Feynman R. The meaning of it all: thoughts of a citizen-scientist. Addison-Wesley; 1998.
[143] Firestein S. Ignorance: how it drives science. Oxford University Press; 2012.
[144] Livio M. Brilliant blunders: from Darwin to Einstein – colossal mistakes by great scientists that changed our understanding of life and the universe. Simon & Schuster; 2013.
[145] Peels R, Blaauw M. The epistemic dimensions of ignorance. Cambridge University Press; 2016.
[146] Boudry M, Pigliucci's M. Science unlimited? The challenges of scientism. University of Chicago Press; 2017.
[147] de Ridder J, Peels R, van Woudenberg R. Scientism: prospects and problems. Oxford University Press; 2018.
[148] Haack S. Scientism and its discontents. Rounded Globe; 2017.
[149] Chemla K, Keller E. Culture without culturalism: the making of scientific knowledge. Duke University Press; 2017.
[150] Cartwright N. Nature's capacities and their measurement. Oxford University Press; 1989.
[151] Groff R, Greco J. Powers and capacities in philosophy: the new Aristotelianism. Routledge; 2012.
[152] Ellis B. Scientific essentialism. Cambridge University Press; 2001.
[153] Simpson W, Koons R, Teh N. Neo-Aristotelian perspectives on contemporary science. Routledge; 2018.

[154] Groff R. Critical realism, post-positivism and the possibility of knowledge. Routledge; 2004.
[155] Evenden M. Critical realism in the personal domain: Spinoza and explanatory critique of the emotions. J Crit Realism 2012;11:163–87.
[156] Landy D. Hume's science of human nature: scientific realism, reason, and substantial explanation. Routledge; 2018.
[157] Smart J. Philosophy and scientific realism. Routledge and Kegan Paul; 1963.
[158] Gironi F, Psillos S. Realist turns: a conversation with Stathis Psillos. Speculations 2012;3:367–427.
[159] Lakoff G, Johnson M. Philosophy in the flesh: the embodied mind and its challenge to Western thought. Basic Books; 1999.
[160] Clark A. Being there: putting brain, body, and world together again. MIT Press; 1997.
[161] Clark A, Chalmers D. The extended mind. Analysis 1998;58:10–23.
[162] Feyerabend P. Against method: outline of an anarchistic theory of knowledge. New Left Books; 1975.
[163] Russell D. Anything goes. Soc Stud Sci 1983;13:437–64.
[164] Murray T. Contributions of embodied realism to ontological questions in critical realism and integral theory. In: Bhaskar R, Esbjörn-Hargens S, Hedlund N, Hartwig M, editors. Metatheory for the twenty-first century: critical realism and integral theory in dialogue. Routledge; 2015.
[165] Göhner J, Jung E-M. Susan Haack: reintegrating philosophy. Springer; 2016.
[166] Bashour B, Muller H. Contemporary philosophical naturalism and its implications. Routledge; 2014.
[167] Clark K. The Blackwell companion to naturalism. Wiley-Blackwell; 2016.
[168] Strawson P. Skepticism and naturalism: some varieties. Columbia University Press; 1985.
[169] Flanagan O. Varieties of naturalism. In: Clayton P, editor. The Oxford handbook of religion and science. Oxford University Press; 2008.
[170] De Caro M, Macarthur D. Naturalism and normativism. Columbia University Press; 2010.
[171] Christias D. Towards a reformed liberal and scientific naturalism. Dialectica 2019;73:507–34.
[172] Annas J. The morality of happiness. Oxford University Press; 1995.
[173] Alexander J. Non-reductionist naturalism: Nussbaum between Aristotle and Hume. Res Publica; 2005;11:157–83.
[174] Dupré J. The disorder of things: metaphysical foundations of the disunity of science. Harvard University Press; 1993.
[175] Cartwright N. The dappled world: a study of the boundaries of science. Cambridge University Press; 1999.
[176] Kitcher P. Preludes to pragmatism: towards a reconstruction of philosophy. Oxford University Press; 2012.
[177] Scheffler I. Worlds of truth: a philosophy of knowledge. Wiley-Blackwell; 2009.
[178] Carroll S. The big picture: on the origins of life, meaning, and the universe itself. Dutton; 2016.
[179] Rosch E. Principles of categorization. In: Rosch E, Lloyd B, editors. Cognition and categorization. Lawrence Erlbaum; 1978.
[180] Russell B. Vagueness. Australas J Psych Philos 1923;1:84–92.
[181] Williamson T. Vagueness. Routledge; 1994.
[182] Keefe R, Smith P. Vagueness: a reader. Bradford Books; 1997.
[183] Sorenson R. Vagueness and contradiction. Oxford University Press; 2001.
[184] Ciprut J. Indeterminacy: the mapped, the navigable, and the uncharted. MIT Press; 2009.

[185] Scheffler I. Beyond the letter: a philosophical inquiry into ambiguity, vagueness and metaphor in language. Routledge; 2010.
[186] Dietz R. Vagueness and rationality in language use and cognition. Springer; 2019.
[187] Maher C. The Pittsburgh school of philosophy: Sellars, McDowell, Brandom. Routledge; 2012.
[188] Price H. Naturalism without mirrors. Oxford University Press; 2011.
[189] Macarthur D. Naturalizing the human or humanizing nature: science, nature, and the supernatural. Erkenntnis 2004;61:29–51.
[190] Leary D. Instead of erklären and verstehen: William James on human understanding. In: Feest U, editor. Historical perspectives on erklären and verstehen. Springer; 2010.
[191] Sayer A. Realism and social science. Sage; 2010.
[192] DeVries W. Wilfrid Sellars. Acumen Publishing; 2005.
[193] Sellars W. Philosophy and the scientific image of man. In: Colodny R, editor. Frontiers of science and philosophy. University of Pittsburgh Press; 1962.
[194] van Niekerk A. Beyond the erklären-verstehen dichotomy. S Afr J Philos 1989;8:189–213.
[195] Flanagan O. The problem of the soul: two visions of mind and how to reconcile them. Basic Books; 2002.
[196] Murphy D. Psychiatry in the scientific image. MIT Press; 2006.
[197] McDowell J. Mind and world. Cambridge University Press; 1994.
[198] Williams M. Naturalism, realism and pragmatism. Philos Exch 2007;37:57–71.
[199] Manicas P. Rescuing Dewey. Lexington Books; 2008.
[200] Magnus P. Scientific enquiry and natural kinds: from planets to mallards. Palgrave Macmillian; 2012.
[201] De Caro M. Naturalism and realism. In: Do Carmo J, editor. Companion to naturalism. NEPFIL; 2015.
[202] Giladi P. Responses to naturalism: critical perspectives from idealism and pragmatism. Routledge; 2019.
[203] Eames S. Pragmatic naturalism: an introduction. Southern Illinois University Press; 1977.
[204] Magnus P. John Stuart: Mill on taxonomy and natural kinds. HOPOS 2015;5:269–80.
[205] Hacking I. A tradition of natural kinds. Philos Stud 1991;61:109–26.
[206] Churchland P. Conceptual progress and word/world relations: in search of the essence of natural kinds. Can J Philos 1985;15:1–17.
[207] Sabbarton-Leary N, Beebee H. The semantics and metaphysics of natural kinds. Routledge; 2010.
[208] Campbell J, O'Rourke M, Slater M. Carving nature at its joints. MIT Press; 2011.
[209] Kendig C. Natural kinds and classification in scientific practice. Routledge; 2016.
[210] Brinkmann S. Psychology as a moral science: perspectives on normativity. Springer; 2011.
[211] Hesse M. The structure of scientific inference. University of California Press; 1974.
[212] Cooper R. Classifying madness. Springer; 2005.
[213] Boyd R. Realism, anti-foundationalism and the enthusiasm for natural kinds. Philos Stud 1991;61:127–48.
[214] Hey J. Genes, categories, and species: the evolutionary and cognitive cause of the species problem. Oxford University Press; 2001.
[215] Wilkins J. Species: the evolution of an idea. 2nd ed. CRC Press; 2018.
[216] Sellars W. Truth and correspondence. In: Sellars W, editor. Science, perception, and reality. Routledge and Kegan Paul; 1963.
[217] O'Shea J. Wilfrid Sellars: naturalism with a normative turn. Polity; 2007.

[218] Searle J. How to derive "ought" from "is". Philos Rev 1964;73:43–58.
[219] Glüer K, Wikforss A. Against content normativity. Mind 2009;118:31–70.
[220] Putnam H. The content and appeal of naturalism. In: Macarthur D, editor. Naturalism in question. Harvard University Press; 2004.
[221] Cohon R. Hume's morality: feeling and fabrication. Oxford University Press; 2008.
[222] Ayer A. Language, truth, and logic. Gollancz; 1936.
[223] Schlick M. Problems of ethics. Prentice-Hall; 1939.
[224] Feigl H. Naturalism and humanism. Am Q 1949;1:135–48.
[225] Putnam H. The collapse of the fact/value dichotomy and other essays. Harvard University Press; 2002.
[226] Wolf M, Koons J. The normative and the natural. Palgrave Macmillan; 2016.
[227] Kincaid H, Dupré J, Wylie A. Value-free science?: ideals and illusions. Oxford University Press; 2007.
[228] Charland L, Zachar P. Fact and value in emotion. John Benjamins; 2008.
[229] Gorski P. Beyond the fact/value distinction: ethical naturalism and the social sciences. Sociology 2013;50:543–53.
[230] Elqayam S. From is to ought: the place of normative models in the study of human thought. Frontiers Media; 2016.
[231] Alexandrova A. A philosophy for the science of well-being. Oxford University Press; 2017.
[232] Okrent M. Nature and normativity: biology, teleology, and meaning. Routledge; 2018.
[233] Flyvbjerg B, Landman T, Schram S. Real social science: applied phronesis. Cambridge University Press; 2012.
[234] Kinesella E, Pitman A. Phronesis as professional knowledge: practical wisdom in the professions. Sense Publishers; 2012.
[235] Churchland P. Neurophilosophy: towards a unified science of the mind/brain. Bradford; 1989.
[236] Churchland P. Brain-wise: studies in neurophilosophy. MIT Press; 2002.
[237] Churchland P. A neurocomputational perspective: the nature of mind and the structure of science. MIT Press; 1989.
[238] Churchland P. Neurophilosophy at work. Cambridge University Press; 2007.
[239] Roskies A. Neuroethics for the new millenium. Neuron 2002;35:21–3.
[240] Buechler S. Psychoanalytic approaches to problems in living: addressing life's challenges in clinical practice. Routledge; 2019.
[241] Angier T. Technē in Aristotle's ethics: crafting the moral life. Continuum; 2010.
[242] Ghaemi S. The rise and fall of the biopsychosocial model: reconciling art and science in psychiatry. Johns Hopkins University Press; 2010.
[243] MacIntyre A. The unconscious: a conceptual analysis. Routledge; 1958.
[244] Hook S. Psychoanalysis, scientific method, and philosophy. New York University Press; 1959.
[245] Gipps R, Lacewing M. The Oxford handbook of psychoanalysis and philosophy. Oxford University Press; 2018.
[246] Ellenberger H. The discovery of the unconscious: the history and evolution of dynamic psychiatry. Basic Books; 1981.
[247] Tallis F. Hidden minds: a history of the unconscious. Arcade Publishing; 2002.
[248] Davidson D. Paradoxes of irrationality. In: Wollheim R, Hopkins J, editors. Philosophical essays on Freud. Cambridge University Press; 1982.

[249] Bouveresse J. Wittgenstein reads Freud: the myth of the unconscious. Princeton University Press; 1995.
[250] Borthmus M. Life conduct in modern times: Karl Jaspers and psychoanalysis. Springer; 2006.
[251] Crews F. Freud: the making of an illusion. Metropolitan Books; 2017.
[252] Forrester J. Dispatches from the Freud wars: psychoanalysis and its passions. Harvard University Press; 1997.
[253] Wilson S, Zarate O. The Freud wars: a graphic illustration. Icon Books; 1999.
[254] Dufresne T. Killing Freud – twentieth century culture and the death of psychoanalysis. Continuum; 2003.
[255] Borch-Jacobsen M, Shamdasani S. The Freud files: an inquiry into the history of psychoanalysis. Cambridge University Press; 2012.
[256] Roazen P. The trauma of Freud. Transaction Publishers; 2002.
[257] Schwartz J. Cassandra's daughter: a history of psychoanalysis. Penguin; 1999.
[258] Wilson M. DSM-III and the transformation of American psychiatry: a history. Am J Psychiatr 1993;150:399–410.
[259] Luhrmann T. Of two minds: the growing disorder in American psychiatry. Knopf; 2000.
[260] Rosenthal N. The gift of adversity: the unexpected benefits of life's difficulties, setbacks, and imperfections. TarcherPerigree; 2013.
[261] MacKinnon R. The psychiatric interview in clinical practice. 3rd ed. American Psychiatric Association Publishing; 2016.
[262] Klein D. False suffocation alarms, spontaneous panics, and related conditions: an integrative hypothesis. Arch Gen Psychiatry 1993;50:306–17.
[263] Thornhill C. Karl Jaspers: politics and metaphysics. Routledge; 2002.
[264] Schwartz M, Mokalewicz M, Wiggins O. Karl Jaspers: the icon of modern psychiatry. Isr J Psychiatry Relat Sci 2017;54:4–7.
[265] Brown M. John Dewey's logic of science. HOPOS 2012;2:258–306.
[266] Hansen N. Patterns of discovery: an inquiry into the conceptual foundations of science. Cambridge University Press; 1958.
[267] Rabinow P, Sullivan W. Interpretive social science: a reader. University of Chicago; 1979.
[268] Hiley R, Bohman J, Shusterman R. The interpretive turn: philosophy, science, culture. Cornell University Press; 1991.
[269] Held B. Psychology's interpretive turn: the search for truth and agency in theoretical and philosophical psychology. American Psychological Association; 2007.
[270] Cooper R, Blashfield R. The myth of Hempel and the DSM-III. Stud Hist Philos Biol Biomed Sci 2018;70:10–9.
[271] Landau W. Clinical neuromythology and other arguments and essays. Futura; 1998.
[272] Tallis R. Why the mind is not a computer: a pocket lexicon of neuromythology. Societas; 2004.
[273] Brigandt I, Love A. Reductionism in biology. In: Zalta E, editor. The Stanford encyclopedia of philosophy. Stanford University; 2017.
[274] Lipowsky Z. Psychiatry: mindless or brainless, both or neither. Can J Psychiatr 1989;34:249–54.
[275] Rees D, Rose S. The new brain sciences: perils and prospects. Cambridge University Press; 2004.
[276] Weisberg D, Keil F, Goodstein J, Rawson E, Gray J. The seductive allure of neuroscience explanations. J Cogn Neurosci 2008;20:470–7.

[277] Satel S, Lilienfeld S. Brainwashed: the seductive appeal of mindless neuroscience. Basic Books; 2013.
[278] Tallis R. Aping mankind: neuromania, Darwinitis and the misrepresentation of humanity. Routledge; 2014.
[279] Morris A. Zuckerman versus Marais: a primatological collision. S Afr J Sci 2009;105:238–40.
[280] Tobias P. Conversion in paleo-anthropology: the role of Robert Broom, Sterkfontein and other factors in Australopithecine acceptance. In: Tobias P, Raath M, Moggi-Cecchi J, Doyle G, editors. Humanity from African naissance to coming millenia. Firenze University Press, Witwatersrand University Press; 2001.
[281] Derricourt R. The enigma of Raymond Dart. Int J Afr Hist Stud 2009;42:257–82.
[282] Stringer C, Andrews P. Genetic and fossil evidence for the origin of modern humans. Science 1988;239:1263–8.
[283] Meldman M, Hatch B. In vivo desensitization of an airplane phobia with penthranization. Behav Res Ther 1969;7:213–4.
[284] Poppen R. Joseph Wolpe. Sage; 1996.
[285] Poppen R. In: O'Donohue W, Henderson D, Hayes S, Fisher J, Hayes L, editors. A history of the behavioral therapies: founders' personal histories. Context Press; 2001.
[286] Winter A. Cats on the couch: the experimental production of animal neurosis. Sci Context 2016;29:77–105.
[287] van der Horst F. John Bowlby – from psychoanalysis to ethology: unraveling the roots of attachment theory. Wiley; 2011.
[288] Rilling M. John Watson's paradoxical struggle to explain Freud. Am Psychol 2000;55:301–12.
[289] Wakefield J. Little Hans and attachment theory: Bowlby's hypothesis reconsidered in light of new evidence from the Freud Archives. Psychoanal Study Child 2007;62:61–91.
[290] Harris B. Whatever happened to little Albert. Am Psychol 1979;34:151–60.
[291] Harris B. Letting go of Little Albert: disciplinary memory, history, and the uses of myth. J Hist Behav Sci 2011;47:1–17.
[292] Kornfeld A. Mary Cover Jones and the Peter case: social learning versus conditioning. J Anxiety Disord 1989;3:187–95.
[293] Eysenck H. Decline and fall of the Freudian Empire. Viking; 1985.
[294] Freud S. A general introduction to psychoanalysis. Boni and Liveright; 1920.
[295] Wolpe J. Behavior therapy according to Lazarus. Am Psychol 1984;39:1326–7.
[296] Wolpe J. Cognition and causation in human behavior and its therapy. Am Psychol 1978;33:437–46.
[297] Wandersman A. Humanism and behaviorism: dialogue and growth. Pergammon; 1976.
[298] Stein D, Bass J, Hofmann S. Global mental health and psychotherapy: adapting psychotherapy for low-and middle-income countries. Academic Press; 2019.
[299] Williams B. Ethics and the limits of philosophy. Oxford University Press; 1985.
[300] Nussbaum M. Aristotle on human nature and the foundation of ethics. In: Altham J, Harrison R, editors. World, mind, and ethics: essays on the ethical philosophy of Bernard Williams. Cambridge University Press; 1995.
[301] Carroll L. Alice's adventures in Wonderland. Macmillan; 1865.
[302] Stone M, Stein D. Essential papers on obsessive-compulsive disorder. New York University Press; 1997.
[303] Kendler K, Munoz R, Murphy G. The development of the Feighner criteria: a historical perspective. Am J Psychiatry 2010;167:134–42.
[304] Rosenhan D. On being sane in insane places. Science 1973;4070:250–8.

[305] Spitzer R. On pseudoscience in science, logic in remission, and psychiatric diagnosis: a critique of Rosenhan's "On being sane in insane places". J Abnorm Psychol 1975;84:442–52.
[306] Cahalan S. The great pretender: the undercover mission that changed our understanding of madness. Grand Central Publishing; 2019.
[307] Whitehead A. Science and the modern world. The Macmillan Company; 1925.
[308] Hyman S. The diagnosis of mental disorders: the problem of reification. Annu Rev Clin Psychol 2010;6:155–79.
[309] Kendler K. The phenomenology of major depression and the representativeness and nature of DSM criteria. Am J Psychiatr 2016;173:771–80.
[310] Kessler R, Merikangas K, Berglund P, Eaton W, Koretz D, Walters E. Mild disorders should not be eliminated from the DSM-V. Arch Gen Psychiatry 2003;60:1177–222.
[311] Smith J. Epic measures: one doctor: seven billion patients. Harper Wave; 2017.
[312] Prince M, Patel V, Saxena S, Maj M, Maseiko J, Phillips M, et al. No health without mental health. Lancet 2007;370:859–77.
[313] Patel V. Rethinking mental health care: bridging the credibility gap. Intervention 2014;12:15–20.
[314] Bhaskar R. Plato etc. problems of philosophy and their resolution. Verso; 1994.
[315] White F. Plato's essentialism: a reply. Australas J Philos 1988;66:403–13.
[316] Shields C. Aristotle. 2nd ed. Routledge; 2013.
[317] Presutti F. Realism and ontology in Spinoza, Abelard and Deleuze. J Br Soc Phenomenol 2011;42:209–22.
[318] Hacohen M. Karl Popper, the Vienna Circle, and Red Vienna. J Hist Ideas 1998;59:711–34.
[319] Nussbaum M. Human functioning and social justice: in defense of Aristotelian essentialism. Political Theory 1992;20:202–46.
[320] Stein D, McLaughlin K, Koenen K, Atwoli L, Friedman M, Hill E, et al. DSM-5 and ICD-11 definitions of posttraumatic stress disorder: investigating "narrow" and "broad" approaches. Depress Anxiety 2014;31:494–505.
[321] Stein D, Scott K, de Jonge P, Kessler R. Epidemiology of anxiety disorders: from surveys to nosology and back. Dialogues Clin Neurosci 2017;19:127–36.
[322] Stein D. Is disorder X in category or spectrum Y? General considerations and application to the relationship between obsessive-compulsive disorder and anxiety disorders. Depress Anxiety 2008;25:330–5.
[323] Nesse R, Stein D. Towards a genuinely medical model for psychiatric nosology. BMC Med 2012;10:5.
[324] Grinker R. A struggle for eclectism. Am J Psychiatr 1964;121:451–7.
[325] Engel G. The need for a new medical model: a challenge for biomedicine. Science 1977;196:129–36.
[326] Bolton D, Gillett G. The biopsychosocial model of health and disease. Springer; 2019.
[327] Pilgrim D. The biopsychosocial model in health research: its strengths and limitations for critical realists. J Crit Realism 2015;14:164–80.
[328] Hladký J, Havlíček J. Was Tinbergen an Aristotelian? Comparison of Tinbergen's four whys and Aristotle's four causes. Hum Ethol Bull 2013;28:3–11.
[329] Mayr E. Cause and effect in biology. Science 1961;134:1501–6.
[330] Haig D. Proximate and ultimate causes: how come? and what for? Biol Philos 2013;28:781–6.
[331] Mitchell S. Biological complexity and integrative pluralism. Cambridge University Press; 2003.
[332] Kellert S, Longino H, Waters C. Scientific pluralism. University of Minnesota Press; 2006.
[333] De Vreese L, Weber E, Van Bouwel J. Explanatory pluralism in the medical sciences: theory and practice. Theor Med Bioeth 2010;31:371–90.

[334] Pedersen N, Wright C. Truth and pluralism: current debates. Oxford University Press; 2013.
[335] Braillard P-A, Malaterre C. Explanation in biology: an inquiry into the diversity of explanatory patterns in the life sciences. Springer; 2015.
[336] Mantzavinos C. Explanatory pluralism. Cambridge University Press; 2016.
[337] Vandenbroucke J, Broadbent A, Pearce N. Causality and causal inference in epidemiology: the need for a pluralistic approach. Int J Epidemiol 2016;45:1776–86.
[338] Ruphy S. Scientific pluralism reconsidered: a new approach to the (dis)unity of science. University of Pittsburgh Press; 2017.
[339] Viet W. Model pluralism. Philos Soc Sci 2019;50:91–114.
[340] Susser M. Causal thinking in the health sciences: concepts and strategies. Oxford University Press; 1973.
[341] Lewis D. Causation. J Philos 1973;70:556–67.
[342] Kendler K. The dappled nature of causes of psychiatric illness: replacing the organic-functional/hardware-software dichotomy with empirically based pluralism. Mol Psychiatry 2012;17:377–88.
[343] Beebee H, Hitchcock C, Price H. Making a difference: essays on the philosophy of causation. Oxford University Press; 2017.
[344] Ghaemi S. Existence and pluralism: the rediscovery of Karl Jaspers. Psychopathology 2007;40:75–82.
[345] Kendler K. The structure of psychiatric science. Am J Psychiatr 2014;171:931–8.
[346] Eisenberg L. Disease and illness distinctions between professional and popular ideas of sickness. Cult Med Psychiatry 1977;1:9–23.
[347] Kleinman A. Rethinking psychiatry: from cultural category to personal experience. Free Press; 1988.
[348] Hofmann B. Disease, illness, and sickness. In: Solomon M, Simon J, Kincaid H, editors. The Routledge companion to philosophy of medicine. Routledge; 2016.
[349] Osler W. The rise and fall of the biopsychosocial model: reconciling art and science in psychiatry. Johns Hopkins University Press; 2010.
[350] Schwartz M, Wiggins O. Science, humanism, and the nature of medical practice: a phenomenological view. Perspect Biol Med 1985;28:331–61.
[351] Shorter E. History of psychiatry: from the era of the asylum to the age of Prozac. Wiley; 1997.
[352] Harrington A. Mind fixers: psychiatry's troubled search for the biology of mental illness. Norton; 2019.
[353] O'Neill O. Autonomy and trust in bioethics. Cambridge University Press; 2002.
[354] Kotzee B. Virtue epistemology and clinical medical judgement. In: Battaly H, editor. The Routledge handbook of virtue epistemology. Routledge; 2018.
[355] Kotzee B, Paton A, Conroy M. Towards an empirically informed account of phronesis in medicine. Perspect Biol Med 2016;59:337–50.
[356] Dreyfus H. What computer's can't do: the limits of artificial intelligence. MIT Press; 1972.
[357] Carver C. Explaining the brain: what a science of the mind-brain could be. Oxford University Press; 2007.
[358] Panksepp J, Biven L. The archeology of mind: neuroevolutionary origins of human emotions. Norton; 2012.
[359] Silver D, Hubert T, Schrittwieser J, Antonoglou I, Lai M, Guez A, et al. A general reinforcement learning algorithm that masters chess, shogi, and Go through self-play. Science 2018;6419:1140–4.
[360] Löwel S, Singer W. Selection of intrinsic horizontal connections in the visual cortex by correlated neuronal activity. Science 1992;5041:209–12.

[361] Hebb D. Organization of behavior. Wiley & Sons; 1949.
[362] Crane T, Patterson S. History of the mind-body problem. Routledge; 2001.
[362a] Robb D. The properties of mental causation, Philosoph Q 2003;47:178–94.
[363] Woodward J. Making things happen: a theory of causal explanation. Oxford University Press; 2003.
[364] Craver C, Bechtel W. Top-down causation without top-down causes. Biol Philos 2007;22:547–63.
[365] Murphy N, Ellis G, O'Connor T. Downward causation and the neurobiology of free will. Springer; 2009.
[366] Van Gulick R. Understanding consciousness – have we cut the Gordian knot or not? (Integration, unity, and the self). Philos Exch 2016;45:2.
[367] Levine J. Materialism and qualia: the explanatory gap. Pac Philos Q 1983;64:354–61.
[368] Nagel T. What is it like to be a bat? Philos Rev 1974;83:435–56.
[369] Chalmers D. Facing up to problem of consciousness. J Conscious Stud 1995;2:200–19.
[370] Flanagan O. Science of the mind. MIT Press; 1991.
[371] McGinn C. The mysterious flame: conscious minds in a material world. Basic Books; 1999.
[372] Humphrey N. How to solve the mind-body problem. Imprint Academic; 2000.
[373] Tye M. Phenomenal concepts: a new perspective on the major puzzles of consciousness. MIT Press; 2009.
[374] Dennett D. The mystery of David Chalmers. J Conscious Stud 2012;19:86–95.
[375] Lavazza A, Robinson H. Contemporary dualism: a defense. Routledge; 2014.
[376] Lorenz H. The brute within: appetitive desire in Plato and Aristotle. Oxford University Press; 2006.
[377] Reuter M, Svensson F. Mind, body, and morality: new perspectives on Descartes and Spinoza. Routledge; 2019.
[378] LeBuffe M. From bondage to freedom: Spinoza on human excellence. Oxford University Press; 2010.
[379] Pitson A. Hume and the mind/body relation. Hist Philos Q 2000;17:277–95.
[380] Ryle G. The concept of mind. University of Chicago Press; 1949.
[381] Tanney J. Rules, reason, and self-knowledge. Harvard University Press; 2013.
[382] Fogel S. The mind is what your brain does for a living. Greenleaf Book Group Press; 2014.
[383] Polanyi M. Life's irreducible structure. Science 1968;160:1308–12.
[384] O'Connor T. Emergent properties. Am Philos Q 1994;31:91–104.
[385] Beckermann A, Flohr H, Kim J. Emergence or reduction? Essays on the prospects of nonreductive physicalism. de Gruyter; 1992.
[386] Bedau M, Humphreys P. Emergence: contemporary readings in philosophy and science. MIT Press; 2008.
[387] Corradini A, O'Connor T. Emergence in science and philosophy. Routledge; 2010.
[388] Sawyer R. Emergence in psychology: lessons from the history of non-reductionist science. Hum Dev 2002;45:2–28.
[389] Smuts J. Holism and evolution. The Macmillan Company; 1926.
[390] Shelley C. Jan Smuts and personality theory: the problem of Holism in psychology. In: Valsiner J, editor. Striving for the whole: creating theoretical syntheses. Routledge; 2008.
[391] Debrock G. Process pragmatism: essays on a quiet philosophical revolution. Rodopi; 2003.
[392] Henning B, Myers W, John J. Thinking with Whitehead and the American pragmatists: experience and reality. Lexington Books; 2015.
[393] Matthews E. Merleau-Ponty's body-subject and psychiatry. Int Rev Psychiatry 2004;16:190–8.
[394] Cassirer E. An essay on man. Yale University Press; 1944.

[395] Henderson E. Homo symbolicus: a definition of man. Man and World 1971;4:131–50.
[396] Henshilwood C, d'Errico F. Homo symbolicus: the dawn of language, imagination and spirituality. John Benjamins; 2011.
[397] Frankl V. Man's search for meaning: an introduction to logotherapy. Beacon; 1946/1992.
[398] May R. The meaning of anxiety. Ronald Press; 1950.
[399] Yalom I. Existential psychotherapy. Basic Books; 1980.
[400] Payne M. Narrative therapy. 2nd ed. Sage; 2006.
[401] Charon R, DasGupta S, Hermann I, Marcus E, Rivera C, Spencer D, et al. The principles and practice of narrative medicine. Oxford University Press; 2017.
[402] Bruner J. Actual minds, possible worlds. Harvard University Press; 1986.
[403] Kagan J. In defense of minds. Yale University Press; 2006.
[404] Midgley M. The myths we live by. Routledge; 2003.
[405] Rosenberg A. How history gets things wrong: the neuroscience of our addiction to stories. MIT Press; 2018.
[406] Kelly M, Heath I, Howick J, Greenhalgh T. The importance of values in evidence-based medicine. BMC Med Ethics 2015;16:69.
[407] Fulford K. Values-based practice: a new partner to evidence-based practice and a first for psychiatry. Mens Sana Monogr 2008;6:10–21.
[408] Hutchinson P, Read R. Reframing health care: philosophy for medicine and human flourishing. In: Loughlin M, editor. Debates in values-based practice: arguments for and against. Cambridge University Press; 2014.
[409] Berg H. Virtue ethics and integration in evidence-based practice in psychology. Front Psychol 2020;11:258.
[410] Ellis A. On Joseph Wolpe's espousal of cognitive-behavior therapy. Am Psychol 1979;34:98–9.
[411] Yakeley J. Psychoanalysis in modern mental health practice. Lancet Psychiatry 2018;5:443–50.
[412] Erickson M. Rethinking Oedipus: an evolutionary perspective of incest avoidance. Am J Psychiatr 2006;150:411–6.
[413] Stein D, Solms M, van Honk J. The cognitive-affective neuroscience of the unconscious. CNS Spectr 2006;11:580–3.
[414] Kandel E. Biology and the future of psychoanalysis: a new intellectual framework for psychiatry revisited. Am J Psychiatr 1999;156:505–24.
[415] Bowlby J. Attachment and loss. Vol. 1: Attachment. Basic Books; 1969.
[416] Wallin D. Attachment in psychotherapy. GuIlford Press; 2007.
[417] Spiegel D. The developing mind: how relationships and the brain interact to shape who we are. 2nd ed. Guilford Press; 2012.
[418] Cassidy J, Shaver P. Handbook of attachment: theory, research, and clinical applications. Guilford Press; 2016.
[419] McCulloch W, Pitts W. A logical calculus of the ideas immanent in nervous activity. Bull Math Biophys 1943;5:115–33.
[420] Sellars W. Empiricism and the philosophy of mind. In: Scriven M, Feyerabend P, Maxwell G, editors. Minnesota studies in the philosophy of science. University of Minneapolis Press; 1956.
[421] Putnam H. The nature of mental states. In: Mind, nature and reality: philosophical papers. Cambridge University Press; 1975.
[422] Rescorla M. The computational theory of mind. In: Zalta E, editor. The Stanford encyclopedia of philosophy. Stanford University; 2015.
[423] Medler D. A brief history of connectionism. Neural Comput Surv 1998;1:18–73.

[424] Stein D, Jacques L. Neural networks and psychopathology. Cambridge University Press; 1998.
[425] Gordon J, Redish A. Computational psychiatry. MIT Press; 2016.
[426] Norman D. Cognition in the head and in the world: an introduction to the special issue on situated action. Cogn Sci 1993;17:1–6.
[427] Millikan R. The son and the daughter: on Sellars, Brandom, and Millikan. In: Millikan R, editor. Language: a biological model. Oxford University Press; 2005.
[428] Dotov D, Nie L, de Wit M. Understanding affordances: history and contemporary development of Gibson's central concept. Avant 2012;3:28–39.
[429] Christensen B. The second cognitive revolution: a tribute to Rom Harré. Springer; 2019.
[430] Gallagher S. Enactivist interventions: rethinking the mind. Oxford University Press; 2017.
[431] Newen A, De Bruin L, Gallagher S. The Oxford handbook of 4E cognition. Oxford University Press; 2018.
[432] Johnson M. Embodied mind, meaning, and reason: how our bodies give rise to understanding. University of Chicago Press; 2017.
[433] Kihlstrom J. The cognitive unconscious. Science 2008;237:1445–52.
[434] Kihlstrom J, Mulvaney S, Tobias B, Tobis I. The emotional unconscious. In: Eich E, Kihlstrom J, Bower G, Forgas J, Niedenthal P, editors. Cognition and emotion. Oxford University Press; 2000.
[435] Kahneman D. Thinking fast and slow. Farrar, Straus and Giroux; 2011.
[436] Weinberger J, Stoycheva V. The unconscious: theory, research, and clinical implications. Guilford; 2020.
[437] Horowitz M. Cognitive psychodynamics: from conflict to character. Wiley; 1998.
[438] Stein D. Cognitive science and the unconscious. American Psychiatric Press; 1997.
[439] Stein D, Young J. Cognitive science and clinical disorders. Academic Press; 1992.
[440] Riso L, du Toit P, Stein D, Young J. Cognitive schemas and core beliefs in psychological problems: a scientist-practitioner guide. American Psychological Association; 2007.
[441] Wachtel P. Psychoanalysis and behavior therapy. Basic Books; 1977.
[442] Arkowitz H, Messer S. Psychoanalytic therapy and behavior therapy: is integration possible. Plenum Press; 1984.
[443] Castonguay L, Eubanks C, Goldfried M, Muran J, Lutz W. Research on psychotherapy integration: building on the past, looking to the future. Psychother Res 2015;25:365–82.
[444] Rucker R. Wetware. Avon Books; 1988.
[445] Bray D. Wetware: a computer in every living cell. Yale University Press; 2009.
[446] John P, Bishop M. Views into the Chinese Room: new essays on Searle and artificial intelligence. Oxford University Press; 2002.
[447] Searle J. The rediscovery of mind. MIT; 1992.
[448] de Freitas Araujo S. Searle's new mystery, or, how not to solve the problem of consciousness. Riv Internazionale Filos Psicol 2013;1:1–12.
[449] Núñez R. What brain for God's-eye?: biological naturalism, ontological objectivism, and Searle. J Conscious Stud 1995;2:149–66.
[450] French R. Peeking behind the screen: the unsuspected power of the standard Turing test. J Exp Theor Artif Intell 2000;12:331–40.
[451] French R. Dusting off the Turing test. Science 2012;336:164–5.
[452] Chalmers D. The conscious mind. Oxford University Press; 1996.
[453] Dennett D. Consciousness explained. Little, Brown and Co; 1991.
[454] Harnad S. Why and how we are not zombies. J Conscious Stud 1995;1:164–7.

[455] Brooks R, Hassabis D, Bray D, Shashua A. Turing centenary: is the brain a good model for machine intelligence? Nature 2012;482:462–3.
[456] Weizenbaum J. Computer power and human reason: from judgment to calculation. WH Freeman & Co; 1976.
[457] Rogers C. Client-centred therapy: its current practice, implications and theory. Constable; 1951.
[458] Lewis-Williams D. The mind in the cave: consciousness and the origins of art. Thames & Hudson; 2002.
[459] Evans H. Music, medicine and embodiment. Lancet 2010;375:886–7.
[460] Schubert E, Canazza S, De Pol G, Rodà A. Algorithms can mimic human piano performance: the deep blues of music. J New Music Res 2017;46:175–86.
[461] Osbeck L, Held B. Rational intuition: philosophical roots, scientific investigations. Cambridge University Press; 2014.
[462] Depaul M, Ramsey W. Rethinking intuition: the psychology of intuition and its role in philosophical inquiry. Rowman & Littlefield; 1998.
[463] Rothenberg D. Why birds sing: a journey into the mystery of bird song. Basic Books; 2006.
[464] Bekoff M. The emotional lives of animals: a leading scientist explores animal joy, sorrow, and empathy—and why they matter. New World Library; 2007.
[465] Ackerman J. The genius of birds. Penguin; 2016.
[466] Hutchinson P. Shame, placebo and world-taking cognitivism. In: George S, Jung P, editors. Cultural ontology of the self in pain. Springer; 2016.
[467] Maund B. Color. In: Zalta E, editor. The Stanford encyclopedia of philosophy. Stanford University; 2019.
[468] Hardin C. Color for philosophers. Hackett; 1988.
[469] Hilbert D. Color and color perception. Stanford University Press; 1987.
[470] Thompson E. Colour vision: a study in cognitive science and the philosophy of perception. Routledge; 1994.
[471] Mizrahi V. Color objectivism and color pluralism. Dialectica 2006;283–306.
[472] Hardin C. Qualia and materialism: closing the explanatory gap. Philos Phenomenol Res 1987;48:281–98.
[473] Ludlow P, Nagasawa Y, Stoljar D. There's something about Mary: essays on phenomenal consciousness and Frank Jackson's knowledge argument. MIT Press; 2004.
[474] Anscombe G. Causality and determination: an inaugural lecture. Cambridge University Press; 1971.
[475] Kendler K. Toward a philosophical structure for psychiatry. Am J Psychiatr 2005;162:433–40.
[476] de Muijnck W. Dependencies, connections, and other relations: a theory of mental causation. Springer; 2003.
[477] Mackie J. The cement of the universe: a study of causation. Oxford University Press; 1974.
[478] Hutto D. Folk psychological narratives: the sociocultural basis of understanding reasons. MIT Press; 2008.
[479] Griffin D. Animal minds: beyond cognition to consciousness. University of Chicago Press; 2001.
[480] Andrews K, Beck J. The Routledge handbook of philosophy of animal minds. Routledge; 2017.
[481] Carruthers P. Human and animal minds: the consciousness questions laid to rest. Oxford University Press; 2020.

[482] King M, Wilson A. Evolution at two levels in humans and chimpanzees. Science 1975;188:107–16.
[483] Stone J, Davenport R. Functional neurological symptoms. Clin Med (London) 2013;13:80–3.
[484] Spitzer R, First M, Williams J, Kendler K, Pincus H, Tucker G. Now is the time to retire the term "organic mental disorders". Am J Psychiatr 1992;149:240–4.
[485] Bell V, Wilkinson S, Greco M, Hendrie C, Mills B, Deeley Q. What is the functional/organic distinction actually doing in psychiatry and neurology. Wellcome Open Res 2020;5:138.
[486] Stein D. Is there a "mosquito net" for anxiety and mood disorders? Curr Psychiatry Rep 2009;11:264–5.
[487] Fuchs T, Schlimme J. Embodiment and psychopathology: a phenomenological perspective. Curr Opin Psychiatry 2009;6:570–5.
[488] Röhricht F, Gallagher S, Geuter U, Hutto D. Embodied cognition and body psychotherapy: the construction of new therapeutic environments. Sensoria J Mind Brain Cult 2014;56:11–20.
[489] Gallagher S, Varga S. Social cognition and psychopathology: a critical overview. World Psychiatry 2015;14:5–14.
[490] Kirmayer L, Gómez-Carrillo A. Agency, embodiment and enactment in psychosomatic theory and practice. Med Humanit 2019;45:169–82.
[491] de Haan S. Enactivism and psychiatry. Cambridge University Press; 2020.
[492] Stein D. Cognitive embodiment and anxiety disorders. Philos Psychiatry Psychol 2020;27:53–5.
[493] Head H, Holmes G. Sensory disturbances from cerebral lesions. Brain 1911–1912;34:102–254.
[494] Shokur S, O'Doherty J, Winans J, Bleuler H, Lebedev M, Nicolelis M. Expanding the primate body schema in sensorimotor cortex by virtual touches of an avatar. Proc Natl Acad Sci U S A 2013;110:15121–6.
[495] Dahln B. Critique of the schema construct. Scand J Educ Res 2000;45:278–300.
[496] Gallagher S. How the body shapes the mind. Oxford University Press; 2005.
[497] Gibbs R. Embodiment and cognitive science. Cambridge University Press; 2005.
[498] Stein D. Schemas in the cognitive and clinical sciences: an integrative construct. J Psychother Integr 1992;2:45–63.
[499] Plant K, Stanton N. The explanatory power of schema theory: theoretical foundations and future applications in ergonomics. Ergonomics 2005;56:1–15.
[500] Simon B. Mind and madness in ancient Greece: the classical roots of modern psychiatry. Cornell University Press; 2000.
[501] Cohen E, Zinaich S. Philosophy, counseling, and psychotherapy. Cambridge Scholars Publishing; 2013.
[502] James W. What is an emotion? Mind 1884;9:188–205.
[503] Roll E. The brain and emotion. Oxford University Press; 1999.
[504] Damasio A, Carvalho G. The nature of feelings: evolutionary and neurobiological origins. Nat Rev Neurosci 2013;14:143–52.
[505] Schmitter A. 17th and 18th century theories of emotions. In: Zalta E, editor. The Stanford encyclopedia of philosophy. Stanford University; 2016.
[506] Hume D. A treatise of human nature: being an attempt to introduce the experimental method of reasoning into moral subjects. John Noon; 1739.
[507] Nussbaum M. Upheavals of thought: the intelligence of emotions. Cambridge University Press; 2003.
[508] Sartre J-P. Sketch for a theory of emotions. Methuen; 1939/1962.
[509] Bakewell S. At the existentialist café: freedom, being, and apricot cocktails with Jean-Paul Sartre, Simone de Beauvoir, Albert Camus, Martin Heidegger, Maurice Merleau-Ponty and others. Other Press; 2016.

[510] Scarantino A, de Sousa R. Emotion. In: Zalta E, editor. The Stanford encyclopedia of philosophy. Stanford University; 2018.
[511] Prinz J. Gut reactions: a perceptual theory of emotion. Oxford University Press; 2004.
[512] Graver M. Stoicism and emotion. Common Knowledge; 2009.
[513] Collier M. Hume's science of emotions: feeling theory without tears. Hume Stud 2011;37:3–18.
[514] Fodor J. Hume variations. Oxford University Press; 2003.
[515] Reed P, Vitz R. Hume's moral philosophy and contemporary psychology. Routledge; 2018.
[516] Hall E. Aristotle's way: how ancient wisdom can change your life. Penguin; 2019.
[517] Ravven H. Spinoza's anticipation of contemporary affective neuroscience. Conscious Emot 2003;4:257–90.
[518] Homiak M. Hume's ethics: ancient or modern? Pac Philos Q 2000;81:215–23.
[519] Cochran M. The Cambridge companion to Dewey. Cambridge University Press; 2010.
[520] Kemp C. Dewey's Darwin and Darwin's Hume. Pluralist 2017;12:1–26.
[521] Barandiaran X, Di Paulo E. A genealogical map of the concept of habit. Front Hum Neurosci 2014;8:522.
[522] Sapiro G. Habitus: history of a concept. In: Wright J, editor. International encyclopedia of the social & behavioral sciences. 2nd ed. Elsevier; 2015.
[523] Ravaisson F. Of habit. Continuum; 1838/2008.
[524] Sparrow T, Hutchinson A. A history of habit: from Aristotle to Bourdieu. Lexington; 2013.
[525] Carlisle C. On habit. Routledge; 2014.
[526] McDowell J. Virtue and reason. The Monist 1979;62:331–50.
[527] Nussbaum M. The fragility of goodness: luck and ethics in Greek tragedy and philosophy. Cambridge University Press; 1986.
[528] Hazony Y. The philosophy of Hebrew scripture. Cambridge University Press; 2012.
[529] Vitz P. The use of stories in moral development. Am Psychol 1990;45:709–20.
[530] Snow N. Virtue as social intelligence: an empirically grounded theory. Routledge; 2009.
[531] Slote M. Education and human values: reconciling talent with an ethics of care. Routledge; 2012.
[532] Annas J, Narvaez D, Snow N. Developing the virtues: integrating perspectives. Oxford University Press; 2016.
[533] Stichter M. The skilfulness of virtue: improving our moral and epistemic lives. Cambridge University Press; 2018.
[534] Hursthouse R. On virtue ethics. Oxford University Press; 1999.
[535] Pappas J. The concept of measure and the criterion of sustainability. The St John's Review 2015;56:74–9.
[536] Price A. Emotions in Plato and Aristotle. In: Goldie P, editor. Oxford handbook of philosophy of emotion. Oxford University Press; 2009.
[537] Reeve C. Aristotelian education. In: Rorty A, editor. Philosophers on education: new historical perspectives. Routledge; 1998.
[538] Kraut R. Aristotle's ethics. In: Zalta E, editor. The Stanford encyclopedia of philosophy. Stanford University; 2018.
[539] Solomon R. Not passion's slave: emotions and choice. Oxford University Press; 2003.
[540] MacIntrye A. Confucian ethics: a comparative study of self, autonomy, and community. Cambridge University Press; 2004.
[541] Angle S, Slote M. Virtue ethics and Confucianism. Routledge; 2013.
[542] Puett M, Gross-Loh C. The path: what Chinese philosophers can teach us about the good life. Simon & Schuster; 2016.

[543] Robinson D. The deep ecology of rhetoric in Mencius and Aristotle: a somatic guide. State University of New York; 2016.
[544] Slote M. Between psychology and philosophy: East-West themes and beyond. Palgrave Macmillan; 2020.
[545] Niedenthal P. Embodying emotion. Science 2007;316:1002–5.
[546] Hutchinson P. Shame and philosophy: an investigation in the philosophy of emotions and ethics. Palgrave Macmillan; 2008.
[547] Seneca L. Moral letters to Lucillius. 65 AD.
[548] Spinoza. Ethics. 1677.
[549] Oring E. The Jokes of Sigmund Freud: a study in humor and Jewish identity. Jason Aronson; 2007.
[550] Gellner E. The psychoanalytic movement: the cunning of unreason. Northwest University Press; 1985.
[551] Madvin G, Markel G. Finding happiness with Aristotle as your guide: action strategies based on 10 timeless ideas. iUniverse; 2012.
[552] Robertson D. The philosophy of cognitive-behavioural therapy (CBT): Stoic philosophy as rational and cognitive psychotherapy. Karnac Books; 2010.
[553] Pies R. The Judaic foundations of cognitive-behavioral therapy: Rabbinic and Talmudic underpinnings of CBT and REBT. iUniverse; 2010.
[554] Mitchell S. Relational concepts in psychoanalysis: an integration. Harvard University Press; 1988.
[555] Friedman R, Downey J, Alfonso C. On the birth of psychodynamic psychiatry. Psychiatr Clin N Am 2018;41:177–82.
[556] Darwin C. The expression of the emotions in man and animals. John Murray; 1872.
[557] Rose S, Bisson J, Churchill R, Wessely S. Psychological debriefing for preventing post traumatic stress disorder. Cochrane Database Syst Rev 2002;2, CD000560.
[558] Stein D. Psychiatric aspects of the Truth and Reconciliation Commission in South Africa. Br J Psychiatry 1998;173:455–8.
[559] Nussbaum M. Anger and forgiveness: resentment, generosity, justice. Oxford University Press; 2016.
[560] Allais L. Wiping the slate clean: the heart of forgiveness. Philos Public Aff 2008;36:33–68.
[561] van Niekerk A. Politico-philosophical perspectives on reconciliation. Dutch Reform Theol J 2010;51:274–8.
[562] Oelofsen R. Afro-communitarian forgiveness and the concept of reconciliation. S Afr J Philos 2015;34:368–78.
[563] Kabat-Zinn J, Davidson R. The mind's own physician: a scientific dialogue with the Dalai Lama on the healing power of meditation. New Harbinger Publications; 2011.
[564] Seppälä E, Simon-Thomas E, Brown S, Worline M, Daryl C, James D. The Oxford handbook of compassion science. Oxford University Press; 2017.
[565] Safina C. Beyond words: what animals think and feel. Henry Holt; 2015.
[566] de Waal F. Mama's last hug: animal emotions and what they tell us about ourselves. Thorndike Press; 2019.
[567] LeDoux J, Brown R. A higher order theory of emotional consciousness. Proc Natl Acad Sci U S A 2016;217:E2016–25.
[568] Berridge K, Kringelbach M. Pleasure systems in the brain. Neuron 2015;86:646–64.
[569] Salzman C, Fusi S. Emotion, cognition, and mental state representation in amygdala and prefrontal cortex. Annu Rev Neurosci 2010;33:173–202.
[570] Pessoa L. The cognitive-emotional brain: from interactions to integration. MIT Press; 2013.
[571] Thagard P. Hot thought: mechanisms and applications of emotional cognition. Bradford Book; 2006.

[572] Elliot R, Zahn R, Deakin J, Anderson I. Affective cognition and its disruption in mood disorders. Neuropsychopharmacology 2011;36:153–82.
[573] Kret M, Bocanegra B. Adaptive hot cognition: how emotion drives information processing and cognition steers affective processing. Front Psychol 2016;7:1920.
[574] Kozhevnikovx M. Cognitive styles in the context of modern psychology: toward an integrated framework of cognitive style. Psychol Bull 2007;133:464–81.
[575] Nessy R. Good reasons for bad feelings: insights from the frontier of evolutionary psychiatry. Penguin; 2019.
[576] Panksepp J, Panksepp J. Seven sins of evolutionary psychology. Evol Cogn 2000;6108:131.
[577] Gould S, Lewontin R. The spandrels of San Marco and the Panglossian paradigm: a critique of the adaptationist programme. Proc R Soc Lond B 1979;205:581–98.
[578] Sahlins M. The use and abuse of biology: an anthropological critique of sociobiology. University of Michigan Press; 1976.
[579] Segerstråle U. Defenders of the truth: the battle for science in the sociobiology debate. Oxford University Press; 2000.
[580] Midgley M. The solitary self: Darwin and the selfish gene. Routledge; 2016.
[581] Mischel W. The marshmallow test: mastering self control. Little, Brown, Spark; 2014.
[582] Mullainathan S, Shafir E. Scarcity: why having too little means so much. Penguin; 2013.
[583] Stein D, Moeller F. The man who turned bad. CNS Spectr 2005;10:88–90.
[584] Sloan E, Hall K, Moulding R, Bryce S, Mildred H, Staiger P. Emotion regulation as a transdiagnostic treatment construct across anxiety, depression, substance, eating and borderline personality disorders: a systematic review. Clin Psychol Rev 2017;57:141–63.
[585] Bateson P, Martin P. Design for a life: how behavior and personality develop. Simon & Schuster; 2000.
[586] Ridley M. Nature via nurture: genes, experience, and what makes us human. HarperCollins; 2003.
[587] LeDoux J. Synaptic self: how our brains become who we are. Penguin; 2003.
[588] Gill M. The British moralists on human nature and the birth of secular ethics. Cambridge University Press; 2009.
[589] Musolf G. John Dewey's social psychology and neopragmatism: theoretical foundations of human agency. Soc Sci J 2001;38:277–95.
[590] Buhle J, Silvers J, Wager T, Lopez R, Onyemekwu C, Kober H, et al. Cognitive reappraisal of emotion: a meta-analysis of human neuroimaging studies. Cereb Cortex 2014;24:2981–90.
[591] Cozolino L. The neuroscience of psychotherapy: healing the social brain. 2nd ed. Norton; 2010.
[592] Erdelyi M. Psychoanalysis: Freud's cognitive psychology. WH Freeman; 1985.
[593] Wilson T. Strangers to ourselves: discovering the adaptive unconscious. Belknap Press; 2002.
[594] Vedentam S. The hidden brain: how our unconscious minds elect presidents, control markets, wage wars, and save our lives. Spiegel & Grau; 2010.
[595] Bargh J. Before you know it: the unconscious reasons we do what we do. Atria Books; 2017.
[596] Festinger L, Riecken H, Schachter S. When prophecy fails. Harper Torchbooks; 1956.
[597] Gilovich T. How we know what isn't so: the fallibility of human reason in everyday life. Free Press; 1993.
[598] Piatelli-Palmarini M. Inevitable illusions: how mistakes of reason rule our minds. Wiley; 1994.
[599] Sternberg R. Why smart people can be so stupid. Yale University Press; 2002.
[600] Fine C. A mind of its own: how your brain distorts and deceives. Norton; 2006.
[601] Travis C, Aaronson E. Mistakes were made (but not by me): why we justify foolish beliefs, bad decisions, and hurtful acts. Houghton Mifflin Harcourt; 2007.

[602] Ariely D. Predictably irrational: the hidden forces that shape our decisions. HarperCollins; 2008.
[603] Brafman O, Brafman R. Sway: the irresistible pull of irrational behavior. Doubleday; 2008.
[604] Chabris C, Simon D. The invisible gorilla: and other ways our intuitions deceive us. MJF Books; 2010.
[605] Halinan J. Why we make mistakes: how we look without seeing, forget things in seconds, and are all pretty sure we are way above average. Broadway Books; 2010.
[606] Dunning D. Self-insight: roadblocks and detours on the path to knowing thyself. Psychology Press; 2012.
[607] Banaji M, Greenwald A. Blindspot: hidden biases of good people. Delacorte Press; 2013.
[608] Epley N. Mindwise: why we misunderstand what others think, believe, feel, and want. Knopf; 2014.
[609] Nisbett R. Mindware: tools for smart thinking. Farrar, Straus and Giroux; 2015.
[610] Mercier H, Sperber D. The enigma of reason. Harvard University Press; 2017.
[611] Mark K. Theses on Feuerbach. 1845.
[612] Dolan P. Happiness by design: change what you do, not how you think. Hudson Street Press; 2014.
[613] Bruner J. The Freudian conception of man and the continuity of nature. Daedalus 1958;87:77–84.
[614] Bergmann M. The origins and organization of unconscious conflict: the selected works of Martin S. Bergmann. Routledge; 2017.
[615] Piaget J. The relation of affectivity to intelligence in the mental development of the child. Bull Menninger Clin 1962;26:129–37.
[616] Furth H. Knowledge as desire: an essay on Freud and Piaget. Columbia University Press; 1987.
[617] Barratt L. Are emotions natural kinds? Perspect Psychol Sci 2006;1:28–58.
[618] Izard C. Basic emotions, natural kinds, emotion schemas, and a new paradigm. Perspect Psychol Sci 2007;2:260–80.
[619] Darwin C. On the origin of species by means of natural selection. John Murray; 1859.
[620] Gould S. Full house: the spread of excellence from Plato to Darwin. Harmony Books; 1996.
[621] Sober E. Evolution, population thinking, and essentialism. Philos Sci 1980;47:350–83.
[622] Dupré J. Processes of life: essays in the philosophy of biology. Oxford University Press; 2012.
[623] Dennett D. Darwinism and the overdue demise of essentialism. In: Smith D, editor. How biology shapes philosophy: new foundations for naturalism. Cambridge University Press; 2017.
[624] Boyd R. Homeostasis, species, and higher taxa. In: Wilson R, editor. Species: new interdisciplinary essays. MIT Press; 1999.
[625] Greenberg L. Emotions, the great captains of our lives: their role in the process of change in psychotherapy. Am Psychol 2012;67:697–707.
[626] Horowitz M. Person schemas and maladaptive interpersonal schemas. University of Chicago Press; 1991.
[627] Bentham J. An introduction to the principles of morals and legislation. T Payne and Son; 1789.
[628] Peale N. The power of positive thinking: a practical guide to mastering the problems of everyday living. Prentice Hall; 1952.
[629] Byrne R. The Secret. Atria; 2006.
[630] Burkeman O. The antidote: happiness for people who can't stand positive thinking. Faber and Faber; 2012.

[631] Bloom H. An elegy for the canon. In: Morrissey L, editor. Debating the canon: a reader from Addison to Nafisi. Palgrave Macmillan; 2005.
[632] Critchley S. Tragedy, the Greeks, and us. Pantheon; 2019.
[633] Leo R. Nil volentibus arduum, Baruch Spinoza, and the reason of tragedy. In: Hoxby B, editor. Darkness visible: tragedy in the enlightenment. Ohio State University Press; forthcoming.
[634] Galgut E. Poetry and the pity. Hume's account of tragic pleasure. Br J Aesthet 2001;4:411–24.
[635] Jacques R. The tragic world of John Dewey. J Value Inq 1991;25:249–61.
[636] Russel B. A free man's worship. Thomas Bird Mosher; 1923.
[637] Pascal B. Man's greatness comes from knowing he is wretched. In: Pensées; 1670.
[638] Jenkins M. Williams, Nietzsche, and Pessimism. J Nietzsche Stud 2012;43:315–25.
[639] van Niekerk A. Death, meaning and tragedy. S Afr J Philos 1999;18:408–26.
[640] Tabensky P. The positive function of evil. Palgrave Macmillan; 2009.
[641] McMahon D. Happiness: a history. Atlantic Monthly Press; 2005.
[642] Bortolotti L. Philosophy and happiness. Palgrave Macmillan; 2009.
[643] Cahn S, Vitrano S. Happiness: classic and contemporary readings in philosophy. Oxford University Press; 2012.
[644] David S, Boniwell I, Ayers A. Oxford handbook of happiness. Oxford University Press; 2012.
[645] Fletcher G. The Routledge handbook of philosophy of well-being. Routledge; 2015.
[646] Rabbås Ø, Emilsson E, Fossheim H, Tuominen M. The quest for happiness: ancient philosophers on happiness. Oxford University Press; 2015.
[647] Bosman P. Ancient routes to happiness. Classical Association of South Africa; 2017.
[648] Goldman A. Life's values: pleasure, happiness, well-being, and meaning. Oxford University Press; 2018.
[649] Kahneman D, Wakker P, Sarin R. Back to Bentham? Explorations of experienced utility. Q J Econ 1997;112:375–405.
[650] Long A. Epictetus: a Stoic and Socratic guide to life. Oxford University Press; 2002.
[651] Irvine W. Guide to the good life: the ancient art of Stoic joy. Oxford University Press; 2008.
[652] Robertson D. Stoicism and the art of happiness. McGraw-Hill Education; 2014.
[653] Pigliucci M. How to be a Stoic: using ancient philosophy to live a modern life. Basic Books; 2017.
[654] Marcano A. More than happiness: Buddhist and Stoic Wisdom for a sceptical age. Icon Books; 2018.
[655] Sherman N. Stoic warriors: the ancient philosophy behind the military mind. Oxford University Press; 2007.
[656] Erasmus D. The Praise of Folly. 1511.
[657] Du Châtelet E. Reflections on happiness. 1796.
[658] Bok S. Exploring happiness: from Aristotle to brain science. Yale University Press; 2010.
[659] Nozick R. Anarchy, State, and Utopia. Basic Books; 1974.
[660] Feldman F. Pleasure and the good life: concerning the nature, variety, and plausibiity of hedonism. Oxford University Press; 2004.
[661] Bramble B. The experience machine. Philos Compass 2016;11:136–45.
[662] Hindriks F, Douven I. Nozick's experience machine: an empirical study. Philos Psychol 2018;31:278–98.
[663] Wittgenstein L. Notebooks, 1914–1916. Blackwell; 1958.
[664] Balaska M. The notion of happiness in early Wittgenstein: towards a non-contentful account of happiness. S Afr J Philos 2014;33:407–15.
[665] Angier T. Happiness: overcoming the skill model. Int Philos Q 2015;55:5–23.
[666] Haybron D. The pursuit of unhappiness: the elusive psychology of well-being. Oxford University Press; 2008.

[667] Feldman F. What is this thing called happiness? Oxford University Press; 2010.
[668] Bostock D. Pleasure and activity in Aristotle's ethics. Phronesis 1988;33:251–72.
[669] Aristotle. Eudemian ethics.
[670] Parfit D. Reasons and persons. Oxford University Press; 1984.
[671] Seligman M. Authentic happiness: using the new positive psychology to realize your potential for lasting fulfillment. Free Press; 2002.
[672] Kekeş J. Enjoyment: the moral significance of styles of life. University of Oxford Press; 2008.
[673] Martin M. Happiness and the good life. Oxford University Press; 2012.
[674] Russell B. The conquest of happiness. Liveright; 1930.
[675] Kisner M. Spinoza on human freedom: reason, autonomy and the good life. Cambridge University Press; 2013.
[676] Immerware J. Hume's essays on happiness. Hume Stud 1989;15:307–24.
[677] Walker M. Reconciling the Stoic and the Sceptic: Hume on philosophy as a way of life and the plurality of happy lives. Br J Hist Philos 2013;21:879–901.
[678] Stack S. Education and the pursuit of happiness: John Dewey's sympathetic character. J Thought 1996;31:25–35.
[679] Hadot P. Philosophy as a way of life: spiritual exercises from Socrates to Foucault. Blackwell Publishers; 1995.
[680] Bentall R. A proposal to classify happiness as a psychiatric disorder. J Med Ethics 1992;18:94–8.
[681] Harris J, Birley J, Fulford K. A proposal to classify happiness as a psychiatric disorder. Br J Psychiatry 1993;162:539–42.
[682] Schneiderman S. The last psychoanalyst. CreateSpace; 2014.
[683] Skinner B. Walden Two. Hackett Publishing Company; 1948.
[684] Pinker S. The blank slate: the modern denial of human nature. Penguin; 2002.
[685] Locke J. An essay concerning human understanding. Thomas Bassett; 1690.
[686] Freud S. Mourning and melancholia. Standard Edition, Hogarth; 1917/1957.
[687] Alloy L, Abramson L. Depressive realism: four theoretical perspectives. In: Alloy L, editor. Cognitive processes in depression. Guilford; 1988.
[688] Haaga D, Beck A. Perspectives on depressive realism: implications for cognitive theory of depression. Behav Res Ther 1995;33:41–8.
[689] Moore M, Fresco D. Depressive realism: a meta-analytic review. Clin Psychol Rev 2012;32:496–509.
[690] Taylor S, Brown J. Illusion and well-being: a social psychological perspective on mental health. Psychol Bull 1988;103:193–210.
[691] Norem J. The positive power of negative thinking. Basic Books; 2002.
[692] Segerstrom S. Breaking Murphy's law: how optimists get what they want from life and pessimists can too. The Guilford Press; 2006.
[693] Sharot T. The optimism bias: a tour of the irrationally positive brain. Pantheon; 2011.
[694] Tierney J, Baumeister R. The power of bad: how the negativity effect rules us and how we can rule it. Penguin; 2019.
[695] Waller B. The sad truth, optimism, pessimism, and pragmatism. Ratio 2003;16:189–97.
[696] Bortolotti L. The epistemic innocence of irrational beliefs. Oxford University Press; 2020.
[697] Rosenberg K, Feder L. Behavioral addictions: criteria, evidence, and treatment. Academic Press; 2014.
[698] Petry N. Behavioral addictions: DSM-5® and beyond. Oxford University Press; 2015.

[699] Stein D, Billieux J, Bowden-Jones H, Grant J, Fineberg N, Higuchi S, et al. Balancing validity, utility, and public health considerations in disorders due to addictive behaviors. World Psychiatry 2018;17:363–4.
[700] Csikszentmihalyi M. Flow: the psychology of optimal experience. HarperCollins; 1991.
[701] Bok D. The politics of happiness: what government can learn from the new research on well-being. Princeton University Press; 2010.
[702] Skidelsky R, Skidelsky E. How much is enough?: money and the good life. Other Press; 2012.
[703] Radcliff B. The political economy of happiness. Cambridge University Press; 2013.
[704] Clark A, Flèche S, Layard R, Powdthavee N, Ward G. The origins of happiness: the science of well-being over the life course. Princeton University Press; 2018.
[705] Jain M, Sharma G, Mahendru M. Can I sustain my happiness? A review, critique and research agenda for economics of happiness. Sustainability 2019;11:6375.
[706] Helliwell J, Layard R, Sachs J, De Neve J. World happiness report. Sustainable Development Solutions Network; 2020.
[707] Schroeder K. Politics of gross national happiness: governance and development in Bhutan. Palgrave Macmillan; 2018.
[708] Kahneman D, Diener E, Schwarz N. Well-being: foundations of hedonic psychology. Russell Sage Foundation; 2003.
[709] Kesebir P, Diener E. In pursuit of happiness: empirical answers to philosophical questions. Perspect Psychol Sci 2008;3:117–25.
[710] Myers D, Diener E. The scientific pursuit of happiness. Perspect Psychol Sci 2018;13:218–25.
[711] Ryff C, Singer B. Know thyself and become what you are: a eudaimonic approach to psychological well-being. J Happiness Stud 2008;9:13–39.
[712] Greene J, Morrison I, Seligman M. Positive neuroscience. Oxford University Press; 2016.
[713] Jeste D, Palmer B. Positive psychiatry: a clinical handbook. American Psychiatric Association Publishing; 2015.
[714] Fava G, Guidi J. The pursuit of euthymia. World Psychiatry 2020;19:40–50.
[715] Peterson C. Character strengths and virtues: a handbook and classification. Oxford University Press; 2004.
[716] Peterson C. Flourish: a visionary new view of happiness and well-being. Atria; 2012.
[717] Kristjánsson K. Virtues and vices in positive psychology: a philosophical critique. Cambridge University Press; 2013.
[718] Ivtzan I, Lomas T, Hefferon K, Worth P. Second wave positive psychology: embracing the dark side of life. Routledge; 2016.
[719] Brown N, Lomas T, Eiroa-Orosa F. The Routledge international handbook of critical positive psychology. Routledge; 2017.
[720] Montero B. Thought in action: expertise and the conscious mind. Oxford University Press; 2016.
[721] Mills J. Autobiography. Longmans, Green, Reader, and Dyer; 1873.
[722] Eggleston B. The Happiness paradox. In: La Follette H, editor. The international encyclopedia of ethics. Blackwell; 2013.
[723] Gruber J, Mauss I, Tamir M. Dark side of happiness? How, when, and why happiness is not always good. Perspect Psychol Sci 2001;6:222–33.
[724] Whippman R. America the anxious: how our pursuit of happiness is creating a nation of nervous wrecks. St. Martin's Press; 2016.
[725] Killingsworth M, Gilbert D. A wandering mind is an unhappy mind. Science 2010;330:932.

[726] Rodski S. The neuroscience of mindfulness: the astonishing science behind how everyday hobbies help you relax. HarperCollins; 2019.
[727] Van Dam N, van Vugt M, Vago D. Mind the hype: a critical evaluation and prescriptive agenda for research on mindfulness and meditation. Perspect Psychol Sci 2017;13:36–61.
[728] Kreplin U, Farias M, Brazil I. The limited prosocial effects of meditation: a systematic review and meta-analysis. Sci Rep 2018;8:2403.
[729] Ratnayake S, Merry D. Forgetting ourselves: epistemic costs and ethical concerns in mindfulness exercises. J Med Ethics 2018;44:567–74.
[730] Purser R. McMindfulness: how mindfulness became the new capitalist spirituality. Repeater; 2019.
[731] Dworkin R. Artificial happiness: the dark side of the new happy class. Carroll & Graf; 2006.
[732] Ehrenreich B. Bright-sided: how positive thinking is undermining America. Metropolitan Books; 2009.
[733] Davies W. The happiness industry: how the government and big business sold us well-being. Verso; 2015.
[734] Frawley A. Semiotics of happiness: rhetorical beginnings of a public problem. Bloomsbury Academic; 2015.
[735] McKenzie J. Deconstructing happiness: critical sociology and the good life. Routledge; 2016.
[736] Doran P. Political economy of attention, consumerism and mindfulness: reclaiming the mindful commons. Routledge; 2017.
[737] Cabanas E, Illouz E. Manufacturing happy citizens: how the science and industry of happiness control our lives. Polity; 2019.
[738] Forbes D. Mindfulness and its discontents: education, self, and social transformation. Fernwood; 2019.
[739] Anand P. What human flourishing is and how we can promote it. Oxford University Press; 2016.
[740] Quiggin J. What happiness conceals. In: Nelson C, Pike D, Ledvinka G, editors. On happiness: new ideas for the twenty-first century. UWA Publishing; 2015.
[741] Francis J. Subversive virtue: ascetism and authority in the second-century pagan world. Pennsylvania State University Press; 1994.
[742] Robinson T, Berridge C. The incentive sensitization theory of addiction: some current issues. Philos Trans Royal Soc Lond B Biol Sci 2008;363:3137–46.
[743] Kringelbach M. The pleasure center: trust your animal instincts. Oxford University Press; 2008.
[744] Bloom P. How pleasure works: the new science of why we like what we like. Norton; 2010.
[745] DiSalvo D. What makes your brain happy and why you should do the opposite. Prometheus; 2011.
[746] Linden D. The compass of pleasure: how our brains make fatty foods, orgasm, exercise, marijuana, generosity, vodka, learning, and gambling feel so good. Viking Penguin; 2011.
[747] Burnett D. The happy brain: the science of where happiness comes from and why. Guardian Faber; 2018.
[748] Foucault M. The use of pleasure: a history of sexuality. Vintage; 1990.
[749] Berridge K. Wanting and liking: observations from the neuroscience and psychology laboratory. Inquiry 2009;52:378.
[750] Hume D. Dialogues concerning natural religion. 1779.
[751] Stephens D, Krebs J. Foraging theory. Princeton University Press; 1986.

[752] Todd P, Hills T. In: Robbins T, editor. Cognitive search: evolution, algorithms, and the brain. MIT Press; 2012.
[753] Ainslie G. Breakdown of will. Cambridge University Press; 2001.
[754] Brown N, Rohrer J. Easy as (happiness) pie? A critical evaluation of a popular model of the determinants of well-being. J Happiness Stud 2019;21:1285–301.
[755] Luhmann M, Intelisano S. Hedonic adaptation and the set point for subjective well-being. In: Diener E, Oishi S, Tay L, editors. Handbook of well-being. DEF Publishers; 2018.
[756] Easterlin R. Does economic growth Improve the human lot? Some empirical evidence. In: David P, Reder M, editors. Nations and households in economic growth: essays in honor of Moses Abramovitz. Academic Press; 1974.
[757] Easterlin R. Paradox lost. Rev Behav Econ 2017;4:311–39.
[758] Easterbrooks G. The progress paradox: how life gets better while people feel worse. Random House; 2004.
[759] Svenson B, Wolfers J. Economic growth and subjective wellbeing: reassessing the Easterlin paradox. Brooking Papers on Economic Activity 2008;1:1–87.
[760] Little B. Me, myself and us: the science of personality and the art of well-being. PublicAffairs; 2014.
[761] Pluess M. Genetics of psychological well-being: the role of heritability and genetics in positive psychology. Oxford University Press; 2015.
[762] Baselmans B, Bartels M. A genetic perspective on the relationship between eudaimonic and hedonic well-being. Sci Rep 2018;8:14610.
[763] Geda Y. A purpose "driven" life: is it potentially neuroprotective? Arch Neurol 2010;67:1010–1.
[764] Nesse R, Berridge K. Psychoactive drug use in evolutionary perspective. Science 1997;278:63–6.
[765] Natterson-Horowitz B, Bowers K. Wildhood: the astounding connections between human and animal adolescents. Scribner; 2019.
[766] Everitt B, Robbins T. Drug addiction: updating actions to habits to compulsions ten years on. Annu Rev Psychol 2016;67:23–50.
[767] Graybiel A. Habits, rituals, and the evaluative brain. Annu Rev Neurosci 2008;31:359–87.
[768] Smith K, Graybiel A. Habit formation. Dialogues Clin Neurosci 2016;18:33–43.
[769] Shaffer J. Neuroplasticity and clinical practice: building brain power for health. Front Psychol 2016;7:1118.
[770] Duhigg C. The power of habit: why we do what we do in life and business. Random; 2012.
[771] Dean J. Making habits, breaking habits: why we do things, why we don't, and how to make any change stick. Da Capo; 2013.
[772] Deary V. How we are: book one of the how to live trilogy. Farrar, Straus and Giroux; 2014.
[773] Rubin G. Better than before: mastering the habits of our everyday lives. Crown; 2015.
[774] Clear J. Atomic habits: an easy & proven way to build good habits and break bad ones. Avery; 2018.
[775] Wood W. Good habits, bad habits: the science of making positive changes that stick. Farrar, Straus, Giroux; 2019.
[776] Wrangham R. Catching fire: how cooking made us human. Basic Books; 2009.
[777] Marean C, Bar-Matthews M, Bernatchez J, Fisher E, Goldberg P, Herries A, et al. Early human use of marine resources and pigment in South Africa during the Middle Pleistocene. Nature 2007;449:905–8.

[778] Crawford M, Bloom M, Broadhurst C, Schmidt W, Cunnane S, Galli C, et al. Evidence for the unique function of docosahexaenoic acid during the evolution of the modern hominid brain. Lipids 1999;34S:39–47.
[779] Stetka B. By land or by sea: how did early humans access key brain-building nutrients. Sci Am Mind 2016;27:31.
[780] Horrobin D. The madness of Adam and Eve: how schizophrenia shaped humanity. Transworld; 2002.
[781] Power M, Schulkin J. The evolution of obesity. Johns Hopkins University Press; 2009.
[782] Gordon E, Ariel-Donges A, Bauman V, Merlo L. What is the evidence for food addiction? A systematic review. Nutrients 2018;10, E477.
[783] Chiva-Blanch G, Badimon L. Benefits and risks of moderate alcohol consumption on cardiovascular disease: current findings and controversies. Nutrients 2019;12, E108.
[784] Nutt D. Alcohol alternatives—a goal for psychopharmacology? J Psychopharmacol 2006;20:318–20.
[785] Marx W, Moseley G, Berk M, Jacka F. Nutritional psychiatry: the present state of the evidence. Proc Nutr Soc 2017;76:427–36.
[786] Thaler R, Sunstein C. Nudge: improving decisions about health, wealth, and happiness. Yale University Press; 2008.
[787] Babor T, Caetano R, Casswell S, Edwards G, Giesbrecht N, Graham K, et al. Alcohol: no ordinary commodity—research and public policy. Oxford University Press; 2010.
[788] Bekoff M, Byers J. Animal play: evolutionary, comparative, and ecological perspectives. Cambridge University Press; 1998.
[789] Brown S. Play: how it shapes the brain, opens the imagination, and invigorates the soul. Avery; 2009.
[790] Restak R. The playful brain: the surprising science of how puzzles improve your mind. Riverhead Books; 2010.
[791] Balcombe J. The exultant ark: a pictorial tour of animal pleasure. University of California Press; 2011.
[792] Panksepp J. Beyond a joke: from animal laughter to human joy? Science 2005;308:62–3.
[793] Vrticka P, Black J, Reiss A. The neural basis of humour processing. Nat Rev Neurosci 2013;14:860–8.
[794] Martin R, Ford T. The psychology of humor: an integrative approach. Academic Press; 2018.
[795] Chan Y, Hsu W, Liao Y, Chen H, Tu C, Wu C. Appreciation of different styles of humor: an fMRI study. Sci Rep 2018;8:15649.
[796] Liu X, Chen Y, Ge J, Mao L. Funny or angry?: Neural correlates of individual differences in aggressive humor processing. Front Psychol 2019;10:1849.
[797] Scott S, Lavan N, Chen S, McGettigan C. The social life of laughter. Trends Cogn Sci 2014;18:618–20.
[798] Critchley S. On humour. Routledge; 2011.
[799] Gordon M. Humor, laughter and human flourishing: a philosophical exploration of the laughing animal. Springer; 2014.
[800] Morreall J. Philosophy of humor. In: Zalta E, editor. The Stanford encyclopedia of philosophy. Stanford University; 2016.
[801] Bergson H. Laughter: an essay on the meaning of the comic. Franklin Classics; 1900/2018.
[802] Cohen T. Jokes: philosophical thoughts on joking matters. University of Chicago Press; 1999.

[803] Cathcart T, Klein D. Plato and a Platypus walk into a bar – understanding philosophy through jokes. Penguin; 2007.
[804] Holt J. Stop me if you've heard this: a history and philosophy of jokes. Norton; 2008.
[805] Gimbel S. Isn't that clever: a philosophical account of humor and comedy. Routledge; 2017.
[806] Eagleton T. Humour. Yale University Press; 2019.
[807] Wiseman R. Quirkology: how we discover the big truths in small things. Basic Books; 2007.
[808] Mosak H. Ha, Ha and Aha: the role of humor in psychotherapy. Routledge; 1987.
[809] Shimamura A. Experiencing art: in the brain of the beholder. Oxford University Press; 2013.
[810] Ehrenreich B. Dancing in the streets: a history of collective joy. Metropolitan Books; 2007.
[811] Stark E, Vuust P, Kringelbach M. Music, dance, and other art forms: new insights into the links between hedonia (pleasure) and eudaimonia (well-being). Prog Brain Res 2018;237:129–52.
[812] De Botton A. Pleasures and sorrows of work. Vintage International; 2010.
[813] Dennett D. Philosophy as the Las Vegas of rational inquiry. Free Inquiry 2017; Aug/Sep: 22–3.
[814] Hurley M, Dennett D, Adams R. Inside jokes: using humor to reverse-engineer the mind. MIT Press; 2011.
[815] Scheffler I. In praise of the cognitive emotions and other essays in the philosophy of education. Routledge; 1991.
[816] Warren J. The pleasures of reason in Plato, Aristotle, and the Hellenistic Hedonists. Cambridge University Press; 2014.
[817] Skilbeck A. Dewey on seriousness, playfulness and the role of the teacher. Educ Sci 2017;7:16.
[818] Gopnik A, Meltzoff A. Words, thoughts, and theories. MIT Press; 1997.
[819] Livio M. Why?: what makes us curious. Simon & Schuster; 2017.
[820] Schlick M. On the meaning of life. In: Klemke E, Cahn S, editors. The meaning of life. 4th ed. Oxford University Press; 2017.
[821] Suits B. The grasshopper: games, life and utopia. Broadview Press; 2005.
[822] Bogost I. Play anything: the pleasure of limits, the uses of Boredom, and the Secret of Games. Basic Books; 2016.
[823] Rumpf H, Achab S, Billieux J, et al. Including gaming disorder in the ICD-11: the need to do so from a clinical and public health perspective. J Behav Addict 2018;7:556–61.
[824] Cappuccio M. Handbook of embodied cognition and sport psychology. MIT Press; 2019.
[825] Liebenberg L. The art of tracking: the origin of science. New Africa Books; 2012.
[826] Lieberman D, Bramble D. Endurance running and the evolution of Homo. Nature 2004;432:345–52.
[827] McDougall C. Born to run: a hidden tribe, superathletes, and the greatest race the world has never seen. Vintage; 2011.
[828] Lieberman D. The story of the human body. Allen Lane; 2013.
[829] Fontes E, Okano A, De Guio F, Schabort E, Min L, Basset F, et al. Brain activity and perceived exertion during cycling exercise: an fMRI study. Br J Sports Med 2015;49:556–60.
[830] Hobson J. Dreaming: an introduction to the science of sleep. Oxford University Press; 2002.
[831] Walker M. Why we sleep: unlocking the power of sleep and dreams. Scribner; 2017.
[832] Young D. How to think about exercise. The School of Life; 2014.
[833] Priest G, Young D. Philosophy and the martial arts: engagement. Routledge; 2014.
[834] Allen B. Striking beauty: a philosophical look at the Asian martial arts. Columbia University Press; 2015.

[835] Gottschall J. The professor in the cage: why men fight and why we like to watch. Penguin Press; 2015.
[836] Chatterjee A, Thomas A, Smith S, Aguirre G. The neural response to facial attractiveness. Neuropsychology 2009;23:135–43.
[837] Zeki S, Nash J. Inner vision: an exploration of art and the brain. Oxford University Press; 1999.
[838] Dutton D. The art instinct: beauty, pleasure, and human evolution. Bloomsbury Press; 2008.
[839] Chatterjee A. The aesthetic brain: how we evolved to desire beauty and enjoy art. Oxford University Press; 2013.
[840] Starr G. Feeling beauty: the neuroscience of aesthetic experience. MIT Press; 2013.
[841] Kandel E. Reductionism in art and brain science: bridging the two cultures. Columbia University Press; 2016.
[842] Tallis R. The limitations of a neurological approach to art. Lancet 2008;372:19–20.
[843] Noë A. Strange tools: art and human nature. Hill and Wang; 2015.
[844] Bratman G, Hamilton J, Gretchen C. The impacts of nature experience on human cognitive function and mental health. Ann N Y Acad Sci 2012;1249:118–36.
[845] Louv R. Last child in the woods: saving our children from nature-deficit disorder. Algonquine Books; 2005.
[846] Bratman G, Hamilton P, Hahn K, Dally G, Gross J. Nature experience reduces rumination and subgenual prefrontal cortex activation. Proc Natl Acad Sci U S A 2015;112:8567–72.
[847] Todes D. Ivan Pavlov: a Russian life in science. Oxford University Press; 2014.
[848] Ridley M. The red queen: sex and the evolution of human nature. Perennial; 1993.
[849] Etcoff N. Survival of the prettiest: the science of beauty. Anchor; 1999.
[850] Miller G. The mating mind: how sexual choice shaped the evolution of human nature. Doubleday; 2000.
[851] Rothenberg D. Survival of the beautiful: art, science, and evolution. Bloomsbury Press; 2011.
[852] Buss D. The evolution of desire: strategies of human mating, revised. Basic Books; 2016.
[853] Prum R. The evolution of beauty: how Darwin's forgotten theory of mate choice shapes the animal world—and us. Doubleday; 2017.
[854] Ryan M. A taste for the beautiful: the evolution of attraction. Princeton University Press; 2018.
[855] Davies S. The artful species: aesthetics, art, and evolution. Oxford University Press; 2012.
[856] Hamermesh D. Beauty pays: why attractive people are more successful. Princeton University Press; 2011.
[857] Maestripieri D, Henry A, Nickels N. Explaining financial and prosocial biases in favour of attractive people: interdisciplinary perspectives from economics, social psychology, and evolutionary psychology. Behav Brain Sci 2017;40:1–56.
[858] Walker C, Papadopoulos L. Psychodermatology: the psychological impact of skin disorder. Cambridge University Press; 2005.
[859] de Botton A. Essays on love. Pan Macmillan; 2014.
[860] Gooley T. How to connect with nature. The School of Life; 1994.
[861] Sternberg E. Healing spaces: the science of place and well-being. Harvard University Press; 2009.
[862] Williams F. The nature fix: why nature makes us happier, healthier, and more creative. Norton; 2017.

[863] Cooper D. Gardening – philosophy for everyone: cultivating wisdom. John Wiley & Sons; 2011.
[864] Cooper D. Convergence with nature: a Daoist perspective. UIT Cambridge; 2012.
[865] Hardy S. Mother nature: maternal instincts and how they shape the human species. Ballantine Books; 2000.
[866] Blum D. Love at Goon Park: Harry Harlow and the science of affection. 2nd ed. Basic Books; 2011.
[867] Insel T. Neurobiological Basis of Social Attachment inability to form normal social attachments. Am J Psychiatr 1997;154:726–35.
[868] Barash D, Lipton J. Making sense of sex: how genes and gender influence our relationships. Island Press; 1997.
[869] Lewis T, Amini F, Lannon R. A general theory of love. Random House; 2000.
[870] Ryan C, Jethá C. Sex at dawn: how we mate, why we stray, and what it means for modern relationships. Harper Perennial; 2012.
[871] Saxon L. Sex at dusk: lifting the shiny wrapping from sex at dawn. CreateSpace; 2012.
[872] Fisher H. Anatomy of love: a natural history of mating, marriage, and why we stray. Norton; 2016.
[873] Wright R. The moral animal: why we are, the way we are: the new science of evolutionary psychology. Pantheon; 1994.
[874] de Botton A. How to think more about sex. Picador; 2012.
[875] Gamble C, Gowlett J, Dunbar R. Thinking big: how the evolution of social life shaped the human mind. Thames & Hudson; 2018.
[876] Hare B, Woods V. Survival of the friendliest: understanding our origins and rediscovering our common humanity. Random House; 2020.
[877] Cheney D, Seyfarth R. Baboon metaphysics: the evolution of a social mind. University of Chicago Press; 2007.
[878] Lieberman M. Social: why our brains are wired to connect. Crown; 2013.
[879] Morris D. The soccer tribe. Jonathan Cape; 1981.
[880] Arendt H. The life of the mind. Harcourt Brace Jovanovich; 1981.
[881] Storr A. Solitude: a return to the self. Free Press; 1988.
[882] Rufus A. Party of one: the loners' manifesto. Da Capo Press; 2003.
[883] Cacioppo J, Patrick W. Loneliness: human nature and the need for social connection. Norton; 2009.
[884] Cain S. Quiet: the power of introverts in a world that can't stop talking. Broadway Books; 2013.
[885] Maitland S. How to be alone. Picador; 2014.
[886] Batchelor S. The art of solitude. Yale University Press; 2020.
[887] Grayling A. https://www.theguardian.com/lifeandstyle/2009/jul/05/a-c-grayling-this-much-i-know; July 5, 2009.
[888] Decety J, Cacioppo J. The Oxford handbook of social neuroscience. Oxford University Press; 2011.
[889] Schutt R, Seidman L, Keshavan M. Social neuroscience: brain, mind, and society. Harvard University Press; 2015.
[890] Cacioppo S, Cacioppo J. Introduction to social neuroscience. Princeton University Press; 2020.
[891] Metz T. Toward an African moral theory. J Polit Philos 2007;15:321–41.
[892] Bell D. Communitarianism and its critics. Oxford University Press; 1993.

[893] Kayange G. Restoration of ubuntu as an autocentric virtue-phronesis theory. S Afr J Philos 2020;39:1–12.
[894] Caluori D. Thinking about friendship: historical and contemporary philosophical perspectives. Palgrave Macmillan; 2013.
[895] Lucash F. Spinoza on friendship. Philosophia 2012;40:305–17.
[896] Grayling A. Friendship. Yale University Press; 2013.
[897] Scruton R. Sexual desire: a philosophical investigation. Weidenfeld and Nicolson; 1986.
[898] Solomon R, Higgins K. The philosophy of (erotic) love. University Press of Kansas; 1991.
[899] Badhwar N. Friendship: a philosophical reader. Cornell University Press; 1993.
[900] Frankfurt H. The reasons of love. Princeton University Press; 2004.
[901] Jollimore T. Love's vision. Princeton University Press; 2011.
[902] Nehamas A. On friendship. Basic Books; 2016.
[903] Grau C, Smuts A. The Oxford handbook of philosophy of love. Oxford University Press; 2017.
[904] Denworth, L. Friendship: the evolution, biology, and extraordinary power of life's fundamental bond. Norton; 2020.
[905] Mill J. On Liberty. Essays on Politics and Society. University of Toronto Press; 1977.
[906] Lyubomirsky S. The how of the happiness: a new approach to getting the life you want. Penguin; 2009.
[907] Buss D. The evolution of happiness. Am Psychol 2000;55:15–23.
[908] Nesse R. Natural section and the elusiveness of happiness. Philos Trans Royal Soc Lond B Biol Sci 2004;359:1333–47.
[909] Darwin C. Autobiography of Charles Darwin. John Murray; 1887.
[910] Schopenhauer A. The world as will and representation. Kegan Paul; 1819/1918.
[911] Baggini J. Oliver Burkeman's the Antidote www.theguardian.com; June 22, 2012.
[912] Nietzsche F. Twilight of the Idols. 1889.
[913] Nietzsche F. Will to power. 1901.
[914] Dweck C. Mindset: the new psychology of success. Ballantine; 2007.
[915] Schulz K. Being wrong: adventures in the margin of error. Ecco; 2010.
[916] Tugend A. Better by mistake: the unexpected benefits of being wrong. Riverhead Trade; 2012.
[917] Holiday R. The obstacle is the way: the timeless art of turning trials into triumph. Portolio; 2014.
[918] Lewis S. The rise: creativity, the gift of failure, and the search for mastery. Simon & Schuster; 2014.
[919] Dalio R. Principles: life and work. Simon & Schuster; 2017.
[920] Billieux J, Flayelle M, Rumpf H-J, Stein D. High involvement versus pathological involvement in video games: a crucial distinction for ensuring the validity and utility of gaming disorder. Curr Addict Rep 2019;6:323–30.
[921] Kraus S, Krueger R, Briken P, First M, Stein D, Kaplan M, et al. Compulsive sexual behaviour disorder in the ICD-11. World Psychiatry 2018;17:109–10.
[922] Saad G. The evolutionary bases of consumption. Routledge; 2007.
[923] Preston S, Kringelbach M, Knutson B. The interdisciplinary science of consumption. MIT Press; 2014.
[924] Hood B. Possessed: why we want more than we need. Oxford University Press; 2019.
[925] Nussbaum M. Virtue ethics: a misleading category. J Ethics 1999;3:163–201.
[926] Cloninger C. Feeling good: the science of well-being. Oxford University Press; 2004.
[927] Sacks J. To heal a fractured world: the ethics of responsibility. Schocken; 2005.
[928] Wolf S. Moral saints. J Philos 1982;79:419–39.

[929] Smith R. The joy of pain: schadenfreude and the dark side of human nature. Oxford University Press; 2013.

[930] van Dijk W, Ouwerkerk J. Schadenfreude: understanding pleasure at the misfortune of others. Cambridge University Press; 2014.

[931] Smith T. Schadenfreude: the joy of another's misfortune. Little, Brown Spark; 2018.

[932] Schopenhauer A. On human nature. George Allen & Unwin; 1897.

[933] Dalrymple T. Mankind is stuffed. Taki's Magazine 2016.

[934] Dewey J. Democracy and education: an introduction to the philosophy of education. Macmillan; 1916.

[935] Harari Y. 21 lessons for the twenty-first century. Random House; 2018.

[936] Russell B. What I believe. Kegan Paul; 1925.

[937] Tabensky P. Happiness: personhood, community, purpose. Aldershot; 2003.

[938] Tessman L. Burdened virtues: virtue ethics for liberatory struggles. Oxford University Press; 2005.

[939] Lilienfeld S, Arkowitz H. Facts & fictions in mental health: can positive thinking be negative? Sci Am Mind 2011;22:64–5.

[940] Brown N, Sokal A, Friedman H. The complex dynamics of wishful thinking: the critical positivity ratio. Am Psychol 2013;68:801–13.

[941] Cavanagh C, Larking K. A critical review of the "Undoing Hypothesis": do positive emotions undo the effects of stress? Appl Psychophysiol Biofeedback 2018;43:259–73.

[942] Held B. Stop smiling, start kvetching: a 5-step guide to creative complaining. St. Martin's Griffin; 2001.

[943] Pearsall P. The last self help book you'll ever need: repress your anger, think negatively, be a good blamer, & throttle your inner child. Basic Books; 2005.

[944] Kleinman A. What really matters: living a moral life amidst uncertainty and danger. Oxford University Press; 2006.

[945] Wilson E. Against happiness: in praise of melancholia. Sarah Crichton; 2008.

[946] Burton N. The art of failure: the anti self-help guide. Acheron; 2010.

[947] Bruckner P. Perpetual euphoria: on the duty to be happy. Princeton University Press; 2011.

[948] Taleb N. Antifragile: things that gain from disorder. Random House; 2012.

[949] Critchley S. How to stop living and start worrying: conversations with Carl Cederström. Polity; 2013.

[950] Hamilton C. How to deal with adversity. The School of Life; 2014.

[951] Parrott W. The positive side of negative emotions. Guilford; 2014.

[952] Bennett M, Bennett S. F*ck feelings: one shrink's practical advice for managing all life's impossible problems. Simon & Schuster; 2015.

[953] McGonigal K. The upside of stress: why stress is good for you (and how to get good at it). Ebury Digital; 2015.

[954] Manson M. The subtle art of not giving a F*ck: a counterintuitive approach to living a good life. HarperOne; 2016.

[955] Brinkmann S. Stand firm: resisting the self-improvement craze. Polity; 2017.

[956] Fowers B, Richardson R, Slife B. Frailty, suffering, and vice: flourishing in the face of human limitations. American Psychological Association; 2017.

[957] Keyes C, Haidt J. Flourishing: positive psychology and the life well-lived. American Psychological Association; 2003.

[958] David S. Emotional agility: get unstuck, embrace change, and thrive in work and life. Avery; 2016.

[959] Webb C. How to have a good day: harness the power of behavioral science to transform your working life. Currency; 2016.

[960] Lazarus A, Fay A. I can if I want to. Warner Books; 1975.
[961] Young J. Reinventing your life: the breakthrough program to end negative behavior and feel great again. Plume; 1994.
[962] Ellis A. How to make yourself happy and remarkably less disturbable. Impact; 1999.
[963] Horowitz M. A course in happiness. Penguin; 2008.
[964] Hayes S. A liberated mind: how to pivot toward what matters. Avery; 2019.
[965] Grinde B. Darwinian happiness: evolution as a guide for living and understanding human behavior. Darwin Press; 2002.
[966] Klein S. The science of happiness: how our brains make us happy – and what we can do to get happier. Scribe; 2006.
[967] Morris D. The nature of happiness. Little Books; 2006.
[968] Nettle D. Happiness: the science behind your smile. Oxford University Press; 2006.
[969] Gilbert D. Stumbling on happiness. Alfred A Knopf; 2006.
[970] Keltner D. Born to be good: the science of a meaningful life. Norton; 2009.
[971] Hanson R. Hardwiring happiness: the new brain science of contentment, calm, and confidence. Harmony; 2013.
[972] Whybrow P. The well-tuned brain: neuroscience and the life well-lived. Norton; 2016.
[973] Geher G, Wedberg N. Positive evolutionary psychology: Darwin's guide to living a richer life. Oxford University Press; 2019.
[974] Haidt J. The happiness hypothesis: finding modern truth in ancient wisdom. Basic Books; 2006.
[975] Camus A. The Myth of Sisyphus. Librairie Gallimard; 1942.
[976] Critchley S. The book of dead philosophers. Vintage; 2009.
[977] Yalom I. The Schopenhauer cure: a novel. Harper; 2005.
[978] Wicks R. Schopenhauer. Blackwell; 2008.
[979] Beiser F. Weltschmerz: pessimism in German philosophy, 1860–1900. Oxford University Press; 2016.
[980] Benatar D. Better never to have been: the harm of coming into existence. Oxford University Press; 2008.
[981] Gray J. The silence of animals: on progress and other modern myths. Farrar, Straus and Giroux; 2013.
[982] Durkheim É. Pragmatism and sociology. Cambridge University Press; 1858/1983.
[983] Gross N. Durkheim's pragmatism lectures: a contextual interpretation. Sociol Theory 1997;15:126–49.
[984] Szasz T. Myth of mental illness: foundations of a theory of personal conduct. Harper & Row; 1961.
[985] Laing R. The divided self: an existential study in sanity and madness. Penguin; 1960.
[986] Cooper D. Psychiatry and anti-psychiatry. Tavistock Publications; 1967.
[987] Foucault M. Madness and civilization: a history of insanity in the age of reason. Librairie Plon; 1961.
[988] Goffman E. Asylums: essays on the social situation of mental patients and other inmates. Doubleday; 1961.
[989] Chapman A. Re-coopering anti-psychiatry: David Cooper, revolutionary critic of psychiatry. Crit Radic Social Work 2016;4:421–32.
[990] Boorse C. On the distinction between disease and illness. Philos Public Aff 1975;5:49–68.
[991] Fulford K. What is (mental) disease?: an open letter to Christopher Boorse. J Med Ethics 2001;27:80–5.
[992] Jaspers K. General psychopathology. Springer-Verlag; 1913.

[993] Canguilhem G. The normal and the pathological. Urzone; 1966.
[994] Pellegrino E, Thomasma D. A philosophical basis of medical practice: towards a philosophy and ethic of the healing professions. Oxford University Press; 1981.
[995] Fulford B. Moral theory and medical practice. Cambridge University Press; 1990.
[995a] Reznek L. The philosophical defence of psychiatry. Routledge; 1991.
[996] Schaffner K. Coming home to Hume: a sociobiological foundation for a concept of "health" and morality. J Med Philos 1999;24:365–75.
[997] Stemsey W. Disease and diagnosis: Value-dependent realism. Springer; 2000.
[998] Sadler J. Descriptions and prescriptions: values, mental disorders, and the DSMs. John Hopkins University Press; 2004.
[999] Thornton T. Essential philosophy of psychiatry. Oxford University Press; 2007.
[1000] Metzl J, Kirkland A. Against health: how health became the new morality. NYU Press; 2010.
[1001] McNally R. What is mental illness? Belknap Press; 2011.
[1002] Graham G. The disordered mind: an introduction to philosophy of mind and mental illness. 2nd ed. Routledge; 2013.
[1003] Wakefield J. Disorder as harmful dysfunction: a conceptual critique of DSM-III-R's definition of mental disorder. Psychol Rev 1992;99:232–47.
[1004] Lillienfeld S, Marino L. Mental disorder as a Roschian concept: a critique of Wakefield's "harmful dysfunction" analysis. J Abnorm Psychol 1995;104:111–20.
[1005] Klein D. A proposed definition of mental illness. In: Spitzer R, Klein D, editors. Critical issues in psychiatric diagnosis. Raven; 1978.
[1006] Wakefield J. Evolutionary versus prototype analyses of the concept of disorder. J Abnorm Psychol 1999;108:374–99.
[1007] Boorse C. A second rebuttal on health. J Med Philos 2014;39:683–724.
[1008] Griffiths P, Matthewson J. Evolution, dysfunction, and disease: a reappraisal. Br J Philos Sci 2018;69:301–27.
[1009] Bayer R. Homosexuality and American psychiatry: the politics of diagnosis. Princeton University Press; 1987.
[1010] Megone C. Mental illness, human function, and values. Philos Psychiatry Psychol 2000;7:45–65.
[1011] Hamilton R. The concept of health: beyond normativism and naturalism. J Eval Clin Pract 2010;16:323–9.
[1012] Jing-Bao N, Gilbertson A, de Roubaix M, Staunton C, van Niekerk A, Tucker J, et al. Healing without waging war: beyond military metaphors in medicine and HIV cure research. Am J Bioeth 2016;16:3–11.
[1013] M Maimonides. Regimen of health.
[1014] Kaye J. History of balance, 1250–1375: the emergence of a new model of equilibrium and its impact on medieval thought. Cambridge University Press; 2014.
[1015] Huber M, et al. How should we define health? BMJ 2011;343:d4163.
[1016] Parsons T. The social system. Free Press; 1951.
[1017] Hall W, Carter A, Forlini C. The brain disease model of addiction: is it supported by the evidence and has it delivered on its promises? Lancet Psychiatry 2015;2:105–10.
[1018] Holton R, Berridge K. Addiction between compulsion and choice. In: Levy N, editor. Addiction and self-control: perspectives from philosophy, psychology and neuroscience. Oxford University Press; 2013.
[1019] Pickard H. Responsibility without blame for addiction. Neuroethics 2017;10:169–80.
[1020] Ross D, Kincaid H, Spurrett D, Collins P. What is addiction? MIT Press; 2010.
[1021] Carter A, Hall W, Illes J. Addiction neuroethics: the ethics of addiction neuroscience research and treatment. Academic Press; 2011.

[1022] Poland J, Graham G. Addiction and responsibility. MIT Press; 2011.
[1023] Levy N. Addiction and self-control: perspectives from philosophy, psychology, and neuroscience. Oxford University Press; 2013.
[1024] Heather N, Segal G. Addiction and choice: rethinking the relationship. Oxford University Press; 2016.
[1025] Pickard H, Ahmed S. The Routledge handbook of philosophy and science of addiction. Routledge; 2019.
[1026] MacLachlan M. Embodiment: clinical, critical and cultural perspectives on health and illness. Open University Press; 2004.
[1027] Pilgrim D. Understanding mental health: a critical realist exploration. Routledge; 2014.
[1028] Leder D. The distressed body: rethinking illness, imprisonment, and healing. University of Chicago Press; 2016.
[1029] Maiese M, Hanna R. The mind-body politic. Palgrave Macmillan; 2019.
[1030] von Wright G. Varieties of goodness. Routledge and Kegan Paul; 1963.
[1031] Bloomfield P. Moral reality. Oxford University Press; 2004.
[1032] Compton A. A study of the psychoanalytic theory of anxiety. I. The development of Freud's theory of anxiety, J Am Psychoanal Assoc 1972;20:3–44.
[1032a] Robertson B. The psychoanalytic theory of depression: II. The major themes. Can J Psychiatr 1979;24:557–74.
[1033] Freud S. Civilization and its discontents. Norton; 1930/1961.
[1034] Rieff P. Freud: the mind of the moralist. Viking; 1959.
[1035] Rieff P. The triumph of the therapeutic. Harper & Row; 1966.
[1036] Webster R. Why Freud was wrong: sin, science, and psychoanalysis. Basic Books; 1995.
[1037] Storr A. Feet of clay: saints, sinners, and madmen: a study of gurus. Free Press; 1996.
[1038] Ehrenreich B, English D. For her own good: two centuries of the experts advice to women. Anchor; 2005.
[1039] Furedi F. Therapy culture: cultivating vulnerability in an uncertain age. Routledge; 2003.
[1040] Illouz E. Saving the modern soul: therapy, emotions, and the culture of self-help. University of California Press; 2008.
[1041] Rosen R. Psychobabble: fast talk and quick cure in the era of feeling. Athenium; 1978.
[1042] Salerno S. Sham: how the self-help movement made America helpless. Crown Publishing; 2005.
[1043] Sommers C, Satel S. One nation under therapy: how the helping culture is eroding self-reliance. St Martin's Press; 2007.
[1044] Tiede T. Self-help nation: the long overdue, entirely justified, delightfully hostile guide to the snake-oil peddlers who are sapping our nation's soul. Monthly; 2001.
[1045] Lewinsohn P, Gotlib I. Behavioral theory and treatment of depression. In: Beckham E, Leber W, editors. Handbook of depression. Guilford Press; 1995.
[1046] Kuhlmann H. Living Walden Two: B. F. Skinner's behaviorist utopia and experimental communities. University of Illinois Press; 2005.
[1047] Beck A, Alford B. Depression: causes and treatment. 2nd ed. University of Pennsylvania Press; 2009.
[1048] Ferrari A, Charlson F, Norman R, Patten S, Freedman G, Murray C, et al. Burden of depressive disorders by country, sex, age, and year: findings from the global burden of disease study 2010. PLoS Med 2013;10:e1001547.
[1049] First M, Wakefield J. Diagnostic criteria as dysfunction indicators: bridging the chasm between the definition of mental disorder and diagnostic criteria for specific disorders. Can J Psychiatr 2013;58:663–9.
[1050] Horwitz A, Wakefield J. The loss of sadness: how psychiatry transformed normal sorrow into depressive disorder. Oxford University Press; 2007.

[1051] Horwitz A, Wakefield J. All we have to fear: psychiatry's transformation of natural anxieties into mental disorders. Oxford University Press; 2012.
[1052] Kendler K, Zachar P, Craver C. What kinds of things are psychiatric disorders? Psychol Med 2011;41:1143–50.
[1053] Zachar P. A metaphysics of psychopathology. MIT Press; 2014.
[1054] Kincaid H, Sullivan J, editors. Classifying psychopathology: mental kinds and natural kinds. Cambridge: MIT Press; 2014.
[1055] Keil G, Keuck L, Hauswald R. Vagueness in psychiatry. Oxford University Press; 2017.
[1056] Lange M. The end of diseases. Philos Top 2007;35:265–92.
[1057] Kendler K. What psychiatric genetics has taught us about the nature of psychiatric illness and what is left to learn. Mol Psychiatry 2013;18:1058–66.
[1058] Maj M. Psychiatric comorbidity: an artefact of current diagnostic systems? Br J Psychiatry 2005;186:182–4.
[1059] Sober E. Ockham's razors: a user's manual. Cambridge University Press; 2015.
[1060] Radden J. The nature of melancholy: from Aristotle to Kristeva. Oxford University Press; 2004.
[1061] Kaczmarek E. How to distinguish medicalization from over-medicalization. Med Health Care Philos 2019;22:119–28.
[1062] Parens E. On good and bad forms of medicalization. Bioethics 2013;27:28–35.
[1063] Stein D, Kaminer D, Zungu-Dirwayi N, Seedat S. Pros and cons of medicalization: the example of trauma. World J Biol Psychiatry 2006;7:2–4.
[1064] Eisenberger N, Lieberman M, Williams K. Does rejection hurt? An fMRI study of social exclusion. Science 2003;302:290–2.
[1065] Stein D, Seedat S, Iversen A, Wessely S. Post-traumatic stress disorder: medicine and politics. Lancet 2007;369:139–44.
[1066] Scarry E. The body in pain: the making and the unmaking of the world. Oxford University Press; 1985.
[1067] van Ommen C, Cromby J, Yen J. The contemporary making and unmaking of Elaine Scarry's the Body in Pain. Subjectivity 2016;9:333–42.
[1068] Ellis B, Del Giudice M. Developmental adaptation to stress: an evolutionary perspective. Annu Rev Psychol 2019;70:111–39.
[1069] Boyce W. The orchid and the dandelion: why some children struggle and all can thrive. Allen Lane; 2019.
[1070] Kagan J. Galen's prophecy: temperament in human nature. Westview Press; 1997.
[1071] Brooks S, Stein D. Psychotherapy and neuroimaging in anxiety and related disorders. Dialogues Clin Neurosci 2015;17:287–93.
[1072] Chmitorz A, Kunzler A, Helmreich I, Tüscher O, Kalisch R, Kubiak T, et al. Intervention studies to foster resilience – a systematic review and proposal for a resilience framework in future intervention studies. Clin Psychol Rev 2018;59:78–100.
[1073] Duckworth A. Grit: why passion and resilience are the secrets to success. Random House; 2017.
[1074] Meckier J. "Great Expectations" and "Self-Help": Dickens frowns on Smiles. J Engl Ger Philol 2001;100:537–54.
[1075] Emerson R. Self-reliance. 1841.
[1076] Berlant L. Cruel optimism. Duke University Press; 2011.
[1077] Cholbi M. Suicide: the philosophical dimensions. Broadview; 2011.
[1078] Stein D. In the aftermath of suicide: any lessons from the sorrow? S Afr Med J 2018;108:12412.

[1079] Stein D, Nesse R. Normal and abnormal anxiety in the age of DSM-5 and ICD-11. Emot Rev 2015;7:223–9.
[1080] Grøn A. The concept of anxiety in Søren Kierkegaard. Mercer University Press; 2008.
[1081] Crocq M-A. A history of anxiety: from Hippocrates to DSM. Dialogues Clin Neurosci 2015;17:319–25.
[1082] Nussbaum M. The monarchy of fear: a philosopher looks at our political crisis. Simon & Schuster; 2018.
[1083] Starkstein S. A conceptual and therapeutic analysis of fear. Palgrave Macmillian; 2018.
[1084] Rozin P, Haidt J, Fincher K. From oral to moral. Science 2009;323:1179–80.
[1085] Douglas M. Purity and danger: an analysis of concepts of pollution and taboo. Routledge and Keegan Paul; 1966.
[1086] Chapman H, Anderson A. Things rank and gross in nature: a review and synthesis of moral disgust. Psychol Bull 2013;139:300–27.
[1087] Nussbaum M. From disgust to humanity: sexual orientation and constitutional law. Oxford University Press; 2010.
[1088] Scruton R. Ethics and welfare: the case of hunting. Philos Psychiatry Psychol 2002;77:543–64.
[1089] Singer P. The expanding circle: ethics, evolution, and moral progress. Princeton University Press; 1981.
[1090] McGinn C. The meaning of disgust. Oxford University Press; 2011.
[1091] Strohminger N, Kumar V. The moral psychology of disgust. Rowman & Littlefield; 2018.
[1092] White G, Mullen P. Jealousy: theory, research, and clinical strategies. Guilford Press; 1989.
[1093] Buss D. The dangerous passion: why jealousy is as necessary as love and sex. Free Press; 2000.
[1094] D'Arms J, Kerr A. Envy in the philosophical tradition. In: Smith R, editor. Envy: theory and research. Oxford University Press; 2008.
[1095] Shapiro G. Nietzsche on envy. Int Stud Philos 1983;15:3–12.
[1096] Pearson G. Aristotle on psychopathology. In: Kontos P, editor. Evil in Aristotle. Cambridge University Press; 2018.
[1097] Simeon D, Hollander E. Self-injurious behaviors: assessment and treatment. American Psychiatric Publishing; 2001.
[1098] Zetterqvist M. The DSM-5 diagnosis of nonsuicidal self-injury disorder: a review of the empirical literature. Child Adolesc Psychiatry Ment Health 2015;9:31.
[1099] Keuthen N, Stein D, Christenson G. Help for hair-pullers: understanding and coping with trichotillomania. New Harbinger; 2001.
[1100] Gibson A. Misanthropy: the critique of humanity. Bloomsbury Academic; 2017.
[1101] Schopenhauer A. On the sufferings of the world. 1851.
[1102] DeBrabander F. Spinoza and the Stoics: power, politics and the passions. Continuum; 2007.
[1103] Miller J. Spinoza and the Stoics. Cambridge University Press; 2015.
[1104] Russell P. Hume's optimism and Williams's pessimism: from 'Science of Man' to genealogical critique. In: Chappell S, van Ackeren M, editors. Ethics beyond the limits: new essays on Bernard Williams' ethics and the limits of philosophy. Routledge; 2019.
[1105] Bloeser C, Stahl T. Hope. In: Zalta E, editor. The Stanford encyclopedia of philosophy. Stanford University; 2017.
[1106] Loemker L. Pessimism and optimism. In: Edwards P, editor. Encyclopedia of philosophy. Macmillan Publishing Co.; 1967.

[1107] Lerner M. The belief in a just world: a fundamental delusion. Springer; 1980.
[1108] Baumeister R. Meanings of life. Guilford; 1991.
[1109] Bowler K. Everything happens for a reason: and other lies I've loved. Random House; 2018.
[1110] Giddens A. Beyond left and right: the future of radical politics. Polity Press; 1994.
[1111] Chomsky N. Optimism over despair: on capitalism, empire, and social change. Penguin; 2017.
[1112] Diamandis P, Kotler S. Abundance: the future is better than you think. Free Press; 2012.
[1113] Kenny C. Getting better: why global development is succeeding and how we can improve the world even more. Basic Books; 2012.
[1114] Norberg J. Progress: ten reasons to look forward to the future. OneWorld; 2016.
[1115] Pinker S. Enlightenment now: the case for reason, science, humanism, and progress. Viking; 2018.
[1116] Ridley M. The rational optimist: how prosperity evolves. Harper Collins; 2010.
[1117] Rosling H. Ten reasons we're wrong about the world – and why things are better than you think. Flatiron; 2018.
[1118] Shermer M. The moral arc: how science leads humanity toward truth, justice, and freedom. Henry Holt; 2015.
[1119] Tallis R. Enemies of hope: a critique of contemporary pessimism, irrationalism, anti-humanism and counter-enlightenment. St Martin's Press; 1997.
[1120] Thomas C. Inheritors of the earth: how nature is thriving in an age of extinction. PublicAffairs; 2017.
[1121] Cioran E. The trouble with being born. Seaver; 1973.
[1122] Dienstag J. Pessimism: philosophy, ethic, spirit. Princeton University Press; 2009.
[1123] Horkheimer M, Adorno T. Dialectic of enlightenment: philosophical fragments. Stanford University Press; 1947/2002.
[1124] Kotkin J. The coming of neo-feudalism: a warning to the global middle class. Encounter Books; 2020.
[1125] Ryan C. Civilized to death: the price of progress. Simon & Schuster; 2019.
[1126] Scruton R. The uses of pessimism and the dangers of false hope. Oxford University Press; 2010.
[1127] Smith J. Irrationality: a history of the dark side of reason. Princeton University Press; 2019.
[1128] Spengler O. The decline of the west. George Allen & Unwin; 1918.
[1129] Tessman L. Moral failure: on the impossible demands of morality. Oxford University Press; 2015.
[1130] Thacker E. Infinite resignation: on pessimism. Repeater; 2018.
[1131] von Wright G. The myth of progress. The tree of knowledge and other essays. EJ Brill; 1993.
[1132] Mann C. The prophet and the wizard: two remarkable scientists and their dueling visions to shape tomorrow's world. Knopf; 2018.
[1133] Illich I. Medical nemesis: the expropriation of health. Pantheon; 1976.
[1134] Ehrenreich B. Natural causes: an epidemic of wellness, the certainty of dying, and killing ourselves to live longer. Twelve; 2018.
[1135] Frances A, Clarkin J. No treatment as the prescription of choice. Arch Gen Psychiatry 1981;38:542–5.
[1136] Bourke J. The story of pain: from prayer to painkillers. Oxford University Press; 2014.
[1137] Nagel T. Mortal questions. Cambridge University Press; 1979.
[1138] Sacks J. Politics of hope. Vintage; 2000.

[1139] Dewey J. Democracy and education. The middle works 1899–1924. Southern Illinois University Press; 1916/1980.
[1140] Dewey J. Reconstruction in philosophy. Holt and Co; 1920.
[1141] Solnit R. Hope in the dark: untold histories, wild possibilities. Nation Books; 2006.
[1142] Popova M. https://www.brainpickings.org/02/09/2015/hope-cynicism; Feb 9, 2015.
[1143] Parfit D. On what matters. Oxford University Press; 2011.
[1144] Aronson R. The dialectics of disaster: a prelude to hope. Verso Editions; 1983.
[1145] van Hooft S. Hope. Acumen; 2011.
[1146] Southwick S, Charney D. Resilience: the science of mastering life's greatest challenges. Cambridge University Press; 2012.
[1147] Hobbes T. Leviathan. Andrew Crooke; 1651.
[1148] Midgley M. Evolution as a religion: strange hopes and stranger fears. Methuen; 1985.
[1149] Li J. Evolutionary progress: Stephen Jay Gould's rejection and its critique. Philosophy Study 2019;9:239–309.
[1150] Shelley M. Frankenstein; or, the modern prometheus. Lackington, Hughes, Harding, Mavor, & Jones; 1818.
[1151] Sarye's Law. http://quoteinvestigator.com.
[1152] Neiman S. Evil in modern thought: an alternative history of philosophy. Princeton University Press; 2015.
[1153] Bernstein R. Radical evil: a philosophical interrogation. Polity; 2002.
[1154] Midgley M. Wickedness: a philosophical essay. Routledge & Kegan Paul; 1984.
[1155] Kekes J. The roots of evil. Cornell University Press; 2005.
[1156] Gaita R. Good and evil: an absolute conception. 2nd ed. Routledge; 2004.
[1157] Grant R. Naming evil: judging evil. University of Chicago Press; 2006.
[1158] Eagleton T. On evil. Yale University Press; 2011.
[1159] Russell L. Evil: a philosophical investigation. Oxford University Press; 2014.
[1160] Meister C, Moser P. The Cambridge companion to the problem of evil. Cambridge University Press; 2017.
[1161] Nys T, de Wijze S. The Routledge handbook of the philosophy of evil. Routledge; 2019.
[1162] Kant I. Groundwork of the metaphysic of morals; 1785.
[1163] Hirst E. The categorical imperative and the golden rule. Philosophy 1934;9:326–35.
[1164] Scruton R. Parfit the perfectionist. Philos Public Aff 2014;89:621–34.
[1165] Berlin I. A message to the 21st century, address in 1994. The New York Review of Books 2014. October 23.
[1166] Heine H. The history of religion and philosophy in Germany. Cambridge University Press; 1834/2007.
[1167] Moore G. The right and the good. Oxford University Press; 1930.
[1168] Beauchamp T, Childress J. Principles of biomedical ethics. 8th ed. Oxford University Press; 2019.
[1169] Clouser K, Gert B. Morality vs principlism. John Wiley and Sons; 1994.
[1170] Savulescu J. Bioethics: why philosophy is essential for progress. J Med Ethics 2015;41:28–33.
[1171] Sinnott-Armstrong W. Moral dilemmas. Basil Blackwell; 1988.
[1172] Sinnott-Armstrong W. Moral skepticisms. Oxford University Press; 2006.
[1173] Joyce R, Garner R. The end of morality: taking moral abolitionism seriously. Routledge; 2019.
[1174] Olson J. Moral error theory: history, critique, defence. Oxford University Press; 2014.

[1175] Megill A. Prophets of extremity: Nietzsche, Heidegger, Foucault, Derrida. University of California Press; 1985.
[1176] Mackie J. Ethics: inventing right and wrong. Penguin; 1991.
[1177] Harrison J. Review. Hume Stud 1982;8:70–85.
[1178] Mackie J. Hume's moral theory. Routledge; 1980.
[1179] Solomon R. Living with Nietzsche: what the great "Immoralist" has to teach us. Oxford University Press; 2003.
[1180] Golomb J, Wistrich R. Nietzsche, godfather of fascism?: on the uses and abuses of a philosophy. Princeton University Press; 2002.
[1181] Leiter B. Nietzsche's naturalism: neither liberal nor illiberal. In: De Caro M, Macarthur D, editors. The Routledge handbook of liberal naturalism. Routledge; forthcoming.
[1182] Swanton C. The virtue ethics of Hume and Nietzsche. Wiley Blackwell; 2015.
[1183] Mill J. A system of logic. Harper & Brothers; 1874.
[1184] Chappell T. Values and virtues: Aristotelianism and contemporary ethics. Oxford University Press; 2006.
[1185] Timpe K, Boyd C. Virtues and their vices. Oxford University Press; 2014.
[1186] Mason E. Value pluralism. In: Zalta E, editor. The Stanford encyclopedia of philosophy. Stanford University; 2018.
[1187] Thompson J. Goodness and advice. Princeton University Press; 2001.
[1188] Anderson E. Value in ethics and economics. Harvard University Press; 1995.
[1189] Berlin I. The crooked timber of humanity. Fontana Press; 1991.
[1190] Harman G, Thomson J. Moral relativism and moral objectivity. Blackwell; 1996.
[1191] Kekes J. Pluralism in philosophy: changing the subject. Cornell University Press; 2000.
[1192] Raz J. The practice of value. Oxford University Press; 2003.
[1193] Slote M. The impossibility of perfection: Aristotle, feminism, and the complexities of ethics. Oxford University Press; 2011.
[1194] Wolf S. The variety of values: essays on morality, meaning, and love. Oxford University Press; 2015.
[1195] Lutz M, Lenman J. Moral naturalism. In: Zalta E, editor. The Stanford encyclopedia of philosophy. Stanford University; 2018.
[1196] Angier T. The Cambridge companion to natural law ethics. Cambridge University Press; 2019.
[1197] Cherry M. The normativity of the natural: human goods, human virtues, and human flourishing. Springer; 2009.
[1198] Boyd R. How to be a moral realist. In: Sayre-McCord G, editor. Essays on moral realism. Cornell University Press; 1988.
[1199] Blackford R. The mystery of moral authority. Palgrave Macmillan; 2016.
[1200] Yong H, Yang X. Moral relativism and Chinese philosophy: David Wong and his critics. State University of New York Press; 2014.
[1201] Wong D. Natural moralities: a defense of pluralistic relativism. Oxford University Press; 2006.
[1202] FitzPatrick W. Teleology and the norms of nature. Garland; 2000.
[1203] Byock I. The best care possible: a physician's quest to transform care through the end of life. Avery; 2012.
[1204] Baggini J. Hume the humane. Aeon; 2018.
[1205] Baier A. A naturalist view of persons. Proc Address Am Philos Assoc 1991;65:5–17.
[1206] MacIntyre A. Dependent rational animals: why human beings need the virtues. Open Court; 1999.

[1207] MacIntyre A. After virtue: a study in moral theory. 2nd ed. University of Notre Dame; 1984.
[1208] Connell S. Aristotle for the modern ethicist. Ancient Philosophy Today 2019;1:192–214.
[1209] De Mesel B. The later Wittgenstein and moral philosophy. Springer; 2018.
[1210] Sharpe M. Stoic virtue ethics. In: van Hooft S, editor. Handbook of virtue ethics. Routledge; 2014.
[1211] Garrett D. Nature and necessity in Spinoza's philosophy. Oxford University Press; 2018.
[1212] Marshall C. Moral realism in Spinoza's Ethics. In: Melamed Y, editor. Spinoza's 'Ethics': a critical guide. Cambridge University Press; 2017.
[1213] Ravven H. The self beyond itself: an alternative history of ethics, the new brain sciences, and the myth of free will. New Press; 2013.
[1214] Gill M. Humean moral pluralism. Oxford University Press; 2014.
[1215] Pigden C. Hume on motivation and virtue. Palgrave Macmillan; 2009.
[1216] Norman A. The machinery of moral progress: interview with Rebecca Newberger Goldstein. Humanist. Sep/Oct: 2014.
[1217] Thompson T. The evolutionary biology of evil. Monist 2002;85:239–59.
[1218] Farber P. The temptations of evolutionary ethics. University of California Press; 1998.
[1219] Flew A. Evolutionary ethics. Macmilllan; 1967.
[1220] Ruse M. Evolutionary ethics and the search for predecessors: Kant, Hume, and all the way back to Aristotle? Soc Philos Policy 1990;8:59–85.
[1221] Angier T. Alasdair MacIntyre's analysis of tradition. Eur J Philos 2011;22:540–72.
[1222] Nussbaum M. Non-relative virtues: an Aristotelian approach. Midwest Stud Philos 1988;13:32–53.
[1223] Carr D, Arthur J, Kristjánsson K. Varieties of virtue ethics. Palgrave MacMillan; 2017.
[1224] Foot P. Natural goodness. Oxford University Press; 2001.
[1225] Midgley M. Heart and mind: the varieties of moral experience. 3rd ed. Routledge; 2016.
[1226] Murdoch I. The sovereignty of good. Routledge and Keegan Paul; 1970.
[1227] Midgley M. Science as salvation: a modern myth and its meaning. Routledge; 1994.
[1228] Hamilton R. Naturalistic virtue ethics and the new biology. In: van Hooft S, editor. Handbook of virtue ethics. Routledge; 2014.
[1229] Fletcher G, Ridge M. Having it both ways: hybrid theories and modern metaethics. Oxford University Press; 2014.
[1230] Midgley M. The owl of Minnerva: a memoir. Routledge; 2005.
[1231] Lamey A. Sympathy and scapegoating in J. M. Coetzee. In: Leist A, Singer P, editors. J.M. Coetzee and ethics: philosophical perspectives on literature. Columbia University Press; 2010.
[1232] Baier A. Hume: The women's moral theorist. In: Kittay E, Meyers D, editors. Women and moral theory. Rowman and Littlefield; 1987.
[1232a] Peperzak A. Ethics as first philosophy: the significance of Emmanuel Levinas for philosophy, literature, and religion. Routledge; 1995.
[1233] Hamington M. Embodied care: Jane Addams, Maurice, Merleau-Ponty and feminist ethics. University of Illinois Press; 2004.
[1233a] Atterton P, Calarco M, Friedman M. Levinas and Buber: dialogue and difference. Duquesne University Press; 2004.
[1234] Slote M. The ethics of care and empathy. Routledge; 2007.
[1235] Thomas A. Virtue ethics and an ethics of care: complementary or in conflict? Eidos 2011;14:132–51.
[1236] Gallagher S. In your face: transcendence in embodied interaction. Front Hum Neurosci 2014;8:495.

[1237] van Niekerk A, Nortjé N. Phronesis and an ethics of responsibility. South Afr J Bioeth Law 2013;6:28–31.
[1238] Appiah K. Cosmopolitanism: ethics in a world of strangers. Norton; 2006.
[1239] Brock G. Global justice: a cosmopolitan account. Oxford University Press; 2009.
[1240] Nussbaum M. The cosmopolitan tradition: a noble but flawed ideal. Harvard University Press; 2019.
[1241] van Hooft S. Cosmopolitanism as virtue. J Global Ethics 2007;3:303–15.
[1242] Seligman M, Railton P, Baumeister R, Sripada C. Homo prospectus. Oxford University Press; 2016.
[1243] Dewey J. Theory of the moral life. Wiley Eastern; 1908.
[1244] Shook J, Solymosi T. Pragmatist neurophilosophy: American philosophy and the brain. Bloomsbury Academic; 2014.
[1245] Solymosi T, Shook J. Neuropragmatism: a neurophilosophical manifesto. Eur J Pragm Am Philos 2013;V:212–34.
[1246] Fesmire S. John Dewey and moral imagination: pragmatism in ethics. Indiana University Press; 2003.
[1247] Keulartz J, Korthals M, Schermer M, Swierstra T. Pragmatist ethics for a technological culture. Springer; 2002.
[1248] Richardson H. Articulating the moral community: toward a constructive ethical pragmatism. Oxford University Press; 2018.
[1249] Serra J. What is and what should pragmatic ethics be? Eur J Pragm Am Philos 2010;11:2.
[1250] McGee G. Pragmatic bioethics. Vanderbilt University Press; 1999.
[1251] Pavarini G, Singh I. Pragmatic neuroethics: lived experiences as a source of moral knowledge. Camb Q Healthc Ethics 2018;27:578–89.
[1252] Aikin S, Talisse R. Pragmatism, pluralism, and the nature of philosophy. Routledge; 2018.
[1253] Putnam H, Putnam R. Pragmatism as a way of life. Harvard University Press; 2017.
[1254] Graham J, Haidt J, Koleva S, Motyl M, Iyer R, Wojcik S, et al. Moral foundations theory: the pragmatic validity of moral pluralism. Adv Exp Soc Psychol 2013;7:55–130.
[1255] Haidt J. The new synthesis in moral psychology. Science 2007;316:998–1002.
[1256] Haidt J. The righteous mind: why good people are divided by politics and religion and politics. Pantheon; 2012.
[1257] Metz T. The virtues of African ethics. In: van Hooft S, editor. Handbook of virtue ethics. Routledge; 2014.
[1258] Tucker M. Two kinds of value pluralism. Utilitas 2016;28:333–46.
[1259] Davis J. The embodiment of virtue: toward a cross-cultural cognitive science. In: Smith J, editor. A history of embodiment. Oxford University Press; 2017.
[1260] Sayre-McCord G. Being a realist about relativism (in ethics). Philos Stud 1991;61:155–76.
[1261] Shermer M. The science of good and evil: why people cheat, gossip, care, share, and follow the golden rule. Holt; 2005.
[1262] van Niekerk J. The virtue of gossip. S Afr J Philos 2005;27:400–12.
[1263] Wattles J. The golden rule. Oxford University Press; 1996.
[1264] Westacott E. The virtues of our vices: a modest defense of gossip, rudeness, and other bad habits. Princeton University Press; 2012.
[1265] Bloomfield P. The virtues of happiness: a theory of the good life. Oxford University Press; 2014.
[1266] Browning H. The natural behavior debate: two conceptions of animal welfare. J Appl Anim Welf Sci 2020;23:325–37.

[1267] Kupper F, De Cock Buning T. Deliberating animal values: a pragmatic-pluralistic approach to animal ethics. J Agric Environ Ethics 2011;24:431–50.
[1268] Cooper D. Animals and misanthropy. Routledge; 2018.
[1269] Sayre-McCord G. Essays on moral realism. Cornell University Press; 1988.
[1270] Bagnoli C. Constructivism in ethics. Cambridge University Press; 2013.
[1271] Nagel T. The view from nowhere. Oxford University Press; 1986.
[1272] Sayre-McCord G. "Good" on twin earth. Philos Issues 1997;8:267–92.
[1273] Nussbaum M. Perfectionist liberalism and political liberalism. Philos Public Aff 2011;39:3–45.
[1274] al-Gharbi M. From political liberalism to para-liberalism, epistemological pluralism, cognitive liberalism & authentic choice. Comp Philos 2016;7:1–25.
[1275] Johnson M. Moral imagination: implications of cognitive science for ethics. University of Chicago Press; 1993.
[1276] Kaag J. Thinking through the imagination: aesthetics in human cognition. Fordham University Press; 2014.
[1277] Cleckley H. The mask of sanity: an attempt to clarify some issues about the so-called psychopathic personality. Mosby; 1941.
[1278] Hare R. Without conscience: the disturbing world of the psychopaths amongst us. Guilford Press; 1999.
[1279] Edens J, Marcus D, Lillienfeld S. Psychopathic, not psychopath: taxometric evidence for the dimensional structure of psychopathy. J Abnorm Psychol 2006;115:131–44.
[1280] Harris G, Rice M, Quinsey V. Psychopathy as a taxon: evidence that psychopaths are a discrete class. J Consult Clin Psychol 1994;62:387–97.
[1281] Kessler R. The categorical versus dimensional assessment controversy in the sociology of mental illness. J Health Soc Behav 2002;43:171–88.
[1282] Barash D, Lipton J. Payback: why we retaliate, redirect aggression and seek revenge. Oxford University Press; 2011.
[1283] Buss D. The murderer next door: why the mind is designed to kill. Penguin; 2005.
[1284] Daly M, Wilson M. Homicide: foundations of human behavior. Routledge; 2017.
[1285] Maestripieri D. Macachiavellian intelligence: how rhesus macaques and humans have conquered the world. University of Chicago Press; 2007.
[1286] Smith D. Why we lie: the evolutionary roots of deception and the unconscious mind. St Martin's Press; 2004.
[1287] Smith D. The most dangerous animal: human nature and the origins of war. St Martin's Press; 2007.
[1288] Wrangham R, Peterson D. Demonic apes: apes and the origins of human violence. Hougton Miflin Harcourt; 1996.
[1289] Pfaff D. The altruistic brain: how we are naturally good. Oxford University Press; 2015.
[1290] Kohn A. The brighter side of human nature: altruism and empathy in everyday life. Basic Books; 1990.
[1291] Bregman R. Humankind: a hopeful history. Bloomsbury; 2020.
[1292] de Waal F. Our inner ape: the best and worst of human nature. Granta; 2014.
[1293] Curtis W. The most invasive species of all. Sci Am 2015;313:32–9.
[1294] Milgram S. Obedience to authority: an experimental view. Harper Row; 1974.
[1295] Zimbardo P. Lucifer effect: understanding how good people turn evil. Random House; 2007.
[1296] Brannigan A, Nicholson I, Cherry F. Introduction to the special issue: unplugging the Milgram machine. Theory Psychol 2015;25:551–63.
[1297] Le Texier T. Debunking the Stanford prison experiment. Am Psychol 2019;74:823–9.
[1298] Arendt H. Eichmann in Jerusalem: a report on the banality of evil. Viking; 1963.

[1299] Lipstadt D. The Eichmann trial. Schocken; 2011.
[1300] Arendt H. The human condition. University of Chicago Press; 1958.
[1301] Benhabib S. Politics in dark times: encounters with Hannah Arendt. Cambridge University Press; 2010.
[1302] Pinnock D. Gang town. Tafelberg; 2016.
[1303] Paulhus D, Williams K. The dark triad of personality: Narcissism, Machiavellianism, and psychopathy. J Res Pers 2002;36:556–63.
[1304] Buckels E, Jones D, Paulhus D. Behavioural confirmation of everyday sadism. Psychol Sci 2013;24:2201–9.
[1305] James A. Assholes: a theory. Doubleday; 2012.
[1306] Schwitzgebel E. Theory of jerks and other philosophical misadventures. MIT Press; 2019.
[1307] Lasch C. The culture of narcissism: American life in an age of diminishing expectations. Norton; 1979.
[1308] Bohart A, Held B, Mendelowitz E, Schneider K. Humanity's dark side: evil, destructive experience, and psychotherapy. American Psychological Association; 2012.
[1309] Storr W. Selfie: how we became so self-obsessed and what it's doing to us. Harry N Abrams; 2018.
[1310] Babiak P, Hare R. Snakes in suits: when psychopaths go to work. HarperCollins; 2006.
[1311] Dutton K. Wisdom of psychopaths: what saints, spies, and serial killers can teach us about success. Doubleday; 2012.
[1312] Konnikova M. The confidence game: why we fall for it … every time. Viking; 2016.
[1313] Cohen G. Deeper into bullshit. In: Buss S, Overton L, editors. Contours of agency: essays on themes from Harry Frankfurt. MIT Press; 2002.
[1314] Lang J. Against obedience: Hannah Arendt's overlooked challenge to social-psychological explanations of mass atrocity. Theory Psychol 2014;24:649–67.
[1315] Goldhagen D. Hitler's willing executioners: ordinary Germans and the Holocaust. Alfred A Knopf; 1996.
[1316] Fenigstein A. Milgram's shock experiments and the Nazi perpetrators: a contrarian perspective on the role of obedience pressures during the Holocaust. Theory Psychol 2015;25:581–98.
[1317] Mastroianni G. Obedience in perspective: psychology and the Holocaust. Theory Psychol 2015;25:657–69.
[1318] Batson C. What's wrong with morality?: a social-psychological perspective. Oxford University Press; 2016.
[1319] Dimsdale J. The anatomy of malice: the enigma of the Nazi war criminals. Yale University Press; 2016.
[1320] Staub E. The psychology of perpetrators and bystanders. Polit Psychol 1985;6:61–85.
[1321] Cesarani D, Levine P. Bystanders to the holocaust: a re-evaluation. Frank Kass; 2002.
[1322] Kaminer D, Stein D. Sadistic personality disorder in perpetrators of human rights abuses. A South African case study. J Pers Disord 2001;15:475–86.
[1323] Marsh A. The caring continuum: evolved hormonal and proximal mechanisms explain prosocial and antisocial extremes. Annu Rev Psychol 2019;70:347–71.
[1324] Glenn A, Kurzban A, Raine A. Evolutionary theory and psychopathy. Aggress Violent Behav 2011;16:371–80.
[1325] Stein D. Philosophy of psychopathy. Perspect Biol Med 1996;39:569–80.
[1326] Prinz J. The emotional construction of morals. Oxford University Press; 2007.
[1327] Maibom H. The mad, the bad, and the psychopath. Neuroethics 2008;1:167–84.
[1328] Miller C, Furr R, Knobel A, Fleeson W. Character: new directions from philosophy, psychology, and theology. Oxford University Press; 2015.

[1329] Goldie P. On personality. Routledge; 2004.
[1330] Miller C. The character gap: how good are we? Oxford University Press; 2017.
[1331] Marsh A. The fear factor: how one emotion connects altruists, psychopaths and everyone in-between. Basic Books; 2017.
[1332] Sonne J, Gash D. Psychopathy to altruism: neurobiology of the selfish–selfless spectrum. Front Psychol 2018;9:575.
[1333] Smith J. Group selection and kin selection. Nature 1964;201:1145–7.
[1334] Hamilton W. Genetical evolution of social behaviour. J Theor Biol 1964;7:1–52.
[1335] Triver R. Natural selection and social theory: selected papers of Robert Trivers. Oxford University Press; 2002.
[1336] Sober E, Wilson D. Unto others: the evolution and psychology of unselfish behavior. Harvard University Press; 1999.
[1337] Barber N. Kindness in a cruel world: the evolution of altruism. Prometheus; 2004.
[1338] Klein S. Survival of the nicest: how altruism made us human and why it pays to get along. Experiment; 2014.
[1339] Kurzban R, Burton-Chellew M, West S. The evolution of altruism in humans. Annu Rev Psychol 2015;66:575–99.
[1340] Oakley B, Knafo A, Madhavan G, Wilson D. Pathological altruism. Oxford University Press; 2011.
[1341] Barash D. Natural selections: natural selections: selfish altruists, honest liars and other realities of evolution. Bellevue Literary Press; 2007.
[1342] van Honk J, Eisenegger C, Terburg D, Stein D, Morgan B. Generous economic investments after basolateral amygdala damage. Proc Natl Acad Sci U S A 2013;110:2506–10.
[1343] Keysers C. The empathic brain: how the discovery of mirror neurons changes our understanding of human nature. Social Brain; 2011.
[1344] Dutton K. Black and white thinking: the burden of a binary brain in a complex world. Farrar, Straus and Giroux; 2021.
[1345] Nagel T. The possibility of altruism. Oxford University Press; 1979.
[1346] Post S, Underwood L, Schloss J, Hurlbut W. Altruism and altruistic love: science, philosophy, and religion in dialogue. Oxford University Press; 2002.
[1347] MacFarquhar L. Strangers drowning: grappling with impossible idealism, drastic choices, and the overpowering urge to help. Penguin; 2015.
[1348] Flescher A, Worthen D. The altruistic species: scientific, philosophical, and religious perspectives of human benevolence. Templeton Press; 2007.
[1349] Kitcher P. The ethical project. Harvard University Press; 2011.
[1350] Nowak M, Coakley S. Evolution, games, and God: the principal of cooperation. Harvard University Press; 2013.
[1351] Singer P. The most good you can do: how effective altruism is changing ideas about living ethically. Yale University Press; 2015.
[1352] MacAskill W. Doing good better: effective altruism and a radical new way to make a difference. Guardian Faber; 2015.
[1353] Coplan A, Goldie P. Empathy: philosophical and psychological perspectives. Oxford University Press; 2011.
[1354] Maibom H. The Routledge handbook of philosophy of empathy. Routledge; 2017.
[1355] Kahn C. Aristotle and altruism. Mind 1981;90:20–40.
[1356] Gill M. Hume's progressive view of human nature. Hume Stud 2000;26:87–108.
[1357] Višňovský E. The pragmatist conception of altruism and reciprocity. Hum Aff 2011;21:437–53.

[1358] Knapp S, Gottlieb M, Handelsman M. Ethical dilemmas in psychotherapy: positive approaches to decision-making. American Psychological Association; 2015.
[1359] MacIntrye A. After virtue. University of Notre Dame Press; 1981.
[1360] Doherty W. Soul searching: why psychotherapy must promote moral responsibility. Basic Books; 1996.
[1361] Wilson T. Redirect: changing the stories we live by. Little, Brown and Company; 2011.
[1362] Banicki K. Philosophy as therapy: towards a conceptual model. Philos Pap 2014;43:17–31.
[1363] Fischer E. How to practice philosophy as therapy: philosophical therapy and therapeutic philosophy. Metaphilosophy 2011;42:49–82.
[1364] Marinoff L. Plato, not Prozac!: applying eternal wisdom to everyday problems. Harper; 1999.
[1365] Pellegrino E, Thomasma D. The virtues in medical practice. Oxford University Press; 1993.
[1366] Fowers B. Virtue and psychology: pursuing excellence in ordinary practices. American Psychological Association; 2005.
[1367] Fröding B. Virtue ethics and human enhancement. Springer; 2013.
[1368] Hamilton R. The frustrations of virtue: the myth of moral neutrality in psychotherapy. J Eval Clin Pract 2013;19:485–92.
[1369] Macaro A. Reason, virtue and psychotherapy. John Wiley & Sons; 2006.
[1370] Martin M. Morality to mental health: virtue and vice in a therapeutic culture. Oxford University Press; 2006.
[1371] Proctor C. Virtue ethics in psychotherapy: a systematic review of the literature. Int J Exist Posit Psychol 2018;8:1–22.
[1372] Radden J, Sadler J. The virtuous psychiatrist: character ethics in psychiatric practice. Oxford University Press; 2010.
[1373] Waring D. The healing virtues: character ethics in psychotherapy. Oxford University Press; 2016.
[1374] Persson I, Savulescu J. Unfit for the future: the need for moral enhancement. Oxford University Press; 2012.
[1375] Spence S. Can pharmacology help enhance human morality? Br J Psychiatry 2008;193:179–80.
[1376] Wiseman H. The myth of the moral brain: the limits of moral enhancement. MIT Press; 2016.
[1377] Tremblay S, Sharika K, Platt L. Social decision-making and the brain: a comparative perspective. Trends Cogn Sci 2017;21:265–76.
[1378] Eddy B. Evolutionary pragmatism and ethics. Lexington Books; 2015.
[1379] Wilson E. Sociobiology: the new synthesis. Belknap; 1975.
[1380] Bloom P. Just babies: the origins of good and evil. Crown; 2014.
[1381] Boehm C. Moral origins: the evolution of virtue, altruism, and shame. Basic Books; 2012.
[1382] Changeux J-P, Ricoer P. What makes us think?: a neuroscientist and a philosopher argue about ethics, human nature, and the brain. Princeton University Press; 2002.
[1383] Changeux J-P. The good, the true, and the beautiful: a neuronal approach. Yale University Press; 2012.
[1384] Churchland P. Braintrust: what neuroscience tells us about morality. Princeton University Press; 2011.
[1385] de Waal F. Good natured: the origins of right and wrong in humans and other animals. Harvard University Press; 1996.
[1386] de Waal F. Primates and philosophers: how morality evolved. Princeton University Press; 2006.
[1387] Decety J, Wheatley T. The moral brain: a multidisciplinary perspective. MIT Press; 2015.
[1388] Dennett D. Darwin's dangerous idea: evolution and the meanings of life. Simon & Schuster; 1995.

[1389] Fowers B. The evolution of ethics: human sociality and the emergence of ethical mindedness. Palgrave Macmillan; 2015.
[1390] Gazzaniga M. The ethical brain. Dana Press; 2005.
[1391] Hauser M. Moral minds moral minds: how nature designed our universal sense of right and wrong. Ecco; 2006.
[1392] Hinde R. Bending the rules: morality in the modern world – from relationships to politics and war. Oxford University Press; 2007.
[1393] Joyce R. The evolution of morality. MIT Press; 2005.
[1394] Mascaro S, Korb K, Nicholson A, Woodberry O. Evolving ethics: the new science of good and evil. Imprint Academic; 2010.
[1395] Midgley M. The ethical primate: humans, freedom and morality. Routledge; 1994.
[1396] Putnam H, Neiman S, Schloss J. Understanding moral sentiments: Darwinian perspectives. Routledge; 2017.
[1397] Ridley M. The origins of virtue: human instincts and the evolution of cooperation. Viking; 1996.
[1398] Ruse M, Richards R. Cambridge handbook of evolutionary ethics. Cambridge University Press; 2017.
[1399] Sinnott-Armstrong W, Miller C. Moral psychology: the evolution of morality: adaptations and innateness. MIT Press; 2007.
[1400] Decety J, Bartal I, Uzefovsky F, Knafo-Noam A. Empathy as a driver of prosocial behaviour: highly conserved neurobehavioural mechanisms across species. Philos Trans Royal Soc Lond B Biol Sci 2016;371:20150077.
[1401] de Waal F, Preston S. Mammalian empathy: behavioural manifestations and neural basis. Nat Rev Neurosci 2017;18:498–509.
[1402] Haidt J. The emotional dog and its rational tail: a social intuitionist approach to moral judgment. Psychol Rev 2001;108:814–34.
[1403] Pizarro D, Bloom P. The intelligence of the moral intuitions: comment on Haidt. Psychol Rev 2001;100:193–8.
[1404] Bazerman M, Tenbrunsel A. Blind spots: why we fail to do what's right and what to do about it. Princeton University Press; 2011.
[1405] Schwitzgebel E, Rust J. The behavior of ethicists. In: Sytsma J, Buckwalter W, editors. A companion to experimental philosophy. Wiley Blackwell; 2016.
[1406] Pellegrino E, Thomasma D. Dubious premises – evil conclusions: moral reasoning at the Nuremberg Trials. Camb Q Healthc Ethics 2000;9:261–74.
[1406a] Edmonds D. Would you kill the fat man?: The trolley problem and what your answer tells us about right and wrong. Princeton University Press; 2013.
[1407] Greene J. From neural 'is' to moral 'ought': what are the moral implications of neuroscientific moral psychology? Nat Rev Neurosci 2003;4:846–9.
[1408] Greene J. The secret joke of Kant's soul. In: Sinnott-Armstrong W, editor. Moral psychology. MIT Press; 2007.
[1409] Berker S. The normative insignificance of neuroscience. Philos Public Aff 2009;37:293–329.
[1410] Awad E, Dsouza S, Kim R, Schultz J, Henrich J, Shariff A, et al. The moral machine experiment. Nature 2018;563:59–64.
[1411] Kontos P. Evil in Aristotle. Cambridge University Press; 2018.
[1412] Midgley M. Wickedness: a philosophical essay. Routledge & Kegan Paul; 1984.
[1413] Shklar J. Ordinary vices. Harvard University Press; 1984.
[1414] Coady C, O'Neill O. Messy morality and the art of the possible. Proc Aristot Soc Suppl Vol 1990;64: 259–79+81–94.
[1415] Hurka T. The best things in life: a guide to what really matters. Oxford University Press; 2010.

[1416] Leisman G, Machado C, Melillo R, Mualem R. Intentionality and "free-will" from a neurodevelopmental perspective. Front Integr Neurosci 2012;36.

[1417] Buchanan A, Powell R. The evolution of moral progress: a biocultural theory. Oxford University Press; 2018.

[1418] Darwin C. The descent of man, and selection in relation to sex. John Murray; 1871.

[1419] Richards R. The romantic conception of life: science and philosophy in the age of Goethe. University of Chicago Press; 2002.

[1420] Reed P. What's wrong with monkish virtues? Hume on the standard of virtue. Hist Philos Q 2012;29:39–56.

[1421] Nozick R. Philosophical explanations. Harvard University Press; 1981.

[1422] FitzPatrick W. Debunking evolutionary debunking of ethical realism. Philos Stud 2015;172:883–904.

[1423] Casebeer W. Natural ethical facts: evolution, connectionism, and moral cognition. Bradford; 2003.

[1424] Hibbing J, Smith K, Alford J. Predisposed: liberals, conservatives, and the biology of political differences. Routledge; 2013.

[1425] Rittel H, Webber M. Dilemmas in a general theory of planning. Policy Sci 1973;4:155–69.

[1426] Stone M, Brucato G. The new evil: understanding the emergence of modern violent crime. Prometheus; 2019.

[1427] Baumeister R. Evil: inside human violence and cruelty. Holt; 2015.

[1428] Baron-Cohen S. The science of evil: on empathy and the origins of cruelty. Basic Books; 2011.

[1429] Benjet C, Bromet E, Karam E, Kessler R, McLaughlin K, Ruscio A, et al. The epidemiology of traumatic event exposure worldwide: results from the World Mental Health Survey Consortium. Psychol Med 2016;46:327–43.

[1430] Desmarais S, Reeves K, Nicholls Telford R, Fiebert M. Prevalence of physical violence in intimate relationships, Part 2: rates of male and female perpetration. Partn Abus 2012;3:170–98.

[1431] van Prooijen J-W, van Lange P. Cheating, corruption, and concealment: the roots of dishonesty. Cambridge University Press; 2018.

[1432] Knutson B. Sweet revenge. Science 2004;305:1246–7.

[1433] Nell V. Cruelty's rewards: the gratifications of perpetrators and spectators. Behav Brain Sci 2006;29:211–57.

[1434] Scott K, Lim C, Hwang I, Adamowski T, Al-Hamzawi A, Bromet E, et al. The cross-national epidemiology of DSM-IV intermittent explosive disorder. Psychol Med 2016;46:3161–72.

[1435] Seneca L. On Anger. 45 AD.

[1436] Bernstein R. Violence: thinking without banisters. Polity Press; 2013.

[1437] Cherry M, Flanagan O. The moral psychology of anger. Rowman & Littlefield; 2018.

[1438] Rae G, Ingala E. The meanings of violence: from critical theory to biopolitics. Routledge; 2018.

[1439] Bell D, Pei W. Just hierarchy: why social hierarchies matter in China and the rest of the world. Princeton University Press; 2020.

[1440] Sue D. Microaggressions in everyday life: race, gender, and sexual orientation. Wiley; 2010.

[1441] Hammels C, Pishva E, Vry D, van den Hove D, Prickaerts J, van Winkel R, et al. Defeat stress in rodents: from behaviour to molecules. Neurosci Biobehav Rev 2015;59:111–40.

[1442] Berger M, Sarnyai Z. "More than skin deep": stress neurobiology and mental health consequences. Stress 2014;18:1–10.

[1443] Paradies Y, Ben J, Denson N, Elias A, Priest N, Pieterse A, et al. Racism as a determinant of health: a systematic review and meta-analysis. PLoS One 2015;10, e0138511.

[1444] Sapolsky R. Behave: the biology of humans at our best and worst. Penguin; 2007.

[1445] Nagel T. Equality and partiality. Oxford University Press; 1991.

[1446] Nussbaum M, Sen A. The quality of life. Clarendon Press; 1993.

[1447] de Botton A. Status anxiety. Hamish Hamilton; 2004.

[1448] Marmot M. Status, anxiety and health or my anxiety is bigger than yours: Review of Status Anxiety. Int J Epidemiol 2005;34:493–6.

[1449] Boehm C. Hierarchy in the forest: the evolution of egalitarian behavior. Harvard University Press; 1999.

[1450] Greene J. Moral tribes: emotion, reason, and the gap between us and them. Penguin Press; 2013.

[1451] MacMullen T. Habits of whiteness: a pragmatic reconstruction. Indiana University Press; 2009.

[1452] Egito J, Nevat M, Shamay-Tsoory S, Osório A. Oxytocin increases the social salience of the outgroup in potential threat contexts. Horm Behav 2020;122:104733.

[1453] Kavaliers M, Choleris E. Out-group threat responses, in-group bias, and nonapeptide involvement are conserved across vertebrates. Am Nat 2017;189:453–8.

[1454] Quammen D. The tangled tree: a radical new history of life. Simon & Schuster; 2018.

[1455] Bowden R, MacFie T, Myers S, Hellenthal G, Nerrient G, Bontrop R, et al. Genomic tools for evolution and conservation in the chimpanzee: pan troglodytes ellioti is a genetically distinct population. PLoS Genet 2012;8, e1002504.

[1456] Soh D. The end of gender: debunking the myths about sex and identity in our society. Threshold Editions; 2020.

[1457] Rippon G. Gender and our brains: how new neuroscience explodes the myths of the male and female minds. Pantheon; 2019.

[1458] Saini A. Inferior: how science got women wrong-and the new research that's rewriting the story. Beacon Press; 2017.

[1459] Jordan-Young R. Brain storm: the flaws in the science of sex differences. Harvard University Press; 2010.

[1460] Baumeister R. Is there anything good about men? How cultures flourish by exploiting men. Oxford University Press; 2010.

[1461] Baron-Cohen S. The essential difference. Basic Books; 2004.

[1462] Tannen D. You just don't understand: men and women in conversation. William Morrow; 1990.

[1463] Zaretsky R. A life worth living: Albert Camus and the quest for meaning. Belknap; 2013.

[1464] Pfaff D. The neuroscience of fair play: why we (usually) follow the golden rule. Dana Press; 2007.

[1465] Stein D, Kaminer D. Forgiveness and psychopathology: psychobiological and evolutionary underpinnings. CNS Spectr 2006;11:87–9.

[1466] Kaminer D, Stein D, Mbanga I, Zungu-Dirwayi N. Forgiveness: toward an integration of theoretical models. Psychiatry 2000;63:344–57.

[1467] de Waal F. Chimpanzee politics: power and sex amongst apes. Johns Hopkins University Press; 1998.

[1468] Hannah S. How to hold a grudge: from resentment to contentment – the power of grudges to transform your life. Scribner; 2019.

[1469] Miller G. Virtue signaling: essays on Darwinian politics & free speech. Cambrian Moon; 2019.

[1470] Tosi J, Warmke B. Grandstanding: the use and abuse of moral talk. Oxford University Press; 2020.
[1471] Sacks J. The great partnership: science, religion, and the search for meaning. Schocken; 2011.
[1472] Landi P, Marazziti D, Rutigliano G, Dell'Osso L. Insight in psychiatry and neurology: state of the art, and hypotheses. Harv Rev Psychiatry 2016;24:214–28.
[1473] Mograbi D, Morris R. Implicit awareness in anosognosia: clinical observations, experimental evidence, and theoretical implications. Cogn Neurosci 2013;4:181–97.
[1474] Pia L, Neppi-Modona M, Ricci R, Berti A. The anatomy of anosognosia for hemiplegia: a meta-analysis. Cortex 2004;40:367–77.
[1475] Goleman D. Vital lies, simple truths: the psychology of self-deception. Simon & Schuster; 1985.
[1476] Tavris C, Aronson E. Mistakes were made but not by me: why we justify foolish beliefs, bad decisions, and hurtful acts. Harcourt; 2007.
[1477] Myslobodsky M. The mythomanias: the nature of deception and self-deception. Psychology Press; 1997.
[1478] Trivers R. The folly of fools: the logic of deceit and self-deception in human life. Basic Books; 2011.
[1479] Booker C. Groupthink: a study in self delusion. Continuum; 2020.
[1480] Eurich T. Insight: the surprizing truth about how others see us, how we see ourselves, and why the answers matter more than we think. Crown; 2017.
[1481] McRaney D. You are now less dumb: how to conquer mob mentality, how to buy happiness, and all the other ways to outsmart yourself. Avery; 2013.
[1482] Mele A. Self-deception unmasked. Yale University Press; 2001.
[1483] McLaughlin B, Rorty A. Perspectives on self-deception. University of California Press; 1988.
[1484] Carnegie D. How to win friends and influence people. Simon & Schuster; 1936.
[1485] Cialdini R. Influence: the psychology of persuasion. revised ed. Harper; 2006.
[1486] Williams G. Huxley's evolution and ethics in sociobiological perspective. Zygone 1988;23:383–407.
[1487] Stein D. The neurobiology of evil: psychiatric perspectives on perpetrators. Ethn Health 2000;5:303–15.
[1488] Smith D. Less than human: why we demean, enslave and exterminate others. St Martin's Press; 2011.
[1489] Bloom P. Against empathy: the case for rational compassion. Ecco; 2016.
[1490] Bauman Z, Donskis L. Liquid evil. Polity Press; 2016.
[1491] Glover J. Humanity: a moral history of the twentieth century. Yale University Press; 2000.
[1492] Angier T, Meister C, Taliaferro C. The history of evil in antiquity, 2000 BCE – 450 CE. Routledge; 2018.
[1493] Chandler J. Neurolaw and neuroethics. Camb Q Healthc Ethics 2018;27:590–8.
[1494] Parens J. Maimonides and Spinoza: their conflicting views of human nature. University of Chicago Press; 2012.
[1495] Seeskin K. From Maimonides to Spinoza: three versions of an intellectual transition. In: Della R, editor. The Oxford handbook of Spinoza. Oxford University Press; 2017.
[1496] Bennett M, Dennett D, Hacker P, Searle J. Neuroscience and philosophy: brain, mind, and language. Columbia University Press; 2009.
[1497] Bickle J. The Oxford handbook of philosophy and neuroscience. Oxford University Press; 2007.
[1498] Choudhury S, Slaby J. Critical neuroscience: a handbook of the social and cultural contexts of neuroscience. Wiley-Blackwell; 2011.
[1499] Moran J. Interdisciplinarity. Routledge; 2002.

[1500] Bernstein J. Transdisciplinarity: a review of its origins, development, and current issues. J Res Pract 2015;11:1–20.
[1501] Nicolescu B. Manifesto of transdisciplinarity. SUNY Press; 2002.
[1502] Choi B, Pak A. Multidisciplinarity, interdisciplinarity and transdisciplinarity in health services research, services, education and policy: 1. Definitions, objectives, and evidence of effectiveness. Clin Invest Med 2006;29:351–64.
[1503] National Research Council. Convergence: facilitating transdisciplinary integration of life sciences, physical sciences, engineering, and beyond. National Academies Press; 2014.
[1504] Sharp P. Convergence: the future of health. National Academies Press; 2014.
[1505] Sharp P. The third revolution: the convergence of the life sciences, physical sciences, and engineering. National Academies Press; 2014.
[1506] Eyre H, Lavretsky H, Forbes M, Raji C, Small G, MGorry P, et al. Convergence science arrives: how does it relate to psychiatry? Acad Psychiatry 2017;41:91–9.
[1507] Horgan J. The end of science: facing the limits of knowledge in the twilight of the scientific age. Little Brown and Company; 1996.
[1508] Sober E. Philosophy of biology. Westview; 1993.
[1509] Sterelny K, Griffiths P. Sex and death: an introduction to philosophy of biology. University of Chicago Press; 1999.
[1510] Hull D, Ruse M. The Cambridge companion to the philosophy of biology. Cambridge University Press; 2007.
[1511] Godfrey-Smith P. Philosophy of biology. Cambridge University Press; 2014.
[1512] Desmond A, Moore J. Darwin's sacred cause: race, slavery and the quest for human origins. Penguin; 2010.
[1513] Polanyi M. The tacit dimension. Doubleday; 1966.
[1514] Carter A, Poston T. A critical introduction to knowledge-how. Bloomsbury Academic; 2017.
[1515] Benson J, Moffett M. Knowing how: essays on knowledge, mind, and action. Oxford University Press; 2012.
[1516] Stanley J. Know how. Oxford University Press; 2011.
[1517] Carruthers P, Stich S, Siegal M. The cognitive basis of science. Cambridge University Press; 2002.
[1518] Fernández ÁN, Magnani L, Salguero-Lamillar F, Barés-Gómez C. Model-based reasoning in science and technology: inferential models for logic, language, cognition and computation. Springer; 2019.
[1519] Giere R. Explaining science: a cognitive approach. University of Chicago Press; 1988.
[1520] Gopnik A. The philosophical baby: what children's minds tell us about truth, love, and the meaning of life. Farrar, Straus and Giroux; 2009.
[1521] Levy A, Godfrey-Smith P. The scientific imagination. Oxford University Press; 2020.
[1522] Magnani L, Nersessian N, Thagard P. Model-based reasoning in scientific discovery. Springer; 1999.
[1523] Morgan M, Morrison M. Models as mediators: perspectives on natural and social science. Cambridge University Press; 1999.
[1524] Thagard P. How scientists explain disease. Princeton University Press; 1999.
[1525] Johnson M. Morality for humans: ethical understanding from the perspective of cognitive science. University of Chicago Press; 2014.
[1526] Ellis G. A mathematical cosmologist reflects on deep ethics: reflections on values, ethics, and morality. Theol Sci 2020;18:175–89.
[1527] Putnam H. What is mathematical truth? In: Mathematics, matter, and method. Cambridge University Press; 1975.

[1528] Bryant L, Srnicek N, Harman G. The speculative turn: continental materialism and realism. re.press; 2011.
[1529] Snow C. The two cultures and the scientific revolution. Cambridge University Press; 1959.
[1530] Snow C. The two cultures: and a second look: an expanded version of the two cultures and the scientific revolution. Cambridge University Press; 1963.
[1531] Wilson E. Consilience: the unity of knowledge. Alfred A Knopf; 1998.
[1532] Gottschall J. Literature, science, and a new humanities. Palgrave Macmillan; 2008.
[1533] Gould S. The hedgehog, the fox, and the magister's pox: mending the gap between science and the humanities. Harmony; 2003.
[1534] Kagan J. The three cultures: natural sciences, social sciences, and the humanities in the 21st century. Cambridge University Press; 2009.
[1535] Gould S. Interdisciplinarity. Routledge; 2002.
[1536] Russell B. Mysticism and logic. Hilbert J 2014;12:780–803.
[1537] James W. Pragmatism: a new name for an old way of thinking. Longmans, Green and Company; 1907.
[1538] Midgley M. What is philosophy for? Bloomsbury; 2018.
[1539] Ross D, Ladyman J, Kincaid H. Scientific metaphysics. Oxford University Press; 2013.
[1540] Maudlin T. The metaphysics within physics. Oxford University Press; 2010.
[1541] Piaget J. Genetic epistemology. Norton; 1971.
[1542] Lorenz K. Behind the mirror: a search for a natural history of human knowledge. Harcourt Brace Jovanovich; 1977.
[1543] McCauley R. The Churchlands and their critics. Wiley-Blackwell; 1996.
[1544] Smith D. How biology shapes philosophy: new foundations for naturalism. Cambridge University Press; 2017.
[1545] Millikan R. Language, thought, and other biological categories: new foundations for realism. MIT Press; 1984.
[1546] Dawkins R. Unweaving the rainbow. Allen Lane; 1998.
[1547] Keats J. Selected letters. Oxford University Press; 2009.
[1548] Wittgenstein L. Culture and value. Blackwell; 1980.
[1549] Josephson-Storm J. The myth of disenchantment. University of Chicago; 2017.
[1550] Overgaard S, Gilbert P, Burwood S. Introduction to metaphilosophy. Cambridge University Press; 2013.
[1551] Edmonds D, Warburton N. Philosophy bites. Oxford University Press; 2010.
[1552] Williams B. Philosophy as a humanistic discipline. Princeton University Press; 2006.
[1553] Quine W. Theories and things. Harvard University Press; 1981.
[1554] Millikan R. Biosemantics. J Philos 1989;86:281–97.
[1555] Papineau D. The poverty of analysis. Proc Aristot Soc Suppl Vol 2009;S83:1–30.
[1556] Wundt W. System der Philosophie. Wilhelm Engelmann; 1897.
[1557] Laplane L, Mantovanic P, Adolphs R, Chang H, Mantovani A, McFall-Ngai M, et al. Why science needs philosophy. Proc Natl Acad Sci U S A 2019;116:3948–52.
[1558] Dewey J. The quest for certainty: a study of the relation of knowledge and action. George Allen & Unwin; 1929.
[1559] Piaget J. Insights and illusions of philosophy. Routledge; 1997.
[1560] Putnam H. Naturalism, realism and normativity. Harvard University Press; 2016.
[1561] Piaget J. Psychology and epistemology: towards a theory of knowledge. Penguin; 1972.
[1562] Weinert F. The scientist as philosopher: philosophical consequences of great scientific discoveries. Springer; 2004.

[1563] Andrea S. Rickert, Heinrich. In: Zalta E, editor. The Stanford encyclopedia of philosophy. Stanford University; 2018.
[1564] Feigl H. The "mental" and the "physical": the essay and a postscript. University of Minnesota Press; 1967.
[1565] Knobe J, Lombrozo T, Nichols S. Oxford studies in experimental philosophy. vol. 1. Oxford University Press; 2015.
[1566] Fischer E, Collins J. Experimental philosophy, rationalism, and naturalism: rethinking philosophical method. Routledge; 2015.
[1567] Sytsma J, Buckwalter W. A companion to experimental philosophy. Wiley Blackwell; 2016.
[1568] Russell B. The problems of philosophy. Henry Holt; 1912.
[1569] Unger P. Empty ideas: a critique of analytic philosophy. Oxford University Press; 2014.
[1570] Read R. Wittgenstein and the illusion of 'progress': on real politics and real philosophy in a world of technocracy. Royal Inst Philos Suppl 2016;78:265–84.
[1571] Swain R. http://quoteinvestigator.com.
[1572] Chalmers D. Why isn't there more progress in philosophy? Philosophy 2015;90:3–31.
[1573] Cappelen H. Disagreement in philosophy: an optimistic perspective. In: D'Oro G, Overgaard S, editors. The Cambridge companion to philosophical methodology. Cambridge University Press; 2017.
[1574] Gallie W. Essentially contested concepts. Proc Aristot Soc 1955;56:167–98.
[1575] Williamson T. Model-building in philosophy. In: Blackford R, Broderick D, editors. Philosophy's future: the problem of philosophical progress. John Wiley; 2017.
[1576] Kekes J. The nature of philosophical problems: their causes and implications. Oxford University Press; 2014.
[1577] Stoljar D. Philosophical progress: in defence of a reasonable optimism. Oxford University Press; 2017.
[1578] Blackford R, Broderick D. Philosophy's future: the problem of philosophical progress. John Wiley; 2017.
[1579] Pigliucci M. The nature of philosophy: how philosophy makes progress and why it matters. City College of New York; 2016.
[1580] Williamson T. The philosophy of philosophy. Blackwell; 2007.
[1581] Sadler J, van Staden W, Fulford K. Oxford handbook of psychiatric ethics. Oxford University Press; 2015.
[1582] Fulford K, Davies M, Gipps R, Graham G, Sadler J, Stanghellini G, et al. Oxford handbook of philosophy and psychiatry. Oxford University Press; 2013.
[1583] Jamison K. An unquiet mind: a memoir of moods and madness. Knopf; 1995.
[1584] Stein D, Harris M, Vigo D, Chiu W, Sampson N, Alonso J, et al. Perceived helpfulness of treatment for post-traumatic stress disorder: findings from the World Mental Health Surveys. Depress Anx 2020;37:972–94.
[1585] Gibson N. Living Fanon: interdisciplinary perspectives. Palgrave Macmillan and the University of Kwa-Zulu Natal Press; 2011.
[1586] van Niekerk A. Is 'decolonization' a legitimate and appropriate value in biomedical research and teaching? S Afr J Bioeth Law 2019;12:4.
[1587] Stein D, Wegener G. Discovery versus implementation research on mental disorders in low- and middle-income countries. Acta Neuropsychiatr 2017;29:191–2.
[1588] Desmond-Hellmann S. Progress lies in precision. Science 2016;353:731.
[1589] Stein D, He Y, Phillips A, Sahakian B, Williams J, Patel V. Global mental health and neuroscience: potential synergies. Lancet Psychiatry 2015;2:178–85.

[1590] Taylor-Robinson D, Kee F. Precision public health – the emperor's new clothes. Int J Epidemiol 2019;48:1–6.
[1591] Maj M. The need for a conceptual framework in psychiatry acknowledging complexity while avoiding defeatism. World Psychiatry 2016;15:1–2.
[1592] Kendler K, First M. Alternative futures for the DSM revision process: iteration v. paradigm shift. Br J Psychiatry 2010;197:263–5.
[1593] Ohlsson H, Kendler K. Applying causal inference methods in psychiatric epidemiology: a review. JAMA Psychiatry; 2020.
[1594] Kendler K. The nature of psychiatric disorders. World Psychiatry 2016;15:5–12.
[1595] Lawlor D, Tilling K, Davey Smith G. Triangulation in aetiological epidemiology. Int J Epidemiol 2016;45:1866–86.
[1596] Flanagan O. Consciousness reconsidered. MIT Press; 1992.
[1597] Becker A. What is real? The unfinished quest for the meaning of quantum physics. Basic Books; 2018.
[1598] Fuchs C. What is real? The unfinished quest for the meaning of quantum physics. Am J Phys 2019;87:317–8.
[1599] Canales J. The physicist and the philosopher: Einstein, Bergson, and the debate that changed our understanding of time. Princeton University Press; 2015.
[1600] Munafò M, Nosek B, Bishop D, Button K, Chambers C, du Sert N, et al. A manifesto for reproducible science. Nat Hum Behav 2017;1:0021.
[1601] Chambers C. The seven deadly sins of psychology: a manifesto for reforming the culture of scientific practice. Princeton University Press; 2017.
[1602] Maj M. Financial and non-financial conflicts of interests in psychiatry. Acta Bioethica 2009;15:2.
[1603] Patel V, Saxena S, Lund C, Thornicroft G, Bungana F, Bolton P, et al. The Lancet Commission on global mental health and sustainable development. Lancet 2018;392:1553–98.
[1604] Mercier H. Not born yesterday: the science of who we trust and what we believe. Princeton University Press; 2020.
[1605] Gorman S, Gorman J. Denying to the Grave: why we ignore the facts that will save us. Oxford University Press; 2017.
[1606] Kang L, Pedersen N. Quackery: a brief history of the worst ways to cure everything. Working Publishing; 2017.
[1607] Offit P. Do you believe in magic? The sense and nonsense of alternative medicine. Harper; 2013.
[1608] Goldeacre B. Bad science: quacks, hacks, and big pharma flacks. McClelland & Stewart; 2010.
[1609] Singh S, Ernst E. Trick or treatment: the undeniable facts about alternative medicine. Norton; 2008.
[1610] Bausell R. Snake oil science: the truth about complementary and alternative medicine. Oxford University Press; 2007.
[1611] Shermer M. Why people believe weird things: pseudoscience, superstition, and other confusions of our time. WH Freeman; 1997.
[1612] Caulfield T. The cure for everything: untangling twisted messages about health, fitness, and happiness. Beacon; 2012.
[1613] Henning B, Scarfe A. Beyond mechanism: putting life back into biology. Lexington Books; 2013.
[1614] Nicolson D. The concept of mechanism in biology. Stud Hist Philos Biol Biomed Sci 2012;43:152–63.

[1615] Craver C. Explaining the brain: mechanisms and the mosaic unity of neuroscience. Clarendon; 2007.
[1616] Bechtel W. Mental mechanisms. Routledge; 2007.
[1617] Damman O. Etiological explanations: illness causation theory. CRC Press; 2020.
[1618] von Wright G. Explanation and understanding. Routledge & Kegan Paul; 1971.
[1619] Pearl J, Mackenzie D. The book of why: the new science of cause and effect. Basic Books; 2018.
[1620] Slater M. Natural kindness. Br Soc Philos Sci 2015;66:375–411.
[1621] Bartlett F. Remembering: a study in experimental and social psychology. Cambridge University Press; 1932.
[1622] Adler M. Ten philosophical mistakes. Macmillan; 1985.
[1623] Wilson E. The meaning of human existence. Norton; 2014.
[1624] Wilson E. On human nature. Harvard University Press; 1978.
[1625] Barash D. Through a glass brightly: using science to see our species as we really are. Oxford University Press; 2018.
[1626] Brown D. Human universals. McGraw-Hill; 1991.
[1627] Hull D. On human nature. In: PSA: Proceedings of the Biennial Meeting of the Philosophy of Science Association 1986;3–13.
[1628] Lewontin R, Rose S, Kamin L. Not in our genes: biology, ideology, and human nature. Haymarket; 2017.
[1629] Bourke J. What it means to be human: historical reflections from the 1800s to the present. Counterpoint; 2011.
[1630] Lewontin R. Biology as ideology: the doctrine of DNA. Anansi; 1991.
[1631] Gross P, Levitt N. Higher superstition: the academic left and its quarrels with science. Johns Hopkins University Press; 1997.
[1632] Scruton R. On human nature. Princeton University Press; 2017.
[1633] Dupré J. Human nature and the limits of science. Oxford University Press; 2002.
[1634] Hannon E, Lewens T. Why we disagree about human nature. Oxford University Press; 2018.
[1635] Kronfeldner M. What's left of human nature?: a post-essentialist, pluralist, and interactive account of a contested concept. MIT Press; 2018.
[1636] Fuentes A, Visala A. Verbs, bones, and brains: interdisciplinary perspectives on human nature. University of Notre Dame Press; 2017.
[1637] Kupperman J. Theories of human nature. Hackett Publishing; 2010.
[1638] Kekes J. The human condition. Oxford University Press; 2010.
[1639] Buller D. Adapting minds: evolutionary psychology and the persistent quest for human nature. MIT Press; 2005.
[1640] Richards J. Human nature after Darwin: a philosophical introduction. Routledge; 2001.
[1641] Midgley M. Beast and man: the roots of human nature. Routledge; 1992.
[1642] Huxley T. Evolution and ethics. Macmillan; 1893.
[1643] Dewey J. Evolution and ethics. Monist 1898;8:321–41.
[1644] Baumeister R. The cultural animal: human nature, meaning, and social life. Oxford University Press; 2005.
[1645] Cavalli-Sforza L, Feldman M. Cultural transmission and evolution. Princeton University Press; 1981.
[1646] Durham W. Coevolution: genes, culture, and human diversity. Stanford University Press; 1991.
[1647] Goldhaber D. The nature-nurture debates: bridging the gap. Cambridge University Press; 2012.
[1648] Lewens T. Cultural evolution: conceptual challenges. Oxford University Press; 2015.
[1649] Prinz J. Beyond human nature: how culture and experience shape the human mind. Norton; 2014.

[1650] van Niekerk A. Biomedical enhancement: makeability or disenchantment? Tydskrift vir Geesteswetenskappe 2012;52:581–95.

[1651] Cuthbert B, Insel T. Toward the future of psychiatric diagnosis: the seven pillars of RDoC. BMC Med 2013;11:126.

[1652] Zwarenstein M. 'Pragmatic' and 'explanatory' attitudes to randomised trials. J R Soc Med 2017;110:208–18.

[1653] Thompson P, Andreassen O, Arias-Vasquez A, Bearden C, Boedhoe P, Brouwer R, et al. ENIGMA and the individual: predicting factors that affect the brain in 35 countries worldwide. NeuroImage 2017;145:389–408.

[1654] Sullivan P, Agrawal A, Bulik C, Andreassen O, Børglum, Breen G, et al. Psychiatric genomics: an update and an agenda. Am J Psychiatr 2018;175:15–27.

[1655] Stein D. An integrative approach to psychiatric diagnosis and research. World Psychiatry 2014;13:51–3.

[1656] Teh R, Valsdottir L, Yeh M, Shen C, Kramer D, Strom J, et al. Parachute use to prevent death and major trauma when jumping from aircraft: randomized controlled. Br Med J 2018;363:k5094.

[1657] Walter H. The third wave of biological psychiatry. Front Psychol 2013;4:582.

[1658] Dimidjian S, Hollon S. How would we know if psychotherapy were harmful? Am Psychol 2010;65:21–33.

[1659] Rosenzweig S. Some implicit common factors in diverse methods of psychotherapy. Am J Orthopsychiatry 1936;6:412–5.

[1660] Frank J. Persuasion and healing: a comparative study of psychotherapy. Johns Hopkins University Press; 1961.

[1661] Frank J. Psychotherapy and the human predicament. Jason Aronson; 1995.

[1662] Mulder R, Murray G, Rucklidge J. Common versus specific factors in psychotherapy: opening the black box. Lancet Psychiatry 2017;4:953–62.

[1663] Wampold B. How important are the common factors in psychotherapy? An update. World Psychiatry 2015;14:270–7.

[1664] Budd R, Huges I. The Dodo Bird Verdict – controversial, inevitable and important: a commentary on 30 years of meta-analyses. Clin Psychol Psychother 2009;16:510–22.

[1665] Andrews J. History of Great Britain. vol. II. T. Cadell and W. Davies; 1796.

[1666] Gardner M. Fads & fallacies in the name of science. 2nd ed. Dover; 1957.

[1667] Laudan L. The demise of the demarcation problem. In: Cohen R, Laudan L, editors. Physics, philosophy and psychoanalysis: essays in honor of Adolf Grünbaum. D. Reidel; 1983.

[1668] Ruse M. But is it science? The philosophical question in the creation/evolution controversy. Prometheus; 1988.

[1669] Weinberg S. Facing up: science and its cultural adversaries. Harvard University Press; 2001.

[1670] Pigliucci M. Nonsense on stilts: how to tell science from bunk. University of Chicago Press; 2010.

[1671] Gordin M. The pseudoscience wars: Immanuel Velikovsky and the birth of the modern fringe. University of Chicago Press; 2012.

[1672] Kitcher P. Abusing science: the case against creationism. MIT Press; 2012.

[1673] Pigliucci M, Boudry M. Philosophy of pseudoscience: reconsidering the demarcation problem. University of Chicago Press; 2013.

[1674] Farha B. Pseudoscience and deception: the smoke and mirrors of paranormal claims. University Press of America; 2014.

[1675] Lilienfeld S, Lynn S, Lohr J. Science and pseudoscience in clinical psychology. 2nd ed. Guilford Press; 2014.

[1676] Storr T. Unpersuadables: adventures with the enemies of science. Harry N Abrams; 2014.
[1677] Kaufman A, Kaufman J. Pseudoscience: the conspiracy against science. MIT Press; 2018.
[1678] McIntyre L. The scientific attitude: defending science from denial, fraud, and pseudoscience. MIT Press; 2019.
[1679] Browne M. Epistemic divides and ontological confusions: the psychology of vaccine skepticism. Hum Vaccin Immunother 2018;14:2540–2.
[1680] Mannheim K. Ideology and utopia. Routledge & Kegan Paul; 1929/1936.
[1681] Merton R. The sociology of science: theoretical and empirical investigations. University of Chicago Press; 1973.
[1682] Popper K. The logic of scientific discovery. Hutchinson; 1959.
[1683] Planck M. Scientific autobiography and other papers. Philosophical Library; 1968.
[1684] Ross D. The philosophy of science at the turn of the millennium. S Afr J Philos 1999;18:91–9.
[1685] Roazen P. Encountering Freud: the politics and histories of psychoanalysis. Transaction Publishers; 1990.
[1686] Derksen A. The seven sins of pseudoscience. J Gen Philos Sci 1993;24:17–42.
[1687] Thagard P. Computational philosophy of science. MIT Press; 1988.
[1688] Putnam H. The 'corroboration' of theories. In: Schilpp P, editor. The philosophy of Karl Popper. Open Court; 1974.
[1689] McHugh P. Try to remember: psychiatry's clash over meaning, memory, and mind. Dana Press; 2008.
[1690] Pendergrast M. The repressed memory epidemic: how it happened and what we need to learn from it. Springer; 2007.
[1691] Crews F. The memory wars: Freud's legacy in dispute. New York Review Books; 1995.
[1692] Moynihan R. Selling sickness: how the world's biggest pharmaceutical companies are turning us all into patients. National Books; 2005.
[1693] Valenstein E. Blaming the brain: the truth about drugs and mental health. Free Press; 2002.
[1694] https://www.madinamerica.com/2015/09/ronald-pies-doubles-down-and-why-we-should-care/.
[1695] Silberman S. NeuroTribes: the legacy of autism and the future of neurodiversity. Avery; 2015.
[1696] Armstrong T. Neurodiversity: discovering the extraordinary gifts of autism, ADHD, dyslexia, and other brain differences. Da Capo; 2010.
[1697] Drescher J, Cohen-Ketttenis P, Reed G. Gender incronguence of childhood in the ICD-11: controversies, proposal, and rationale. Lancet Psychiatry 2016;3:297–304.
[1698] Gleiser M. The island of knowledge: the limits of science and the search for meaning. Basic Books; 2014.
[1699] Flanagan O. The really hard problem: meaning in a material world. MIT Press; 2007.
[1700] Ayer A. The claims of philosophy, 1947. In: Klemke E, editor. The meaning of life. 3rd ed. Oxford University Press; 2008.
[1701] Landau I. Why has the question of the meaning of life arisen in the last two and a half centuries? Philosophy Today 1997;41:263–9.
[1702] Ogden C, Richards I. Meaning of meaning. Harcourt, Brace and World; 1923.
[1703] Eagleton T. The meaning of life: a very short introduction. Oxford University Press; 2008.
[1704] Seachris J. The meaning of life: the analytic perspective. In: Fieser J, Dowden B, editors. Internet encyclopedia of philosophy. 2021.
[1705] Martela F, Steger M. The three meanings of meaning in life: distinguishing coherence, purpose, and significance. J Posit Psychol 2016;11:531–45.

[1706] Hauskeller M. The meaning of life and death: ten classic thinkers on the ultimate question. Bloomsbury; 2019.
[1707] Leach S, Tartaglia J. The meaning of life and the great philosophers. Routledge; 2018.
[1708] Klemke E, Cahn S. The meaning of life. 4th ed. Oxford University Press; 2017.
[1709] Benatar D. Life, death & meaning: key philosophical readings on the big questions. 3rd ed. Rowe and Littlefield; 2016.
[1710] Wong P. The human quest for meaning: theories, research and applications. 2nd ed. Routledge; 2012.
[1711] Seachris J. Exploring the meaning of life: an anthology and guide. Wiley-Blackwell; 2012.
[1712] Friend D. The meaning of life: reflections in words and pictures on why we are here. Little, Brown; 2011.
[1713] Durant W. On the meaning of life. Ray Long and Richard R Smith; 1932.
[1714] Colebrook C. Deleuze and the meaning of life. Continuum International Publishing; 2010.
[1715] Smith E. The power of meaning: finding fulfillment in a world obsessed with happiness. Broadway Books; 2017.
[1716] Pink D. Drive: the surprising truth about what motivates us. Riverhead Books; 2009.
[1717] Landau I. Finding meaning in an imperfect world. Oxford University Press; 2017.
[1718] Einstein A. The world as I see it. John Lane; 1935.
[1719] Weinberg S. The first three minutes. Basic Books; 1977.
[1720] Baier K. The meaning of life. In: Jecker N, editor. Aging and ethics: contemporary issues in biomedicine, ethics, and society. Humana Press; 1992.
[1721] Cahn S, Vitrano C. Happiness and goodness: philosophical reflections on living well. Columbia University Press; 2015.
[1722] Sacks J, https://rabbisacks.org/happiness-found/; 2008.
[1723] Schopenhauer A. Parerga and paralipomena: short philosophical essays. Oxford University Press; 1851/1974.
[1724] Adams D. The hitchhiker's guide to the galaxy. Pocket Books; 1981.
[1725] Python M. The meaning of life. 1983.
[1726] Nietzsche F. The gay science. 1882.
[1727] Gray J. Feline philosophy: cats and the meaning of life. Farrar, Straus and Giroux; 2020.
[1728] Nagel T. The absurd. J Philos 1971;68:716–27.
[1729] Critchley S. There's no theory of everything. In: Catapano P, Critchley S, editors. Modern Ethics in 77 Arguments: A Stone Reader. Liveright; 2017.
[1730] Critchley S. Very little ... almost nothing: death, philosophy, literature. Routledge; 1997.
[1731] Matthews G. Socratic perplexity and the nature of philosophy. Oxford University Press; 1999.
[1732] Nagel T. The view from nowhere. Oxford University Press; 1986.
[1733] Vernon M. After atheism: science, religion and the meaning of life. Palgrave Macmillan; 2007.
[1734] Tartaglia J. Philosophy in a meaningless life: a system of nihilism, consciousness and reality. Bloomsbury Academic; 2016.
[1735] Trisel B. How human life matters in the universe: a reply to David Benatar. J Philos Life 2019:9:1–15.
[1736] Midgley M. Purpose, meaning and Darwinism. Philosophy Now 2009;17:16–9.
[1737] Metz T. Meaning in life: an analytic study. Oxford University Press; 2013.
[1738] Thagard P. The brain and the meaning of life. Princeton University Press; 2010.
[1739] Peterson J. Twelve rules for life: an antidote to chaos. Random House; 2018.
[1740] Kass L. Leading a worth life: finding meaning in modern times. Encounter; 2018.
[1741] Veit W. Existential nihilism: the only really serious philosophical question. J Camus Stud 2018;211–32.

[1742] Yalom I. Love's executioner and other tales of psychotherapy. Basic Books; 1989.
[1743] James W. Principles of psychology. Henry Holt & Co; 1890.
[1744] James W. Is life worth living? Int J Ethics 1895;6:1–24.
[1745] Taylor R. Good and evil. Macmillan; 1970.
[1746] Froese P. On purpose: how we create the meaning of life. Oxford University Press; 2016.
[1747] Little B, Salmela-Aro K, Phillips S. Personal project pursuit: goals, action, and human flourishing. Lawrence Erlbaum; 2007.
[1748] Marino G. The existentialist's survival guide: how to live authentically in an inauthentic age. HarperOne; 2018.
[1749] Ruse M. A meaning to life. Oxford University Press; 2019.
[1750] Messerly J. www.reasonandmeaning.com.
[1751] Messerly J. The meaning of life: religious, philosophical, transhumanist, and scientific perspectives. Darwin and Hume; 2013.
[1752] Hosseini R. Wittgenstein and meaning in life: in search of the human voice. Palgrave Macmillan; 2015.
[1753] Cabrera I. Meaning of life and nonsense in Tractatus: no answer for no question and some fictional illuminations. J Philos Life 2019;9:33–53.
[1754] Wolf S. Meaning in life and why it matters. Princeton University Press; 2010.
[1755] Russell D. Happiness for humans. Oxford University Press; 2012.
[1756] Russell B. A free man's worship; 1903.
[1757] Popova M. www.brainpickings.org/2012/02/27/purpose-work-love.
[1758] Britton K. Philosophy and the meaning of life. Cambridge University Press; 1969.
[1759] de Muijnck W. Good fit versus meaning in life. Symposion 2016;3:309–24.
[1760] May T. A significant life: human meaning in a silent universe. University of Chicago Press; 2016.
[1761] Svensson F. A subjectivist account of life's meaning. De Ethica J Philos Theol Appl Ethics 2017;4:45–66.
[1762] Ting A. The 'subjective attraction' and 'objective attractiveness' of the practice of the rites in the Xunzi. J Philos Life 2019;9:16–32.
[1763] Baggini J. Revealed – the meaning of life. The Guardian; September 24, 2004.
[1764] Tolstoy L. A calendar of wisdom: daily thoughts to nourish the soul. Prentice Hall; 1997.
[1765] Messer S, Winokur M. Some limits to the integration of psychoanalytic and behavior therapy. Am Psychol 2000;35:818–27.
[1766] Ellis A. Misrepresentation of behavior therapy by psychoanalysts. Am Psychol 1981;36:798–9.
[1767] Brandstätter M, Baumann U, Borasio G, Fegg M. Systematic review of meaning in life instruments. Psycho-Oncology 2012;21:1034–52.
[1768] Erikson E, Erikson J. Life cycle completed. Wiley; 1998.
[1769] Maslow A. Motivation and personality. Harper & Brothers; 1954.
[1770] Volkert J, Schulz H, Brütt A, Andreas S. Meaning in life: relationship to clinical diagnosis and psychotherapy outcome. J Clin Psychol 2014;70:528–35.
[1771] Slote M. Human development and human life. Springer; 2016.
[1772] Held B. The negative side of positive psychology. J Humanist Psychol 2004;44:9–46.
[1773] Wong P. Positive psychology 2.0: towards a balanced interactive model of the good life. Canadian Psychol 2011;52:69–81.
[1774] Hayes L. Happiness in valued living: acceptance and commitment therapy as a model for change. In: David S, Boniwell I, Ayers A, editors. Oxford handbook of happiness. Oxford University Press; 2012.

[1775] Mahoney M, Granvold D. Constructivism and psychotherapy. World Psychiatry 2005;4:74–7.
[1776] Metz T. Meaning in life as the aim of psychotherapy: a hypothesis. In: Hicks J, Routledge C, editors. The experience of meaning in life: classical perspectives, emerging themes, and controversies. Springer; 2013.
[1777] Metz T. The proper aim of therapy: subjective well-being, objective goodness, or a meaningful life? In: Russo-Netzer P, Schulenberg S, Batthyany A, editors. Clinical perspectives on meaning: positive and existential psychotherapy. Springer; 2016.
[1778] Murdoch I. Existentialists and mystics: writings on philosophy and literature. Penguin; 1999.
[1779] Norman A. Getting humanism right-side up: A reality-based "mattering map" and alternative humanist manifesto. Humanist; Jan/Feb: 2015.
[1780] Campbell J. The hero with a thousand faces. Pantheon; Jan/Feb:2015.
[1781] Bering J. The belief instinct: the psychology of souls, destiny, and the meaning of life. Norton; 2011.
[1782] Booker C. The seven basic plots: why we tell stories. Continuum; 2004.
[1783] Boyd B. On the origin of stories: evolution, cognition, and fiction. Harvard University Press; 2009.
[1784] Caruso G, Flanagan O. Neuroexistentialism: meaning, morals, and purpose in the age of neuroscience. Oxford University Press; 2018.
[1785] Gottschall J. The storytelling animal: how stories make us human. Houghton Mifflin Harcourt; 2012.
[1786] Gottschall J, Wilson D. The literary animal: evolution and the nature of narrative. Northwestern University Press; 2005.
[1787] Shermer M. The believing brain: from ghosts and gods to politics and conspiracies – how we construct beliefs and reinforce them as truths. Times Books; 2011.
[1788] Storr W. The science of storytelling: why stories make us human and how to tell them better. Abrams Press; 2020.
[1789] Bargh J. Free will is un-natural. In: Baer J, Kaufman B, Baumeister R, editors. Are we free? Psychology and free will. Oxford University Press; 2008.
[1790] Harris S. Free will. Free Press; 2012.
[1791] Wegner D. The illusion of conscious will. Bradford Book; 2003.
[1792] Wittgenstein L. Tractatus logico-philosophicus. Kegan Paul; 1922.
[1793] Sellars W. … this I or he or it (the thing) which thinks …. Proc Am Philos Assoc 1970;44:5–31.
[1794] Midgley M. Are you an illusion? Routledge; 2014.
[1795] von Wright G. Of human freedom. In: von Wright G, editor. In the shadow of Descartes: essays in the philosophy of mind. Springer; 1984/1998.
[1796] Mele A. Free: why science hasn't disproved free will. Oxford University Press; 2014.
[1797] Searle J. Freedom and neurobiology: reflections on free will, language, and political power. Columbia University Press; 2007.
[1798] Dennett D. Freedom evolves. Viking; 2003.
[1799] Unger P. Free will and scientiphicalism. Philos Phenomenol Res 2002;65:1–25.
[1800] Baumeister R. Free will in scientific psychology. Perspect Psychol Sci 2008;3:14–9.
[1801] Kopp S. If you meet the buddha on the road, kill him! Science & Behavior Books; 1972.
[1802] Strawson P. Freedom and resentment. Proc Br Acad 1962;48:1–25.
[1803] Russell P. Hume's lengthy digression: free will in the Treatise. The Cambridge companion to Hume's Treatise. Cambridge University Press; 2015.
[1804] Shook J. Jonathan Edwards' contribution to John Dewey's theory of moral responsibility. Hist Philos Q 2004;21:299–312.

[1805] Bromhall K. Embodied akrasia: James on motivation and weakness of will. William James Stud 2018;14:26–53.
[1806] Uszkai R. Aristotle's account of akrasia: towards a contemporary analogy. Ann Philos 2012;5:85–90.
[1807] Ainslie G. "Free will" as recursive self-prediction: does a deterministic mechanism reduce responsibility? In: Poland J, George G, editors. Addiction and responsibility. MIT Press; 2011.
[1808] Baer J, Kaufman J. Baumeister R. Are we free? The psychology of free will. Oxford University Press; 2008.
[1809] Baumeister R. Self-regulation and self-control: selected works of Roy F. Baumeister. Routledge; 2018.
[1810] Clark A, Kiverstein J, Vierkantagain T. Decomposing the will. Oxford University Press; 2013.
[1811] Elster J. Ulysses and the sirens: studies in rationality and irrationality. Cambridge University Press; 1979.
[1812] Glover J. Responsibility. Routledge & Kegan Paul; 1970.
[1813] Hassin R, Oschner K, Trope Y. Self control in society, mind, and brain. Oxford University Press; 2010.
[1814] Holton R. Willing, wanting, waiting. Oxford University Press; 2009.
[1815] Kane R. The Oxford handbook of free will. 2nd ed. Oxford University Press; 2011.
[1816] Mele A. Surrounding free will: philosophy, psychology, neuroscience. Oxford University Press; 2015.
[1817] Timpe K, Griffith M, Levy N. The Routledge companion to free will. Routledge; 2016.
[1818] Trakakis N, Cohen D. Essays on free will and moral responsibility. Cambridge Scholars Publishing; 2008.
[1819] Milkman K, Rogers T, Bazerman M. Harnessing our inner angels and demons: what we have learned about want/should conflicts and how that knowledge can help us reduce short-sighted decision making. Perspect Psychol Sci 2008;3:324–38.
[1820] Baumeister R, Scher S. Self-defeating behavior patterns among normal individuals: review and analysis of common self-destructive tendencies. Psychol Bull 1988;104:3–22.
[1821] Baumeister R, Tierney J. Willpower: rediscovering the greatest human strength. Penguin; 2011.
[1822] Inzlicht M, Schmeidel B, Macrae C. Why self-control seems (but may not be) limited. Trends Cogn Sci 2014;18:127–33.
[1823] Job V, Walton G, Bernecker K, Dweck C. Beliefs about willpower determine the impact of glucose on self-control. Proc Natl Acad Sci U S A 2013;110:14387–842.
[1824] Friese M, Loschelder D, Gieseler K, Frankenbach J, Inzlicht M. Is ego depletion real? An analysis of arguments. Personal Soc Psychol Rev 2019;23:107–31.
[1825] Watts A. Wisdom of insecurity: a message for an age of anxiety. Pantheon; 1958.
[1826] Ainslie G. Willpower with and without effort. Behav Brain Sci 2020;1–81.
[1827] Miller A. The drama of the gifted child: the search for the true self. Basic Books; 2007.
[1828] Miller M. The true "Drama of the Gifted Child": the phantom Alice Miller – the real person. Amazon; 2018.
[1829] Christensen S, Turner D. Folk psychology and philosophy of mind. Psychology Press; 2017.
[1830] Peels R, van Woudenberg R. The Cambridge companion to common-sense philosophy. Cambridge University Press; 2020.
[1831] Gidden A. Modernity and self-identity: self and society in the late modern age. Polity Press; 1991.

[1832] Perry J. Personal identity. University of California Press; 1975.
[1833] Glover J. The philosophy and psychology of personal identity. Penguin; 1988.
[1834] Bermudez J, Eilan N, Marcel A. The body and the self. MIT Press; 1995.
[1835] Flanagan O. Self expressions: mind, morals, and the meaning of life. Oxford University Press; 1996.
[1836] Radden J. Divided minds and successive selves: ethical issues in disorders of identity and personality. MIT Press; 1996.
[1837] Harré R. The singular self: an introduction to the psychology of personhood. SAGE Publications; 1998.
[1838] Sorabji R. Self: ancient and modern insights about individuality, life, and death. University of Chicago Press; 2006.
[1839] Williams B. Problems of the self. Cambridge University Press; 2008.
[1840] Noë A. Out of our heads: why you are not your brain, and other lessons from the biology of consciousness. Hill and Wang; 2009.
[1841] Baggini J. The ego trick: what does it mean to be you. Granta; 2012.
[1842] Goldie P. The mess inside: narrative, emotion, and the mind. Oxford University Press; 2012.
[1843] Churchland P. Touching a nerve: the self as brain. Norton; 2013.
[1844] Gallagher S. The Oxford handbook of the self. Oxford University Press; 2013.
[1845] Northoff G. Neuro-philosophy and the healthy mind: learning from the unwell brain. Norton; 2016.
[1846] Gabriel M. I am not a brain. Polity; 2017.
[1847] Baumeister R. Public self and private self. Springer-Verlag; 1986.
[1848] Crick F. The astonishing hypothesis: the scientific search for the soul. Charles Scribner; 1993.
[1849] Panksepp J. The periconscious substrates of consciousness: affective states and the evolutionary origins of the SELF. J Conscious Stud 1998;5:566–82.
[1850] Thompson E. Waking, dreaming, being: self and consciousness in neuroscience, meditation, and philosophy. Columbia University Press; 2015.
[1851] Hood B. Self illusion: how the social brain creates reality. Oxford University Press; 2013.
[1852] Damasio A. Self comes to mind: constructing the conscious brain. Pantheon; 2010.
[1853] Metzinger T. The ego tunnel: the science of mind and the myth of the self. Basic Books; 2009.
[1854] Keenan T, Keenan J. The lost self: pathologies of the brain and identity. Oxford University Press; 2005.
[1855] Kircher T, David A. The self in neuroscience and psychiatry. Cambridge University Press; 2003.
[1856] Ramachandran V, Blakeslee S. Phantoms in the brain: probing the mysteries of the human mind. William Morrow; 1998.
[1857] Stein D. What is the self? A psychobiological perspective. CNS Spectr 2007;12:333–6.
[1858] Ersner-Hershfield H, Wimmer G, Knutson B. Saving for the future self: neural measures of future self-continuity predict temporal discounting. Soc Cogn Affect Neurosci 2009;4:85–92.
[1859] Maiese M. Embodied selves and divided minds. Oxford University Press; 2016.
[1860] Pennebaker J, Smyth J. Opening up by writing it down: how expressive writing improves health and eases emotional pain. 3rd ed. Guilford Press; 2016.
[1861] de Montaigne M. Essays. Simon Millanges; 1580.
[1862] Bakewell S. How to live: or a life of Montaigne in one question and twenty answers. Other Press; 2010.
[1863] Hartle A. Michel de Montaigne: accidental philosopher. Cambridge University Press; 2003.

[1864] Barsky A. The paradox of health. N Engl J Med 1988;318:414–8.
[1865] Ainslie G, Monterosso J. A marketplace in the brain? Science 2004;306:421–3.
[1866] Byrne A. Transparency and self-knowledge. Oxford University Press; 2018.
[1867] Cassam Q. Self-knowledge for humans. Oxford Univerity Press; 2017.
[1868] Renz A. Self-knowledge: a history. Oxford University Press; 2016.
[1869] Coliva A. The varieties of self-knowledge. Palgrave Macmillan; 2016.
[1870] Gerler B. Self-knowledge. Routledge; 2011.
[1871] Carruthers P. The opacity of mind: an integrative theory of self-knowledge. Oxford University Press; 2011.
[1872] Heatherington S. Self-knowledge: beginning philosophy right here and now. Broadview; 2007.
[1873] Cassam Q. Self-knowledge. Oxford University Press; 1994.
[1874] Emre M. The personality brokers: the strange history of Myers-Briggs and the birth of personality testing. Doubleday; 2018.
[1875] Leddy T. John Dewey. In: Giovannelli A, editor. Aesthetics: the key thinkers. Bloomsbury Academic; 2012.
[1876] Galgut E. Hume's aesthetic standard. Hume Stud 2012;38:183–200.
[1877] Rice L. Spinoza's relativistic aesthetics. Tijdschrift voor Filosofie 1996;3:476–89.
[1878] Dutton D. Aesthetics and evolutionary psychology. In: Levinson D, editor. The Oxford handbook for aesthetics. Oxford University Press; 2003.
[1879] Shusterman R. The invention of pragmatist aesthetics: genealogical reflections on a nation and a name. In: Malecki W, editor. Practicing pragmatist aesthetics: critical perspectives on the arts. Brill; 2014.
[1880] Fenner D. Varieties of aesthetic naturalism. Am Philos Q 1993;30:353–62.
[1881] Morais I. Aesthetic realism. Palgrave; 2019.
[1882] Goodman N. Languages of art: an approach to a theory of symbols. Bobbs-Merrill; 1968.
[1883] Hitchens C, Dawkins R, Harris S, Dennett D. The four horsemen: the conversation that sparked an atheist revolution. Penguin; 2019.
[1884] Spencer N. Atheists: the origin of the species. Bloomsbury Academic; 2014.
[1885] Bellah R. Religion in human evolution: from the Paleolithic to the axial age. Belknap Press; 2011.
[1886] Boyer P. Religion explained: the evolutionary origins of religious thought. Basic Books; 2001.
[1887] Dennett D. Breaking the spell: religion as a natural phenomenon. Viking; 2006.
[1888] Feierman J. The biology of religious behavior – the evolutionary origins of faith and religion. Praeger; 2009.
[1889] Hinde R. Why gods persist: a scientific approach to religion. Routledge; 1999.
[1890] Rue L. Religion is not about God: how spiritual traditions nurture our biological nature and what to expect when they fail. Rutgers University Press; 2004.
[1891] Torrey E. Evolving brains: emerging gods: early humans and the origins of religion. Columbia University Press; 2017.
[1892] Vallaint G. Spiritual evolution: how we are wired for faith, hope, and love. Harmony; 2009.
[1893] Wright R. The evolution of God. Little, Brown and Company; 2009.
[1894] Midgley M. Evolution as a religion: strange hopes and stranger fears. Metheun; 1985.
[1895] Midgley M. Science as salvation: a modern myth and its meaning. Routledge; 1992.
[1896] Eagleton T. Reason, faith, and revolution: reflections on the God debate. Yale University Press; 2009.
[1897] Gray J. The soul of the marionette: a short enquiry into human freedom. Penguin; 2015.
[1898] Einstein A. Ideas and opinions. Crown Publishers; 1954.
[1899] Viskontas I. Myths exploded: lessons from neuroscience. The Teaching Company; 2017.

[1900] Jarrett C. Great myths of the brain. Wiley-Blackwell; 2014.
[1901] Rauch J. The happiness curve: why life gets better after 50. Thomas Dunne; 2018.
[1902] Jeste D, Oswald A. Individual and societal wisdom: explaining the paradox of human aging and high well-being. Psychiatry 2014;77:317–30.
[1903] Takahashi M, Singh R, Stone J. A theory for the origin of human menopause. Front Genet 2017;7:222.
[1904] Johnstone R, Cant M. The evolution of menopause in cetaceans and humans: the role of demography. Proc Royal Soc B Biol Sci 2010;277:3765–71.
[1905] Sternberg R, Glück J. The Cambridge handbook of wisdom. Cambridge University Press; 2019.
[1906] Fuentes A, Deane-Drummond C. Evolution of wisdom: major and minor keys. University of Notre Dame; 2018.
[1907] Schwartz B, Sharpe K. Practical wisdom: the right way to do the right thing. Riverhead Books; 2011.
[1908] Meeks T, Jeste D. Neurobiology of wisdom: a literature overview. Arch Gen Psychiatry 2009;66:355–65.
[1909] Thiele L. The heart of judgment: practical wisdom, neuroscience, and narrative. Cambridge University Press; 2006.
[1910] Jaspers K. On my philosophy. In: Kaufman W, editor. Existentialism from Dostoyevsky to Sartre. Meridian; 1941/1956.
[1911] Solomon R. Spirituality for the skeptic: the thoughtful love of life. Oxford University Press; 2002.
[1912] Hägglund M. This life: secular faith and spiritual freedom. Pantheon; 2019.
[1913] van Niekerk A. Pragmatism and religion. In: Malachowski A, editor. Cambridge companion to pragmatism. Cambridge University Press; 2013.
[1914] Dawkins R. Unweaving the rainbow: science, delusion and the appetite for wonder. Mariner; 2000.
[1915] Hofstadter D. Popular culture and the threat to rational inquiry. Science 1998;281:512–3.
[1916] Wilczek F. A beautiful question: finding nature's deep design. Penguin; 2015.
[1917] Lightman A. A sense of the mysterious: science and the human spirit. Pantheon; 2005.
[1918] Schneider K. The resurgence of awe in psychology: promise, hope, and perils. Humanist Psychol 2017;45:103–8.
[1919] Gallagher S, Reinerman L, Janz B, Bockelman P, Trempler J. A neurophenomenology of awe and wonder: towards a non-reductionistic cognitive science. MIT Press; 2015.
[1920] Pearsall P. Awe: the delights and dangers of our eleventh emotion. Health Communications; 2008.
[1921] Keltner D, Haidt J. Approaching awe, a moral, spiritual, and aesthetic emotion. Cognit Emot 2003;17:297–314.
[1922] Dettelbach M. Alexander von Humboldt between enlightenment and romanticism. Northeast Nat 2001;8:9–20.
[1923] Leslie J, Kuhn R. The mystery of existence: why is there anything at all. Wiley-Blackwell; 2013.
[1924] Krauss L. A universe from nothing: why there is something rather than nothing. Atria; 2012.
[1925] Holt J. Why does the world exist?: An existential detective story. Liveright; 2012.
[1926] Atkins P. On being: a scientist's exploration of the great questions of existence. Oxford Universit Press; 2011.
[1927] Stoddard J, Afari N. The big book of ACT metaphors: a practitioner's guide to experiential exercises & metaphors in acceptance & commitment therapy. New Harbinger; 2014.

[1928] Stott R, Mansell W, Salkovskis P, Lavender A, Cartwright-Hatton S. Oxford guide to metaphors in CBT: building cognitive bridges. Oxford University Press; 2010.
[1929] Kopp R. Metaphor therapy: using client generated metaphors in psychotherapy. Routledge; 1995.
[1930] Siegelman E. Metaphor and meaning in psychotherapy. Guilford; 1990.
[1931] Jaye A. The golden rule of schmoozing: the authentic practice of treating others well. Sourcebooks; 1998.
[1932] Auerbach E. Mimesis: the representation of reality in western literature. Princeton University Press; 1953.
[1933] Miles J. A biography of God. Knopf; 1995.
[1934] Wiesel E. The trial of God. Random House; 1979.
[1935] Neiman S. Moral clarity: a guide for grown-up idealists. Houghton Mifflin Harcourt; 2009.
[1936] Kushner H. The book of Job: when bad things happened to a good person. Schocken; 2012.
[1937] Camus A. The plague. Vintage International; 1947/1991.
[1938] Metz T. Judaism's distinct perspectives on the meaning of life. forthcoming.
[1939] Kasparov G. How life imitates chess: making the right moves, from the board to the boardroom. Bloomsbury; 2007.
[1940] Greenfeld L. Mind, modernity, madness: the impact of culture on human experience. Harvard University Press; 2013.
[1941] Berger P, Luckmann T. Modernity, pluralism and the crisis of meaning: the orientation of modern man. Bertelsmann Foundation; 1995.
[1942] Thomson G. On the meaning of life. Wadsworth; 2003.
[1943] Fassin D, Rechtman R. The empire of trauma: an inquiry into the condition of victimhood. Princeton University Press; 2009.
[1944] Dalrymple T. Spoilt rotten: the toxic cult of sentimentality. Gibson Square; 2010.
[1945] Haidt J, Lukianoff G. The coddling of the American mind: how good intentions and bad ideas are setting up a generation for failure. Penguin; 2018.
[1946] Campbell B, Manning J. The rise of victimhood culture: microaggressions, safe spaces, and the new culture wars. Palgrave Macmiillan; 2018.
[1947] Furedi F. The culture of fear; risk taking and the morality of low expectations. Continuum; 1997.
[1948] Brodsky J. On grief and reason: essays. Farrar, Straus, and Giroux; 1997.
[1949] Gopnik A. The gardener and the carpenter: what the new science of child development tells us about the relationship between parents and children. Farrar, Straus, and Giroux; 2016.
[1950] Foot P. Mencius. Penguin; 2005.
[1951] Angel M. Why be a parent? The New York Review of Books; November 10, 2016.
[1952] Van Den Abbeele G. Metaphor: from Montaigne to Rousseau. University of Minnesota Press; 1992.
[1953] Solnit R. Wanderlust: history of walking. Penguin; 2001.
[1954] Gros F. Philosophy of walking. Verso; 2014.
[1955] Nicholson G. The lost art of walking: the history, science, philosophy, and literature of pedestrianism. Riverhead; 2008.
[1956] Revel J-F, Ricard M. The monk and the philosopher: a father and son discuss the meaning of life. Schocken; 1997.
[1957] Rodman P. Derech Eretz. https://www.myjewishlearning.com/article/derech-eretz/.
[1958] Cousineau P. Art of pilgrimage: the seeker's guide to making travel sacred. Conari; 1998.
[1959] Bly J. Iron John: a book about men. Addison-Wesley; 1990.

[1960] Pirsig R. Zen and the art of motorcycle maintenance: an inquiry into values. William Morrow; 1974.
[1961] Feldman F. Whole life satisfaction concepts of happiness. Theoria 2008;74:219–38.
[1962] van Niekerk A. Modernity, mortality, and mystery. Philos Today 1999;43:18–34.
[1963] Hannah S. Happiness, a mystery: & 66 attempts to solve it. Profile Books; 2020.
[1964] Hermanowicz J. The culture of mediocrity. Minerva 2013;51:363–87.
[1965] Milo D. Good enough: the tolerance for mediocrity in nature and society. Harvard University Press; 2019.
[1966] Horowitz A. On looking: a walker's guide to the art of observation. Scribner; 2013.
[1967] Thoreaux H. Walking. Ticknor and Fields; 1861.
[1968] Solnit R. A field guide to getting lost. Penguin; 2006.
[1969] Yalom I. Love's executioner and other tales of psychotherapy. HarperPerennial; 1989.
[1970] Fingarette H. Death: philosophical soundings. Open Court; 1999.
[1971] Bradley B, Feldman F, Johansson J. The Oxford handbook of philosophy of death. Oxford University Press; 2020.
[1972] Kessler D. Finding meaning: the sixth stage of grief. Scribner; 2019.
[1973] Kerouac J. On the road. Viking; 1957.
[1974] Isaacson W. Einstein: his life and universe. Simon & Schuster; 2007.
[1975] Head H. Schopenhauer and the Stoics. Pli Warwick J Philos 2016;90–105.
[1976] Merton R. On the shoulders of giants: a Shandian postscript. Harcourt Brace Janovich; 1985.
[1977] Sarte J-P. No exit. Samuel French; 1958.
[1978] Smith H. Aldous Huxley – a tribute. Psychedelic Rev 1964;1:264–5.
[1979] Maner J, DeWall C, Baumeister R, Schaller M. Does social exclusion motivate interpersonal reconnection? J Pers Soc Psychol 2007;92:42–55.
[1980] Barris J. Sometimes always true: undogmatic pluralism in politics, metaphysics, and epistemology. Fordham University Press; 2015.
[1981] Gíslason G, Eddy J. The hygge life: embracing the Nordic art of coziness through recipes, entertaining, decorating, simple rituals, and family traditions. Ten Speed; 2017.
[1982] Wiking M. The little book of hygge: Danish secrets to happy living. Ten Speed; 2017.
[1983] Brantmark N. Lagom: not too little, not too much: the Swedish art of living a balanced, happy life. Harper Design; 2017.
[1984] Brones A. Live Lagom: balanced living, the Swedish way. Ten Speed; 2017.

Index

Note: Page numbers followed by *t* indicate tables.

D

E

F

G

H

I

N

O

P

Printed in the United States
by Baker & Taylor Publisher Services